History of Mathematics in Hungary until the 20th Century

by
Barna Szénássy

W0245817

Springer-Verlag Berlin Heidelberg GmbH

Prof. Barna Szénássy
Lajos Kossuth University
Debrecen 10
Hungary

This book is the translated and enlarged version of the original Hungarian
A MAGYARORSZÁGI MATEMATIKA TÖRTÉNETE
Akadémiai Kiadó, Budapest

Translated by
Judit Pokoly

English text revised by
János Bognár

ISBN 978-3-662-02745-5   ISBN 978-3-662-02743-1 (eBook)
DOI 10.1007/978-3-662-02743-1

41/3140-543210

History of Mathematics
in Hungary
until the 20th Century

Contents

Introduction

This book was first published in 1970 and reprinted four years later, but since then it has sold out. The helpful comments of my colleagues have strengthened my conviction that some changes and corrections were to be done in the third edition. These are summarized below, supplementing a nearly unaltered part of the Preface to the original edition.

Any work on the history of science spanning a considerably long period of time has to satisfy a great number of criteria as the discipline under scrutiny has to be examined as it evolved, embedded in the intricate network of relations in the national and universal history of culture. To compound the problem, the rise of mathematics out of backwardness in Hungary was fraught with relapses instead of leading in a straight line to today's heights.

To begin with, the author of a book on science history encounters the problem of what material to include and how to treat it. To stretch the point a little, one might say that as many authors and reviewers, as many opinions. It is almost impossible to coordinate all the divergent points of view and expectations. A *mathematician*, for example, would like to learn more about the background and circumstances of how the leading ideas of his science were sparked off and developed, and what their use and subsequent effects were like; a *teacher* is far more interested in the questions of education, teaching methods and textbooks; a *historian* would obviously hail a meticulous elaboration of the archival material and a thorough study of the interplay between the forces and conditions of production and mathematical culture.

The list could be continued on end. To satisfy all the different expectations, many heavy tomes should be written about the history of mathematics in Hungary as well.

To a certain extent, all the varying points of view are taken into consideration in this work, but the highlight is on the history of mathematical *research* in Hungary, on the evaluation of significant results and the description of their influence on further development. This programme has influenced the division of the history of Hungarian mathematics into periods and the grouping of the material to be discussed. Unlike, for example, the history of Hungarian literature, arts or the nation, Hungarian mathematics cannot be periodized by well-established criteria. Universal histories of mathematics also greatly differ as to such criteria, so they cannot provide a model for the Hungarian case, which significantly deviates from the development of universal mathematical research, anyway. Some foreign authors, for example, match the history of mathematics in the Middle Ages and modern times against the major currents of thought (e.g. scholasticism, renaissance, baroque, enlightenment, etc.) and break them into periods accordingly. The relative profusion of documentary material on world mathematics offers a suitable basis for such classification. For Hungary, however, this periodization would be forced and meaningless since the scarcity of known documents from old times does not allow us to deduce the culture-historical criteria of the above-mentioned currents of thought. This prompted me to choose a criterion for periodization which seems more suitable to the particular Hungarian development leading to original mathematical results.

The book ends with a few comments on some outstanding contributions at the beginning of our century; an evaluation of subsequent development is a hard task still to be undertaken. Needless to say, the earlier and relatively easily arranged material also posed plenty of problems. The difficulties are well-known, I suppose. Anyone must be familiar with them who has ever read old Latin, German or Hungarian mathematical works in an archaic style full of old terms and obscure symbols, and having deciphered their meaning, tried to establish priorities, decide on originality and evaluate them objectively.

I have studied nearly all the relevant works of our mathematicians in the original, even when the commentary literature would have been sufficient. In many cases the conclusions I have come to differ from the opinions of other scholars. A list of the *most important* sources that I used can be found after each chapter. With some most recently published references included and a few former ones omitted, the bibliographies display the greatest change against the first edition. The omitted writings have not lost their relevance but the latest publications provide rich bibliographies for those looking for further references. To make the reading of the book easier, most of the foreign language (mainly German and Latin) quotations are given in literal translation unless it distorts faithful interpretation.

In one respect I have significantly changed the customary lay-out of histories of mathematics: I attached a separate chapter to the book with a concise summary of the biographies of Hungarian scientists arranged in alphabetic order. The reader quite rightly expects some information in this regard, but it has been my experience that such references are rather disturbing when mixed in the discussion of mathematical results. Following the intentions of one of our late mathematicians of worldwide reputation, I tried to "weigh out" the biographical summaries according to the significance of the scholar — within certain limits, of course (otherwise a whole volume would be added on the life of *János Bolyai* alone!). Over the decades, a great many false biographical data have taken root in the literature which I have tried to rectify, seeking out more reliable sources. Though quite impossible to misunderstand them, the data in parentheses after the names indicate place of birth, year of birth, year of death, and place of death.

To provide the Reader with the necessary geographical information, an index of geographical names has been enclosed at the end of the book. In our century, the frontiers of Central European countries have been modified several times, and as a result the internationally familiar names of several places have changed. When necessary, I indicate after a name of a place located outside Hungary now the letter symbol of the given country, to help the Reader.

8

I received valuable observations from *István Fenyő* and *Ferenc Kárteszi* concerning the first draft of my book. *László Makkai, Gábor Szász* and *László Vekerdi* gave me great help as the literary editors of the book's first edition.

I owe a word of gratitude to *Sándor Gacsályi* and *György Pollák* for their useful comments upon reading the present text. I have carried out almost all the changes they proposed. I herewith thank *Judit Pokoly* for the conscientious translation, *János Bognár* for the professional supervision of the translation, *Piroska Polyánszky* for the meticulous editing work and the publisher *Akadémiai Kiadó* for their endeavours in publishing the book.

Barna Szénássy

I. Data on mathematical culture in Hungary before printing

1. The origin of Hungarian numerals and cuneiform number symbols

The very first data on the arithmetic knowledge of ancient Hungarians are offered by linguistics, archaeology and history.

Absorbed in studying the number names of different peoples, linguists try to conjecture from their findings what their level of culture and kindred relations must have been. Earlier this ambition was fuelled by the mistaken notion that the numerals, just like the names for the parts of the body, belonged to the most ancient layer of a people's word stock. It is now well-known that the development of the concept of number is the result of a long and difficult process of abstraction in the evolution of people, just like in the intellectual development of a child. Linguistic investigations have also proved that originally the names of most *one-digit* numbers denoted commonly used concrete objects and not numeric properties.

The way how natural numbers and the respective number names evolved in different languages remains hidden in the mystery of prehistoric times. Nevertheless, research findings entitle us to propose certain hypotheses. Hungarian linguistic research has also provided weighty arguments for the origin and relations of the Hungarian people as well as the number systems used by our forebears.

For prehistoric man the human hand was a natural tool for counting, and accordingly, in some languages the names of numbers had developed from the names of the fingers. Comparative linguistic studies have proved this conclusively, e.g., for Indo-Germanic languages. For some other peoples the origin of numerals was in some way connected to the personal pronouns *I* and *you*. However, among the basic numerals of Hungarian-related Finno-Ugric languages (Vogul /Manshi/, Ostyak /Chanti, Hanti/, Cheremiss /Mari/, Mordvin /Moksha, Erze/, Zyrian /Komi/, Votyak /Udmurt/, Finnish /Suomi/, Estonian /Eesti/ and Lapp /Sabme/) neither words for the parts of the body nor personal pronouns can be tracked down. Even though the etymological explanation of the origin of our numerals has not yet crystallized, several linguists consider it highly probable that Hungarian *egy* (1) and its equivalents in related languages has developed from the contraction of the demonstrative pronoun *ez itt* ('this here'). On the other hand, it is beyond doubt that in all Finno-Ugric languages the names of the cardinal numbers 2, 3, 4, 5, 6 and 100 go back to common roots. Apparently, this is what our historians consider as some of the most reliable proof for the Hungarian people belonging to the family of the Finno-Ugric peoples regarding their origin, culture and ancient ethnicity.

For the sake of comparison, let us present the equivalents in some related languages:

Hungarian	Vogul	Ostyak	Finnish	
két	kit	kat	kaksi	2
három	churum	chulem	kolme	3
négy	nyila	nyal	neljä	4
öt	at	wet	viisi	5
hat	chot	chut	kuusi	6
száz	sat	sot	sata	100

From the fact that the correspondence can be detected only among the lower numbers with alterations above *six* one may conclude that in the oldest times accessible to linguistic studies our forefathers regarded *six* as a kind of "key number", or to put it a bit more liberally, they used a (non-positional) number system *based on 6*.

In order to avoid misunderstanding, it must be noted that the notion of a *number system* is used here, unlike today, in a highly general sense. We have to distinguish between *non-positional* and *positional* numeral systems. In the course of the spiritual evolution of mankind, non-positional numeral systems were the first to develop, well before the invention of writing. Thus, for instance, in the non-positional number system to *base 5* the numbers 1 through 5 had separate names of their own, a set of 5 items was considered a unit and had a name of its own, 5 such bigger units made up a higher unit (of 25 items), etc. This system of numbers was based on grouping by 5. Most of the examples are of grouping to the base 5, 10 or 12 (dozen, gross). A special but still not positional system is the one in which e.g 4 or 5 are taken as the basis and twice, three-times, etc. the base are the higher key numbers. However, soon after the invention of writing, examples of positional numeral systems can also be found, first in Babylonia about 2000 BC, where the simple grouping system to the base 10 and a sexagesimal system employing the principle of position were mixed. With the same wedge-shaped symbols denoting different numbers and for want of a zero sign, the Babylonian way of writing was not yet perfect. The oldest records of today's positional numeral system to base 10 (with the zero symbol) have survived from India from the 6th—7th centuries AD. The principle of multiplicative grouping can be discerned in the number systems of several primitive peoples even today.

To this supposition, however, we must immediately add that we have no idea at all how it had developed in the life of the ancient Magyars. One thing is certain: the statement (by BERNÁT MUNKÁCSI) that the use of six as a key number in the life of Finno-Ugric peoples demonstrates the influence of Sumerian-Babylonian culture, is false. No data whatever can be gleaned from the highly sophisticated and complex number system developed in Mesopotamia that might support this view.

Several Hungarian papers attest that later on 7 became the key number. PÁL HUNFALVY is one of the exponents of this view which is presumably supported by such expressions in the Hungarian language as: seven wise men, seven tribal chiefs, seven-headed dragon, seven stars, over seven lands, seven come-seven go (meaning: 'many people go by somewhere'), etc. All this seems to verify that the number seven played a salient role in our tales and legends. What can be declared on the basis of linguistic research is that while the names of the numbers 2, 3, . . ., 6 are Finno-Ugric, the word *hét* ('seven') reveals a direct *Ugrian* (Magyar, Vogul and Ostyak) origin. In the latter three languages, the word *hét* (7) denotes not only a number but a period of time consisting of seven days.

Some studies claim that the number *5* was at some time the key number of our ancestors. There are no acceptable arguments verifying this statement, but its denial is also unjustified. The use of five as a key number can be traced back to counting with fingers, which must have been a general practice among ancient Magyars as well.

Through these phases did the ancient Hungarians arrive at a (non-positional) number system *to base 10*. Our philologists, otherwise slightly differing in their opinions, unanimously agree that after prehistory our forebears already used ten as the basic number. According to ELEMÉR MOÓR [7] it was through barter transactions with the Old Iranians that the Finno-Ugric peoples got acquainted with the number system based on *ten* (but not using the principle of position yet). Only the eastern branch of the Finno-Ugrians including our Ugrian forefathers was in direct contact with the Persians. That this number system developed in this way is also proved by the numeral *tíz* (10). It comes from Old Iranian where it meant a *decas* or 'a unit consisting of ten items'. That is why we count like this: *tizen-egy* (11), *tizen-kettő* (12), etc., meaning 'one the unit ten; two on ten' (cf. English *eleven* 'one left over', *twelve* 'two left over', etc.). In the Indo-Germanic languages in which the original meaning of 10 was a *decas*, the way of counting after 10 is the same (e.g. the Balto-Slavic languages and Albanian).

Linguists have been hard put to explain the etymology of the Hungarian number words for 8 and 9. It is certain that both are compounds in which the components can hardly be recognized today. In the *tz* of the old forms *nyoltz* (8) and *kilentz* (9) most certainly the consonants of *tíz* (10) can be discovered, and originally they might have meant 'two out of ten', 'one out of ten'. *Nyol* is a word of Ugrian origin, meaning 'peak, nose, distant place jutting out'; the Finnish *külki* means 'side'. Thus the origin of *nyolc* (8) and *kilenc* (9) may have been connected with counting on fingers, since most probably these two numbers were indicated with a finger placed on the tip and the side of the nose, respectively. Similarly, the word *tíz* (10) is concealed at the end of *harmintz* (30). The numeral *húsz* (20) is also of Finno-Ugric origin, though its etymology is still unsettled. Some scholars are inclined to link the name to the number of fingers and toes.

The origin of the suffices *van, ven* in the names of some multiples of ten (*negyven, ötven, ..., kilencven,* = 40, 50, ..., 90) is still questionable. According to some linguists these words also used to mean ten, so ancient Magyars were likely to have more than one name for this number (probably at different times, though).

It was through the trade transactions with the old Iranians that the words denoting 100 and 1000 struck root in the Finno-Ugric languages. However, the old Iranian word for 1000 disappeared from old Hungarian, since due to phonological changes it became ambiguous. To replace it, the old Hungarians borrowed the word for 1000 from the Alan people[1] between 200—600 AD. Around that time did we take over the word *szám* 'number' indicative of a high level of abstraction from the language of Bulgar-Turkic peoples with a more advanced culture.

What has been said so far suffices to prove that there are still many questions to be clarified about the origin of our number names as well as the numeral systems

---

[1] The Alans belonged to the North-Iranian Sarmatian ethnic group. They had trade connections with the Magyars.

used by our ancestors and only the findings of thorough interdisciplinary research can provide satisfactory explanation.

Though the magnitude of numbers used by the Magyars before they settled in the Carpathian Basin ranged to 10,000, there are no written records of their names from earlier than the period between the 11th and 15th centuries[2] (when orthography was of course different). The word *cifra* from German *Ziffer* meaning *digit* at the beginning and later *zero* through contraction of meaning, first appeared in some documents in the late 15th century.[3] This confirms the fact also borne out by other data that the Hindu-Arabic notation started to spread in Hungary in the 16th century, since the positional notation of numbers requires the use of zero.[4]

The ample material preserved in museums and archives of "cuneiform numbers" or "shepherd's tallies" offers some insight into the mathematical culture of the remote past. The remains date back to around the 12th century, while some linguistic data put this date even earlier: according to them the use of shepherd's tallies was brought along by the conquering Magyars. This is allegedly proved by the fact that certain words (*rovás* 'notch'; *párja, parjál, pariál* — words referring to paired tallies; see later) were borrowed by the Slavic, Neo-Greek and Romanian languages from Hungarian.

The mere fact, however, that JÓZSEF BUDENZ has detected the stems "rav" and "rov" in the Finnish and Cheremiss languages does not allow one to draw definite conclusions as to the beginnings of the *use of runic numbers* since these words may refer to *runic script* as well. The origin of *runic script* used in the earlier centuries in a limited area (mainly among the Magyars in Bukovina) and that of *runic numerals* used nearly everywhere in the country (even as late as the early 20th century) have to be treated separately as they definitely derive from different sources.

The surviving Hungarian relics of runic numerals and the relevant questions have been studied by many scholars (GYULA SEBESTYÉN, KÁLMÁN SZILY SR., OTTÓ HERMAN and others). Their findings inform us that this simple method of accounting used to be essential in the housekeeping, in the assessment and recording of taxes, in trade and other areas. With the spread of literacy it gradually lost significance but it can be spotted even today, especially in the remotest recesses of the country (e.g. in the Hortobágy plain).

In accordance with the different purposes, Hungarian runic number symbols show a great variety. The numbers are cut into long quadratic (sometimes regular hexagonal or octagonal) wooden sticks (or into ax- or whip-handles, etc.). The stick was

---

[2] Data in more detail: *egy* (1) c. 1200 in *Halotti Beszéd* (Death Sermon, first extant written record in Hungarian); *két, kettő* (2) 1245; *három* (3) 1055; *négy, öt, hat* (4, 5, 6) 1211; *hét* (7) in the Anonymous codex, 12th c.; *nyolc* (8) 15th c.; *kilenc* (9) 1378; *tíz* (10) 1240; *húsz* (20) 15th c.; *száz* (100) 1067; *ezer* (1000) 14th c.

[3] The words *nulla* 'nil' (ONADI, 1693), *semmi* 'nothing' and *zérus* 'zero' (MARÓTHI, 1743) appeared later. It must be noted that the word *semmi* was used as a common noun in some of our earlier codices but it was Maróthi who recommended using it for zero. Etymologically, the words *cifra* 'ornate', *cifráz* 'embellish' go back to the fact that figures were also used for decoration.

[4] It was through the book of LEONARDO PISANO (FIBONACCI, 1180?—1250?), *Liber Abaci* (first draft 1202, second 1228) that the Hindu-Arabic number symbols began to spread in Italy and Germany. The other route to Europe was through the Iberian Peninsula via the teachers of the Moorish school in Toledo who translated several Arabic works into Latin. Finally, the crusades had also largely contributed to the dissemination of the mathematical achievements of the East.

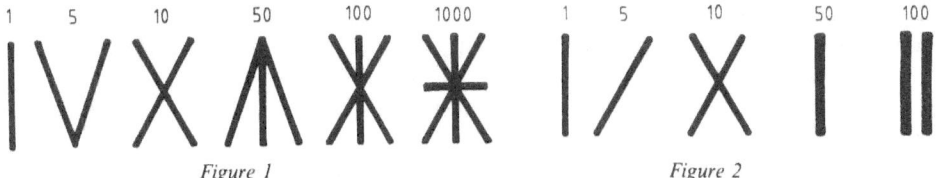

generally cut out of the piece of wood parallel to the network, and for convenience's sake, the marks were engraved perpendicularly or obliquely to the grain. In the course of time, relatively simple wedge-shaped symbols developed, avoiding curves and even marks at right angle (because of the wood grain). The signs formed in the above manner display some kinship with the Roman numerals, and similarly to the latter, are presumably the ideograms of raised fingers. The form of numerals widely varies by age and region. *Figure 1* shows the most commonly used cuneiform symbols, while *Figure 2* shows the numbers the shepherds in Hortobágy are still familiar with. According to GYULA SEBESTYÉN [9] the most common runic number symbols can be traced back to the remote Etruscan script and may have come to Hungary by

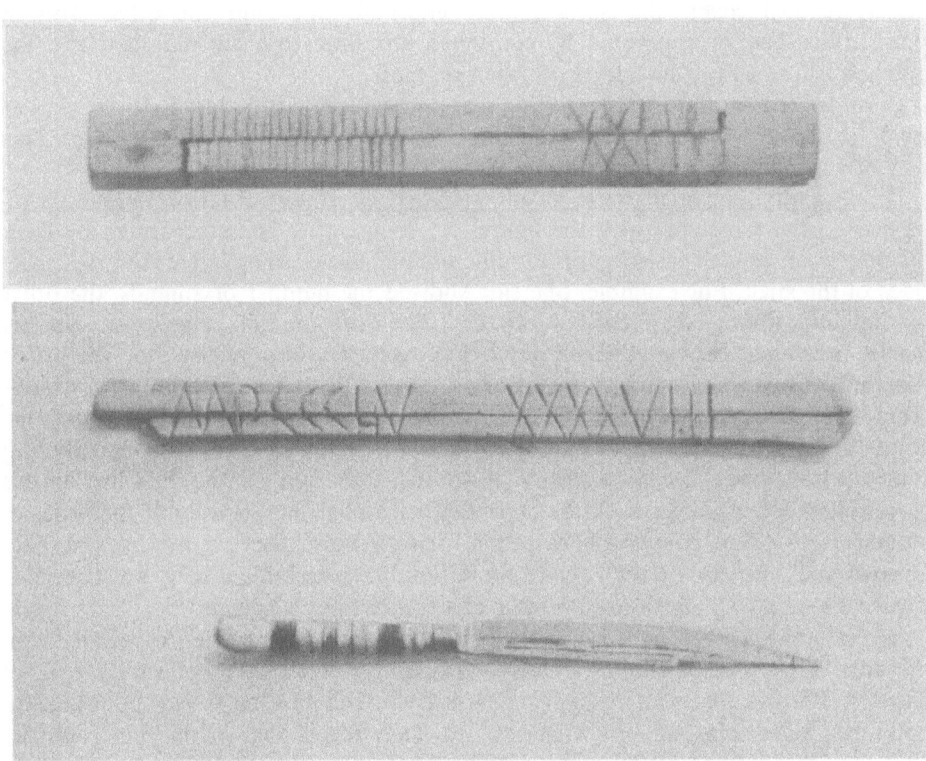

1.a), 1.b) Objects with running number symbols in the Déri Museum, Debrecen

Roman mediation. In another opinion (of Dezső Csallány) both runic script and cuneiform number signs are of *Avar* origin. However, only further studies can give a definitive answer to this question.

On the basis of surviving records, counting with runic numbers can be classified in two aspects.

1. a) *Simple runes:* a shepherd, herdsman or swineherd engraved just as many marks on the squared handle of his whip as many animals he had to tend.

b) *Paired runes:* the marks were engraved, at the same time on two squared sticks of equal size (e.g. the way the top of a wooden pencil box is slipped into place) and the writing was not intelligible unless the two parts were brought into the original position. One part belonged to the owner, the other to the shepherd *(Figure 3)*.

Figure 3

2. a) *Running count:* the notches were engraved one after the other without summation. The writing could be continued any time with any number, and the notched marks had to be added up when read off.

b) In *additive* writing, after reaching a round sum (ten, fifty, hundred) the figures were "closed up" indicating the sum total, followed again by small numbers. This way of writing made addition faster.

If sufficiently long sticks were given, the livestock of several farmers could be recorded on the same side with the figures for each owner being separated by clear dividing lines. In such cases figures for different animals or things were cut in different sides of the stick. For example, one side showed the number of animals, the other the liabilities (flour, salt, bread, money, etc.), the third the milk, cheese, etc. Special marks, or brands, indicated which farmer's property the data referred to. The brand was the distinguishing mark of the farmer, a kind of identification number or monogram, sometimes formed of old letters. At times, some agreed order of listing the animals was observed, and the categories were separated by a dot or a short line. If someone had none of a certain animal, it was also indicated by a dot or a line. In this way, quite a lot of things could be expressed with a few marks only. If the order of animals is, say, bull, cow and heifer, then *Figure 4* reads: Steve Smith has one bull, 16 cows and 3 heifers. *Figure 5* reads: Steve Smith has no bulls, 2 cows and 8 heifers. *Figure 6* says: Steve Smith has no bulls or cows but has 9 heifers.

Several written records are available on how numbers were written, especially from the area of the Principality of Transylvania as a result of the collecting work of Sándor Tóth, a mathematician from Kolozsvár. Indelible signs can be found on buildings, bells, coins, stamps, charters, etc. They reveal that in the first centuries after the Magyars had settled in Pannonia only Roman numerals were used, and as our earliest deeds attest, our forefathers tried to avoid them and write their names

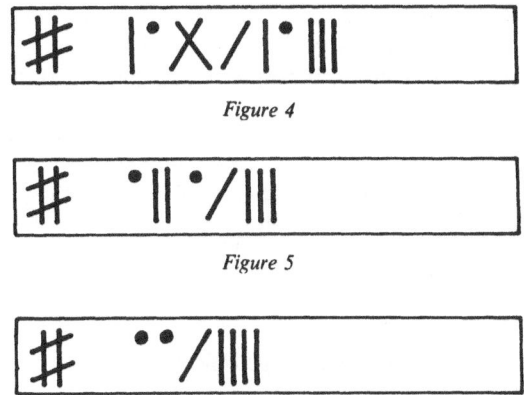

Figure 4

Figure 5

Figure 6

instead. Just like all over Europe, they were reluctant to adopt the Hindu-Arabic notation. In our country, the 15th century was the period of transition. On the inscriptions and documents dating from that time the Roman numerals were often mixed with Arabic signs *(Figure 7)*.[5] At any rate, dates written in the Hindu-Arabic way have been preserved from the beginning of the 15th century. The earliest such data in SÁNDOR TÓTH's collection go back to 1407, as he communicated to me. However, it was a long time before the Hindu-Arabic notation struck root and it also took long for the numerals to develop into their present form, a process accelerated by the invention of printing. *Figure 8* shows the Hindu-Arabic numerals as they first appeared in Hungary, while *Figure 9* illustrates one of the earliest such records based on a book of accounts from Selmecbánya from 1436.

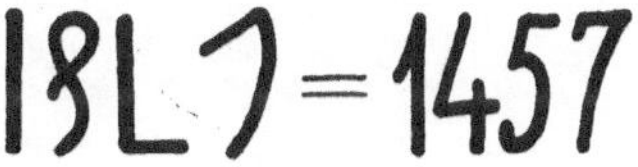

$$I3L7 = 1457$$

Figure 7

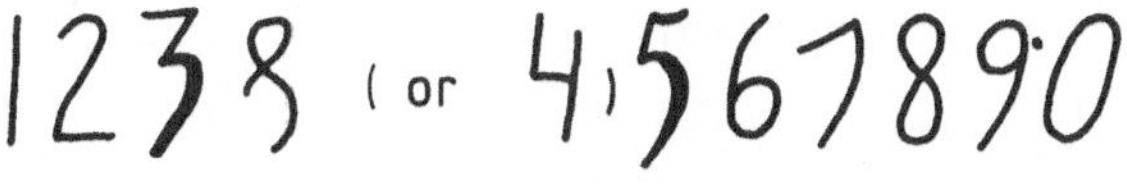

$$1238 \quad (\text{or} \quad 4)567890$$

Figure 8

$$1236$$

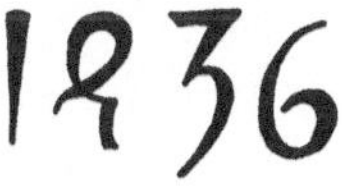

Figure 9

---

[5] According to PÁL LŐVEI, the oldest such written record of "mixed" numerals in Hungary can be found on a gravestone in the National Museum (inventory number: 1878.95.4). It indicates the date of death (1346) with Roman (first three digits) and Hindu-Arabic (last digit) numerals. *Matematikai Lapok*, 33, 1982—1986, pp. 25—26. (In Hung.)

For the sake of completeness, it must be mentioned that epitaphs with Hebrew numerals (the oldest from 1278) and registry entries in Cyrillic script were also found in Hungary, indicative of our cultural relations and historical development.

The book of SÁNDOR T. TÓTH and ÁRPÁD SZABÓ, *The Framework of Hungarian Mathematical Culture*, Gondolat, Budapest, 1988; in Hung.) contains the bibliographic data of manuscripts and incunabula that can be found in archives and libraries of Hungary and the neighbouring countries. They date from the 11th—17th centuries. Although their subject-matter is not explicitly mathematical, one can draw inferences from them as to the current level of mathematics in the given age. This volume of high scholarly merit largely improves our knowledge of the period studied.

Bibliography

[1] *An ethnography of the Magyars.* Vol. II. The material culture of the Magyars. Budapest, no date (193?). (In Hung.)
[2] *Is runic script still used by Hungarians?* — Report submitted to a committee delegated by the 1st Department of the Hungarian Academy of Sciences for deliberation. Budapest, 1903. (In Hung.)
[3] BÁRCZI, GÉZA: *Hungarian etymological dictionary.* Budapest, 1941. (In Hung.)
[4] — *The origin of the Hungarian word stock.* (2nd enlarged ed.) Budapest, 1958. (In Hung.)
[5] LÜKŐ, GÁBOR: Shepherd's runic numerals in Debrecen. *A Debreceni Déri Múzeum Évkönyve*, 1938, pp. 83—89. (In Hung.)
[6] *A history of Hungarian culture.* (Ed.: DOMANOVSZKY, SÁNDOR) Vol. I. Prehistoric and mediaeval culture. Budapest, 1939. (In Hung.)
[7] MOÓR, ELEMÉR: Die Ausbildung des ungarischen Konsonantismus. *Acta Linguistica Acad. Sci. Hung.*, 1952, pp. 1—92, 355—463. Reprint, 1953, Akadémiai Kiadó, Budapest.
[8] ORBÁN, GÁBOR: *Numerals in Finno-Ugric languages.* Bratislava, 1932. (In Hung.)
[9] SEBESTYÉN, GYULA: *Authentic records of the Hungarian runic script.* Budapest, 1915. (In Hung.)

2. Written relics of the beginnings of mathematical culture in Hungary

There is a scarcity of data on the mathematical culture of Hungarians in the first centuries after the Conquest, and the relevant conclusions of the universal history of culture for the period in question can only be applied to the specific Hungarian conditions with great caution. There is no answer to the question as to what mathematical knowledge was needed in everyday life and who were in possession of that knowledge. MASTER GYÖRGY of Hungary wrote the following in the late 15th century: "The fruit of numbers is not only advantageous and useful, but also necessary for everyone, regardless of their post or occupation. First of all for the greatest and most distinguished of the men: kings, princes, magnates and all the dignitaries of the state, as well as for the men who are engaged in the major military or martial arts... Then for the most outstanding scholars of the most sacred theology and the holy canon, and for all prominent people who have devoted themselves to studying pure philosophy; for all the clergy, prelates, monks, priests; and finally, for all respectable merchants and craftsmen."[1]

However, the above-quoted words of MASTER GYÖRGY referred to Western countries of a more advanced culture — as he wrote his work there — and expressed the higher cultural standards of the age of Humanism. In the life of our conquering forebears mathematics can hardly have been such a "prime necessity". Determined by their social class or layer, the Hungarian youths of that period were trained in different ways: the offspring of the royal house were usually trained by priests in the Hungarian or foreign courts, acquiring the knowledge considered necessary to rule a country, arithmetic probably having a very little part in it; for the nobility strategics was more important than arithmetic or Latin; the already existing thin urban layer of merchants and artisans acquired the knowledge they needed within guilds or later in town schools; and the poor living in serfdom had no chance at all to attain any knowledge to improve their miserable lives.

Yet our predecessors must have had some skills in counting, as it was necessary in everyday life. The calculation of the "blood fine" (or ransom to be paid by the murderer in steers or money, a general form of compensation for homicide in the tribal

---

[1] See HÁRS, JÁNOS: *How did Master György of Hungary count in 1499?* Budapest, 1936, pp. 10—11. (In Hung.)

system) also required some arithmetic, for example. The people who settled in Hungary after the Conquest had to pay landrent; the merchants and their customers at royal market places had to pay tax; so a considerable part of the population had to know something about the simple arithmetic operations of selling and buying.

Following the Magyars' conquest, schools were first organized and headed by different religious orders and priests. Schools were founded near the bigger *chapters* as early as the reign of the first Hungarian king ISTVÁN I (reigned 1000—1038), and from the 12th century on religious order, also engaged in teaching and monastic schools started to spread all over the country. The *Benedictine* order was the first to settle in our country, followed by the *Cistercian, Premonstratensian, Augustinian, Franciscan* and *Dominican* orders. In constant contact with the West, they mediated scholasticism to our country. For want of suitable forms of education, many priests and monks had to attend universities abroad. In the late 12th and early 13th centuries Paris University had the greatest appeal to most of the students, and later the ones in Padua and Bologna. We have no reason to doubt that several clergymen in Hungary were highly trained in those days, but the data at our disposal are silent about their qualifications in mathematics.

On grounds of our connections with the West and the word "Dacia" in his name, some scholars have concluded that PETRUS DE DACIA, a Dominican monk featuring in the universal history of mathematics, originated from Hungary. According to LÁSZLÓ MAKKAI, this view is apparently false since the name *Dacia* was not used for Transylvania again before the 15th century Humanism, and in spite of incomplete biographical data we know that Petrus de Dacia was the rector of Paris University as early as 1326. So he must have been of *Danish* extraction. His only book on mathematics that survived was published by CURTZE in Copenhagen (*Petri Philomeni de Dacia in Algorismum vulgarem Johannis de Sacrobosco commentarius.* 1897). In it PETRUS DE DACIA systematically analyzes and comments on JOHANNES DE SACROBOSCO's book *Tractatus de arte numerandi* (see later).

It is impossible to distinguish between primary, secondary and higher education in this period. No doubt the parish and monastic schools and the town schools to be discussed later provided primary education, the collegiate schools mainly in episcopal sees were secondary schools, some of them giving higher education as well. It is certain that church schools following the practice of scholasticism, chose their material from the subjects of the "seven liberal arts", teaching the subjects of *trivium* (grammar, rhetoric, dialectic) in parish and monastic schools, and those of *quadrivium* (arithmetic, geometry, astronomy, music) in the collegiate schools.

The compilation of calendars, a duty of the priests, required some knowledge of arithmetic. On the other hand, the fascinating ornaments in churches and chapters speak of considerable geometrical knowledge. It was undoubtedly necessary for the conquering Magyars to have some arithmetic and geometric knowledge to construct well-defendable fortresses, to dig ditches around them and erect masterly fortifications from the dug-out earth on wooden skeletons. The wealth of finds in architecture and arts also testify to their knowledge in geometry. It is also proved that in the monasteries scribes were busy copying codices. As the catalogues of some libraries reveal, the books of BOËTHIUS, CASSIODORUS, CAPELLA and SACROBOSCO significant for mathematics also reached Hungary.

Born in 480?, the first and one of the most outstanding Latin scholars of the times after Rome's fall, BOËTHIUS was sentenced to death by Emperor THEODORIC in 524. Especially two books on mathematics are associated with his name, one on arithmetic (*Institutiones arithmeticae*, 500?) and a geometry. The latter, however, was proved by research to be the work of an 11th century anonymous author. His arithmetic, the adaptation of a work by NICOMACHUS (around 150 BC) was not very successful either. It was again revised by a professor of Paris University JEAN DE MEURS (JOHANNES DE MURIS 1310?—1360?), who improved it a lot so that it was used even in the 16th century. After the invention of printing, his work was published about 30 times between 1488 and 1570. The selection of the subject-matter for the *quadrivium* also comes from BOËTHIUS, highlighting arithmetic, geometry, astronomy and music for a long time to come. That was his main influence on the development of mathematics. CASSIODORUS FLAVIUS (480?—570?), a high-ranking official during the reign of the Goths in Italy, was a prolific writer. Retiring into a convent in the prime of his life, he wrote his two-volume encyclopaedia *Institutiones divinarum et humanarum litterarum*, used as a textbook in the Middle Ages. CAPELLA MARTIANUS lived in the 5th century in Carthage. Following PLINY, VARRO and other Latin authors he wrote a superficial and high-flown encyclopaedia in nine volumes in the form of a novel, covering the material of the seven liberal arts (*De nuptiis Philologiae et Mercurii*. Latest edition: Leipzig, 1866). In mediaeval monastic schools this encyclopaedia was used as a guide. The arithmetic of JOHANNES DE SACROBOSCO (John of Holywood, around the mid-13th century), *Tractatus de arte numerandi*, acquainted the reader with the Hindu-Arabic computation, greatly contributing to its spread in Europa.

We know very little about teaching methods in ecclesiastical schools. It can be conjectured from later arithmetics that religious and mysthic elements were mixed even in mathematics. Most probably the teachers of the time dictated the material to the pupils who tried to memorize the rules of arithmetic word for word from their notes. With the passing of time the lectures became more polished and extended as required, and our first arithmetics published after the invention of printing were the products of this long process and the influence of foreign examples. Unfortunately, almost all the lecture notes taken down by pupils were lost. However, the two Hungarian sources that are mentioned below provide us with some information.

One of them is a codex now at the library of the Cathedral of Esztergom, *Tractatus in Cantica Canticorum* probably from the first half of the 12th century. The other manuscript, also preserved there, dates from later (1490): it contains the lectures of a baccalaureate teacher, JÁNOS KISVÁRDAI taken down by SÁNDOR SZALKAI, later archbishop of Esztergom, when a pupil in Sárospatak.

Thanks to detailed analyses [2, 9] we are familiar with the contents of both manuscripts: they seem to cover the material of the *trivium*, though not exactly in the old sense. Adjusting to the changing requirements of different times, the subject-matter of the *trivium* kept changing and expanding. The above-cited manuscripts, for example, also include a detailed account of "computus", which is worthy of noting from our point of view.

In order to understand the role of "computus", we have to know that the first day of Easter — as decided by the First Council of Nicaea (325 AD) is the Sunday following the first full moon after the vernal equinox. The importance of this day in the liturgical life of the Church is obvious, since all the other holidays adjust to the date of Easter. At the same time, fixing the date of Easter required a precise astronomic knowledge on solar and lunar motions as well as elementary arithmetical skills. These were indispensable to compute which calendar day (of the *Julian* year) coincided with the full moon determined by the Moon's synodic cycle (c. 29 days and 7 hours). In mathematical terms, the necessary computations were confined to simple arithmetic operations (addition, subtraction and multiplication). The so-called "perpetual calendars" known even today were constructed by competent persons in possession of the necessary data. Some knowledge of arithmetic was essential for their use, the instructions for which were usually given in short poems in Latin.

"Computus[2] as a subject was soon included in school instruction since all clergymen had to be well-versed in the use of church calendars, in the dating of documents and in recording historical data. In addition, they had to be able to perform the computations included in the textbooks of computus by themselves, which required elementary mathematical skills. Therefore, computus studies at that level were not begun before the end of elementary schooling... Thus the mediaeval way of computing the date of Easter can be considered to be the forerunner of teaching mathematics." ([9], p. 387).

While the collegiate, monastic and parochial schools mainly prepared the pupils for priesthood, the *town schools* gradually set up from the 13th century were mostly attended by children from the middle classes. With the growing weight of these classes, the significance of these schools kept increasing, yet they never reached the level of church schools. Some arithmetic and geometry was also taught in them, a fact indirectly verified by the highly developed craftsmanship of goldsmiths, carpet weavers, etc. The only concrete data, however, that has some down to us says that in order to improve the level of teaching arithmetic, "the town obtained a blackboard for the school in Bártfa in 1509 for half a florin, one in Kassa in 1533 for 60 denarius and one in Késmárk in 1540 for 32 denarius." ([3], p. 40).

Our ancestors measured distance even in the pre-Conquest times on the basis of the divisions of the foot and arm. The town of Sopron had a land register as early as 1379, and from the reign of the ÁRPÁD dynasty some written records on surveying estate boundaries have survived. At first, large distances were measured in steps and by the distance a horse covered in one day, but before long different measuring instruments (e.g. measuring chain, measuring rope or stick) had also appeared. But we have information on more sophisticated "engineering" tasks (*mérnök*, the Hungarian word for engineer literally means 'measurer'), and from the time between 1337 and 1725 we have several data on "rodmasters" (Hung. *rudasmester*[3]). The data divulge that the "rodmasters", some of whom must have attained their knowledge abroad, were mostly in the employment of landlors. Also acknowledged financially, they had to carry out jobs in architecture (designing and building), geodesy and agronomy. Later

---

[2] Computus tables are also included in two codices from the age of the ÁRPÁD dynasty: The *Pray* Codex of Benedictine origin (late 12th or early 13th c.) and the Franciscan codex from Németújvár (1470).

[3] LÁSZLÓ BENDEFY gave an account of the life and work of a 16th century Hungarian "rodmaster", BENEDEK SZOMBATHELYI [6]. His study reveals that the Hungarian "engineers" were highly appreciated abroad in old times, too. PÉTER LOSSAI, for example, a surveyor born in Magyaróvár, worked in Lithuania in the early 16th century. Except for the first 57 chapters that were lost, his manuscript on land survey (*De geometricis mensurationibus...*) has survived. He is supposed to have written it as a student of Bologna University in 1498. A fine facsimile edition of the codex with its Latin text, Hungarian translation and 5 pages of notes, was published by the Geodetic and Cartographic Company of Pécs in 1970, edited by ZOLTÁN PORONYI and ALAJOS FLECK.

on they had to be competent in map-making and water engineering as well. "Rodmasters" were slowly replaced by land surveyors.

In recent years our knowledge of currencies and different weights and measures used in Hungary has considerably increased.[4] It is obvious that due to the feudal division of the country there was complete chaos in this respect. Different regions, towns and landlords used any measures they had arbitrarily chosen. In order to stabilize the market, the issue of standardization was brought up at the Diet as early as 1405, which decreed that the units valid in Óbuda should be binding throughout the entire kingdom. The ordinance was, however, completely ignored. Later those concerned tried to have the units that they used officially sanctioned. The second such act was passed by the Diet of 1717—1722, declaring the units used in Pozsony compulsory for the whole country. However, this act failed again and the issue was not finally settled before the second half of the 19th century when the urgent need for uniform trade and market terms made it unavoidable.

Our data on the first Hungarian "universities" are rather insufficient, too. The word "university" must be put in quotes because we can hardly give an exact answer to the question whether the institutions below really represented "studium generale" in the mediaeval meaning of the word and had the right to confer the title "ius ubique docendi".

There is no denying that our country was quick enough to follow the West in setting up institutions of higher education. Around 1250 King Béla IV raised the collegiate school in Veszprém to a higher rank (it was destroyed by the troops of the oligarch Péter Csák in 1276), though in Remig Békefi's opinion it should still be ranked among collegiate schools. The institution in Pécs established by King Louis the Great in 1367 with the assent of the pope was nearer to the idea of a university (we do not know exactly when it ceased to exist). According to some fairly reliable data from that period, medical and natural sciences were also taught there. A scion of the Anjou house in close contact with Italy, Louis the Great was in the position to invite Italian scholars to this university, which must have favourably affected mathematics, Italy being the European centre of science at the time. That might explain why the first printed Hungarian arithmetics dwelt at length on the "Italian division" or the "Italian rule".[5] However, this is only one assumption as these terms, widely used all over Europe, could have got into our terminology in an indirect way as well.

Around 1389 King Sigismund of Luxemburg asked for, and was granted, permission by the pope to promote the collegiate school in *Óbuda* to the rank of university. The date when it closed down is still uncertain. We know from relevant sources that philosophy had prominence in the institute. This in turn suggests that mathematics

---

[4] In addition to Emma Léderer's pioneering study [8], several essays are devoted to this question in *Chapters from the history of Hungarian measures* [7]. Most relevant to this point are: *Some data on the mediaeval history of Hungarian measures* (in Hung.), by István N. Kiss (pp. 5—23) and *Mediaeval Hungarian measures of length and area* (in Hung.), by László Bendefy (pp. 45—97). All these papers underline the diversity of measures, which was also characteristic of currencies. Let it suffice to refer to the immense conversion table for currencies at the end of Maróthi's arithmetic (see later).

[5] A way of notating the division of numbers different from today's. See, e.g. Maróthi, György: *Arithmetica...* Debrecen, 1782, pp. 84—91.

was also taught there, since at the time of scholasticism and humanism that subject was included in the curriculum of philosophy at many places. In 1454 King LADISLAUS V invited PEURBACH[6] to Buda as court astrologer. PEURBACH, one of the most famous mathematicians and astronomers of his time, wrote his work on astronomy entitled *Canones pro compositione et usu gnomonis...* in Hungary.

In Hungary, the influence of the Renaissance on the development of science and arts reached its climax during the reign of King MATTHIAS CORVINUS. In 1467, he founded a university in *Pozsony* with four faculties, mainly to offset the influence of similar institutions in Krakow and Vienna. Although the "Academia Istropolitana" existed only for a short time (probably until 1491), it played an important part in our mathematical life as *Regiomontanus*, one of the greatest astronomers and mathematicians of the age, was among its guest scholars.

Regiomontanus (JOHANNES MÜLLER, 1436—1476) was born to a miller's family near Königsberg. At a young age he went to Vienna where he studied under PEURBACH before he became his associate. He completed PEURBACH's translation of PTOLEMY's Almagest into Latin. Having left Vienna, he lived in Italy for some time and then in 1467 he came to Hungary. During his four-year stay he lectured on astronomy at Pozsony University. In the meantime he wrote his work on astronomy, "Ludus Pannoniensis quem alias vocare libuit tabulam directionum". Unfortunately this work of his has not survived and we have no knowledge of its exact contents either. From Hungary he left for Nurenberg where he established an observatory and a printing press. In 1475 he was called to Rome by pope SIXTUS IV for the reformation of the calendar, where he met his death at an early age. From his correspondence with Italian and German scholars we know that in addition to the geometric solutions of extreme-value problems, he also dealt with questions of number theory. However, his historic contribution to the development of mathematics was the introduction of entire surds and the definition of operations with them, as a result of which it soon became possible to raise and examine the question of solution with radicals for broader and broader classes of equations. As some sources attest, Regiomontanus planned to publish a series of classical Greek mathematical works in his press in Nurenberg but this plan was aborted by his early death. For the computation of distances he introduced the tangent function and compiled the first tangent table. His best-known works on mathematics include: *De doctrina triangulorum* (Venice, 1463), *De quadratura circuli* (Venice, 1463), *De triangulis omnimodis* (Nurenberg, 1533), *Tabulae directionum profectionumque in nativitatibus multum utiles* (Venice, 1585).

In 1473, MATTHIAS established a printing press in Buda preceding several Western countries (England, Austria, Spain). It was also during his reign that the scientific society "Litteraria Sodalitas Danubiana" was set up; it later moved to Vienna (1497) and became the kernel of the would-be scientific academy there.

The above listed scientific institutions and the first universities did not take root in Hungary, and as to their influence, they cannot be compared to the noted old universities in the West (mainly the ones in Paris and Bologna) or even to those in our neighbourhood (Prague, Vienna or Krakow). Apart from the tempestuous history of the 16th—17th centuries, that is the reason why Hungarians failed to contribute original results to mathematical.

---

[6] GEORG PEURBACH (Burbach, Peyerbach, 1423—1461) was born in Burbach near Linz. Having completed his studies in Vienna with a master's degree, he toured Italy and returned to Vienna in 1453. A year later King LADISLAUS V invited him to Buda where he soon wrote his above-mentioned book published in Nurenberg in 1516. The book also includes a sine table for astronomical calculations. Even more famous is his other book on astronomy, *Tractatus Georgii Burbachii super Propositiones Ptolemaei de sinubus et chordis.*

Bibliography

[1] *A history of Hungary*, Vols 1—2. Akadémiai Kiadó, Budapest, 1984. (In Hung.)

[2] BARTHA, DÉNES: *Archbishop Szalkai's musical notes from the time he was a monastic school pupil (1490)*. Budapest, 1934. (In Hung.)

[3] BÉKEFI, REMIG: *A history of public education in Hungary up to 1540*. Budapest, 1906. (In Hung.)

[4] — *The University of Pécs*. Budapest, 1909. (In Hung.)

[5] — *A history of collegiate schools in Hungary up to 1540*. Budapest, 1910. (In Hung.)

[6] BENDEFY, LÁSZLÓ: *The rodmaster Benedek Szombathelyi. Some data to the history of land survey in Hungary*. Tankönyvkiadó, Budapest, 1958. (In Hung.)

[7] *Chapters from the history of Hungarian measures* (Collection of Studies). Ed.: MAKKAI, LÁSZLÓ. Közgazdasági és Jogi Kiadó, Budapest, 1959. (In Hung.)

[8] LÉDERER, EMMA: Old Hungarian capacity measures. *Századok*, 1923, Vol. 57, pp. 123—157, 305—326. (In Hung.)

[9] MÉSZÁROS, ISTVÁN: A Hungarian schoolbook from the first half of the 12th century. *Magyar Könyvszemle*, 1961, Vol. 77, pp. 371—398. (In Hung.)

II. The age of elementary arithmetics

3. Education.
General level of learning

The 16th — and even more the 17th — century is regarded as the time of great discoveries launching a new period in the universal history of mathematics. Achievements in algebra (VIÈTE, FERRO, TARTAGLIA, CARDANO), the discovery of logarithm which precipitated numerical calculations (NAPIER, BRIGGS), deeper number theory (FERMAT), analytic geometry (DESCARTES, FERMAT), and first and foremost, the infinitesimal calculus (NEWTON, LEIBNIZ) had indeed lent salience to these two centuries in the history of mathematics, and of scientific thought in general. Scientific research often provided its practicioners with a full-time occupation by which they could make a living in the Western countries then well on the way to capitalism. Scientific life needed organization: 1560 saw the establishment of a scientific academy in *Naples*, followed by Rome's "Accademia dei Lincei" (1603), *Florence*'s "Accademia del Cimento" (1657), then the *British* Royal Society (1662) and the *French* academy (1666). It is to their credit, too, that several mathematicians could devote their whole lives to scientific investigations.

In view of that, Hungary's performance seems very poor regarding *original* contributions to mathematics. One cannot find a single person in the period who could be called "mathematician" in the true sense of the word, nor can one adduce a single original result in Hungarian mathematics.

Nevertheless, it would be rash to conclude that the Hungarian intellectuals had a mathematical knowledge below the *average* level in Europe. The lack of independent research merely proves that under the less favourable conditions that prevailed in Hungary the need for independent investigations had not arisen yet. Obviously, Hungary's mathematical culture did not grow out of nothing; only, the preparatory period preceding independent research was drawn out here.

One can retrace the slow but gradual progress in the extant printed arithmetics, the available mathematical manuscripts, the dissertations defended at foreign universities, and in the changes in public education. The major turn in the latter came with the Reformation: with a feel for utilizing their limited resources, Protestant towns kept vying with each other in establishing schools of humanistic learning at various levels. That was the time when the renowned colleges of *Sárospatak* (1531), and *Debrecen* (1538) were founded among others. To restore the equilibrium tilted in favour of the Reformed Churches, the archbishop of Esztergom MIKLÓS OLÁH called in the Jesuitic order, mainly to head the seminary at Nagyszombat that he had turned into a central theological school in 1558. Later (1635) PÉTER PÁZMÁNY, archbishop of Esztergom entrusted the new university of Nagyszombat to the care of the Jesuits. In 1657 a new university was set up in Kassa on BENEDEK KISDY's foundation, but it lived a short life and remained insignificant for mathematics.

Inner stability was relatively higher in far more autonomous Transylvania. Transylvania also succeeded in keeping up a certain level of material welfare, and in culture this was the region that offered opportunities for progress for a long time. Especially under the reign of the Princes GÁBOR BETHLEN and GYÖRGY RÁKÓCZI I did Transylvania's cultural and economic life experience prosperity. Outstanding representatives of these few decades of middle-class advancement included JÁNOS APÁCZAI CSERE and FERENC PÁPAI PÁRIZ.

Obviously, an inquiry into the history of mathematical culture must guide one's eye to the institutions of higher education. But we must not judge their work too severely, as the primary, almost exclusive, duty of the professors was the dissemination of existing knowledge, and although there were teachers who excelled in systematizing or methodically elaborating their special subjects, the betterment of science including research into new topics was outside their liabilities.

Taking a closer look at mathematics at the university of Nagyszombat (and the Jesuitic schools in general) and at Protestant colleges, one is immediately made aware of the differences between them both in character and standards. The central objective of the Jesuitic order was to reinforce the Catholic faith in the teeth of rapidly spreading Reformation. This aim was also expressed in the curricula of the university: priority was given to theological and philosophical subjects, while exact sciences and practical skills were of secondary importance. Christian obedience, the unconditional reverence of authority was to mould the youths for the service of the state and the church. Formalism was typical of teaching the neglected sciences. Indeed, until the late 18th century there was no high-level *university* teaching of mathematics in Hungary. Mathematics was merely subsidiary material within the comprehensive subject of philosophy in the Jesuitic curriculum. This position assigned to it by scholasticism could not be easily changed abroad either, because analysis had a penchant for philosophical speculations (on the concepts of zero, the infinite, etc.) in that age. The first Jesuitic curriculum, *Ratio Studiorum*, introduced in 1599, which later became binding for the university of Nagyszombat as well, spelt out that the students of philosophy get acquainted with the introductory parts of EUCLID's geometry and the elementary theorems of plane trigonometry, followed by some lectures on spheric trigonometry and later, on mathematical geography. The philosophical questions of the examination for the bachelor's degree included a few mathematical problems — but only from the early 18th century on. These problems were taken from the areas of arithmetic, algebra or geometry. Some practical problems were also put to the candidates (such as determination of geographic definition of position, calculation of height, certain questions of commercial arithmetic, etc.).

It greatly hindered the efficiency of Jesuitic schools that the members of the order were only allowed to stay at one place for a limited period of time, so they regarded their current posts as temporary, since they were liable to being transferred to another place any time. Those, however, who taught mathematics were in a somewhat more favourable position. Firstly, they had been trained in mathematics at some major university abroad (Vienna, Rome, Graz, Padua, Prague); secondly, their transfer was always seriously considered in view of the difficulties of teaching mathematics. As a result, mathematicians were among the highest-trained university personnel, although mathematics had a relatively subordinate role. And although the common use of the Latin and German languages would have enabled them to apply books published abroad, first of all in Vienna, the Hungarian members of the order strove to supply the schools with textbooks of their own authorship. The first book in this line was HENRIK BERZEVITZY's *Practica Arithmetica*, a multiplication table of which at least 12 editions are extant from the period between 1668 and 1803 (Kolozsvár, Kassa, Pozsony, etc.). The language of the editions varies, ranging from Hungarian, German and Latin to Slovak. It is certain that the teachers and students at Nagyszombat did use the book. It was originally compiled by JULIUS CAESAR OF PADUA (probably a Jesuit of Italian origin); BERZEVITZY wrote and partly translated the few pages added to the multiplication table. The little book contains the products of natural numbers, with an ever richer content in the later editions. The "miniature" edition of 1803, for example, proceeds up to 100 times 1000. The addenda list the weights used in Hungary, the weight units of gold and silver objects, the proportions of the two metals in their alloys. Also included are coinages used in Hungary and the surrounding countries.

The first multiplication tables were soon followed by a 34-page trigonometric table entitled *Canon sinuum, tangentium et secantium ad partes Radii 100.000* (Nagyszombat, 1694). It was most probably compiled by the Jesuitic teachers JÁNOS DUBOVSZKY and FERENC SZÉKELY. The first 15 pages of the table contain the five-place values of the sine, tangent and secant functions for every 6″ of arc. By way of explaining the title, the authors note that the radius of the circle generating the functions was taken as 10^5 in order to avoid the cumbersome calculations with vulgar fractions. Those days the decimal fractions were not in common use yet. In 1737 MIKLÓS JÁNOSI published in Kolozsvár the British Jesuit JACOB GOODEN's (1670—1730) work, *Trigonometria plana et sphaerica*, followed in the next year by the first algebra in Hungary, MIHÁLY LIPSICZ's *Algebra, seu analysis speciosa* (Kassa, 1738).

The Protestant colleges of Debrecen and Sárospatak, the extensive network of subsidiary institutions belonging to them, as well as the six colleges founded in this period (Nagyenyed, Kolozsvár, Marosvásárhely, Székelyudvarhely, Szászváros, Zilah) played a significant role in the history of culture in Hungary. Earlier the young Protestants had favoured the humanistic universities of Krakow and Wittenberg, but from the mid-17th century their interest shifted to Holland and Switzerland. There they embraced the popular Cartesian philosophy and the views of the rationalist philosophers (DESCARTES, SPINOZA, LEIBNIZ) on the universal character of mathematical methods. Returning home, they became the advocates of rationalism and other early tenets of the Enlightenment. The Protestant schools did not prevent their teachers from adducing rationalistic arguments in support of religious doctrine. As a result, the Reformation had democratized the cultural life.

It is easy to answer the question what had attracted the Protestant youths to the universities of the Low Countries, for example. Sweeping progress had been made there over the 17th century in middle-class development, the bourgeois revolution had been fought, a popular-national uprising had cast off the Spanish yoke. These events had laid the groundwork for the development of the doctrines that served as starting point for the philosophy of rationalism. It is only too obvious that middle-class development which was coming to a head in Hungary, and especially in Debrecen, turned to philosophical trends that included in their programme the abolition of the prerogatives of feudal society. Of course, no sooner had the exponents of these tenets started voicing more radical views of social transformation or had their theses confronted with the doctrines of the church than the official protest was launched. At any rate, during this period the old teaching material got imbued with more modern ideas at the colleges of Sárospatak and Debrecen, and through their mediation, at different Protestant schools. It has to be added that at that time Debrecen, for instance, was one of the country's most advanced towns in industry and commerce, and had a well-equipped printing press as well. These and similar reasons account for the fact that the instruction of mathematics was more *life-like* and *practice-orientated* in the *Protestant* colleges as well as at the *Unitarian* college of Kolozsvár and the *Lutheran* colleges in Northern Hungary. The authors of textbooks on mathematics adapted to the needs of everyday life were mostly teachers of these schools.

The social basis of Protestant colleges included the burghers of towns and some strata of poorer nobility who tried to advance their lot through intellectual work (becoming town or county clerks, estate bailiffs, etc.) and needed some experience in solving everyday mathematical problems.

Bibliography

[1] Békefi, Remig: *A history of public education in Hungary up to 1540.* Budapest, 1906. (In Hung.)
[2] Kiss, Áron: *Some data on the history of teaching and education in Hungary.* Budapest, 1874. (In Hung.)
[3] *A history of the College of Debrecen.* (Ed. in cooperation with Nagy, Sándor and Zsigmond, Ferenc, by Révész, Imre.) Debrecen, 1940. (In Hung.)
[4] Szentpétery, Imre: *A history of the Faculty of Arts, 1635—1935.* Budapest, 1935. (In Hung.)

4. Mathematical culture in Hungary as reflected by the first arithmetics and geometries

Two and a half centuries passed between the publication of the arithmetic by MASTER GYÖRGY OF HUNGARY[1] (1499) and that of GYÖRGY MARÓTHI (1743). As far as we know, some 15 textbooks on mathematics-related subjects had appeared by Hungarian authors during this span of time, not counting here the simple multiplication tables and the mathematical dissertations to be discussed separately later. Even if we reckon with the possible reeditions, we hardly get as much as twice fifteen. It is revealing of the slow pace of progress that the publications appeared in 2—300 copies at the beginning. A brief computation will show that no more than some 10,000 arithmetic books were available over two and a half centuries to lay the basis for a country's mathematical culture and its improvement.

Better known books on mathematics published in Hungary up to 1743 include:

First edition	*First and further editions*
1499 MASTER GYÖRGY: Arithmetic	1499 MASTER GYÖRGY
1563 PÜHLER: Geometry	1563 PÜHLER: Geometry
1577 Arithmetic of Debrecen	1577 Arithmetic of Debrecen
	1582 Arithmetic of Debrecen
1591 Arithmetic of Kolozsvár	1591 Arithmetic of Kolozsvár
1655 APÁCZAI: Encyclopaedia	1655 APÁCZAI
1668 CAESAR: Arithmetic	1668 CAESAR
	1671 CAESAR
1674 MENYŐI: Arithmetic	1674 MENYŐI
	1675 MENYŐI
	1678 CAESAR
1687 BERZEVITZY: Arithmetic	1687 BERZEVITZY
1693 ONADI: Arithmetic	1693 MENYŐI, ONADI
1694 Canon sinuum	1694 Canon sinuum, MENYŐI
	1696 CAESAR
	1698 MENYŐI

[1] It shows the bibliographical significance of MASTER GYÖRGY's *Arithmetic* that it was published in facsimile recently (1965) in Nieuwkoop, Holland, with an instructive foreword by A. J. E. M. SMEUR.

	1701	Caesar, Menyői
	1703	Menyői
	1706	Menyői
	1727	Menyői
	1729	Menyői
1737 Jánosi: Trigonometry	1737	Jánosi
1738 Lipsicz: Algebra	1738	Lipsicz
1743 Maróthi: Arithmetic	1743	Maróthi

The published books include quite a few multiplication tables that did not demand much methodological invention or professional competence of their authors. The chapters on mathematics in Apáczai's *Encyclopaedia* will be discussed separately. Some of the works (Master György's *Arithmetic*, the *Arithmetics of Debrecen* and *of Kolozsvár*, the ones by Menyői Tolvaj, Onadi and Maróthi, the *Geometry* of Pühler) permit the student to detect the outlines of the general position of mathematics in Hungary in the studied period.

To avoid boring repetitions, I am not going to discuss all these works separately. With the exception of Kristóf Pühler's *Geometry*, all are books on arithmetic that explicate the basic operations with integers or at most with common fractions, in Hindu-Arabic notation. Decimals are outside the scope of these books, or only passing mention is made of them (e.g. Maróthi). The use of decimal fractions had already spread in Europe but in Hungary they came to stay rather late. Most of our weights, measures and currencies did not belong to the decimal system (which was eventually introduced by Act No. VIII of 1874) and the index number of various units was given in vulgar fraction. The *accounting forint* = 100 *denar*s was decimal, and so was the *tithe*,[2] the collection of which required much written work. But the necessary operations of conversion were carried out with vulgar fractions even in these cases.

Apart from Master György's *Arithmetic* in Latin, all the arithmetics are written in Hungarian. Their efforts to establish the Hungarian terminology of arithmetic had various degrees of success, from laboured, confused, hardly understandable wording (Tolvaj, Onadi) to a most enjoyable and vivid Hungarian phrasing (*Arithmetics of Debrecen* and *of Kolozsvár*, Maróthi). The newly coined Hungarian words are often followed by the old and commonly used Latin equivalents without which reading and understanding these books would hardly be possible.

One of the most conspicuous features of the first Hungarian arithmetics is their emphasis on usefulness, or narrow *practicism*, so to speak. That this was a conscious aim is made explicit at several points: "The fruit ... of numbers is not only beneficial and useful but also necessary for all" ([4], p. 10); "... competence in figures is immensely profitable"; "... this knowledge will be of use for those who aspire at the main and old sciences, especially astronomy... The same knowledge helps us understand the science of geometry as well" ([5], p. 59). "I did not leave out anything that I thought would be necessary for our country", Maróthi stresses ([12], p. 3). Not only such and

---

[2] The "free" peasant class which had evolved beginning with the late 13th century had to deliver one-tenth (decima pars) of all their harvest to the Church.

Scribe primo pro   1500   divisor
qui; fulioꝛ ptibus
nõ dcrminatis p   10000   4350
tes dcrminatas ß
mõ ʒ adde õnes ß   750
mul ʒ erũt 4350
divisor cuius deide   600   3000 mltiplic.
multiplica vnuus   500
cuiſꝗ ptẽ dcrminatam per multiplicatoꝛẽ ʒ diuide
p diuiſoꝛẽ ʒ patebit ꝓ vnuuſcui°q;

    ¶ Octaua ꝛegula de lepoꝛe fugiẽte
Fugit ꝗdã de pariſi°uſus romã et ãbulat ꝗñdie no
uẽ ſtadia Aliuo auẽ pſeꝗt cũ poſt quiꝗ dies i ꝗbus
pãbulauerat fugiẽs ꝗoꝛagiꜩaquiꝗ ſtadia .Et ãbu
lat pſeꝗuẽs ꝗñdie 14 ſtadia In ꝗt ergo dieb°pote it
pſeꝗꝰ ꝏ̃ꝓhẽdere fugiẽtem Si vis ſcire ſcribe nũm
ſtadioꝛũ que fugiẽs perãbulat ꝗñdie et ſimiliter ꝗ
pſecucoꝛ pãbulat ꝗñdie   9   45 diuidẽd°
Deinde cõſidera quãtũ   54   5 ercettus
ercedit perſequẽo ſugi   diuiſoꝛ.
ẽtem Et ꝑ illa ꝑ ꝗt ercedit Diuide diſtãnã intmedi
ã hoc ẽ numeꝛ quẽ pãbulauerat fugiens pnſꝗ per
ſeꝗuẽo inaperet icinciare .

    ¶ Nona regula de ſolutione incerta
Sunt duo hoĩes ementes mille arteſias pro vigin
tiocto ſtuferis ʒ dimidio Sꝫ vnus illoꝛ volt habere
ſercentas arteſias .ſcd° vero ꝗdꝛingẽtas Quãtum
ſoluet igitur ꝑm° pro ſercẽtis arteſhs Et quãtũ ſcd°
pro quadꝛingentis Si vis ſcire dupla vigitiocto ſtu

AZ
ARITHMETI-
KANAK;

Avagy

Az Számlálásnak öt Speci-
esinek rövid Magyar Ré-
gulákban foglaltatott
Mestersége.

Taliter disponente

FRANC: TOLVAJ MENYŐI,
Gyöngyösinen: Sch. Rectore.

*Az Arythmetikát Tanúló Magyarok
kedvekért irattatott és bővebben
ki-botsáttatott.*

LŐTSÉN,

Nyomtattatott 1701 Eszt:

ARITHMETICA,
vagy
SZÁMVETÉSNEK
MESTERSÉGE,

Mellyet írt, és közönséges Haſzonra, fő-
képen a' MAGYAR ORSZÁGON
elő-fordúlható Dolgokra
alkalmaztatni igye-
kezett

MAROTHI GYÖRGY,
D. P.

DEBRETZENBEN,

Nyomtt. MARGITAI JÁNOS által.
1 7 4 3. Eſztendőben.

4. The title-page of Maróthi's *Arithmetic*

egy illyen Táblátska, mellyben fel-légyen téve, hány Krajtzár van mindenik Pénz' Nemében:

Ebből mindenkor	*Garas* —	3 *Krajtzár*
ki-ſzedheted a'	*Peták* —	7 x.
két ſzámot; a'	*Máriás* —	17
melly meg-jelen-	*Húſz-póltrás* —	30
ti a' *Proportiot.*	*Magyar forint* —	50
De nem minde-	*Vonás forint* —	51
niket lehet	*R forint* —	60
oſztán Oſztás-	*Taller* —	90
ſal kiſſebbí-	*Tſáſzár-Taller* -	120
teni.	*Tſáſzár-Aranya* -	219
	Körmötzi-Arany -	252

Ha már illyen Kérdés adná elö magát: 149 *Vonás forint* hány *R forintot* téſzen? E' Táb-láből (ha másképen nem tudnám) ki-ſzedem e' két Pénz' nemére való ſzámot: és látom, hogy

$$V f. \quad R f.$$
$$X$$

60 Vforint téſzen 51 Rfo-   x. 51   60

rintot. Ha már így mondom: 60 Vforint té-ſzen 51 Rforintot: Hát 149 hányat téſzen? E leſzſz a' három ſzám: —— 60 51 149

A' két elsőt el-oſztom 3-ra, a' fellyebb irtt 2-dik Régula ſzerént: leſzſz — 20 17 149

Ezt ha a' Hármas Régulán fel-vetem; a' Sokſzorozás és Oſztás után jö ki 126 P forint. De az Oſztásból fenn-marad 13.

```
          17
         1043
          149
        253  R126
         2    0
          13
```

G 5   Az

5. A page from Maróthi's *Arithmetic*

similar quotations but the contents of the arithmetics also prove that the authors kept wide masses of the population in mind when pursuing their endeavours.

The numbers in these books are written with the Hindu-Arabic symbols, and the names of lesser numbers are already uniformly used. Divergences occur in designating the higher powers of ten. For example:

In MASTER GYÖRGY	in MARÓTHI
10^6 thousand times thousand, or cuentus	million
10^9 thousand times thousand times thousand, or million	thousand million
10^{12} thousand times, etc. thousand, or summa	bimillion
10^{15} thousand times, etc. thousand, or draga[3]	—

As we see, MASTER GYÖRGY uses not only the complicated Latin[4] phrases but also the names used at the time in Spain (probably upon the influence of his source, PEDRO CIRUELO's arithmetic published in Paris in 1495). In MARÓTI we already find those terms which were more usual all over Europe, although partly different from the present-day forms.

It is intriguing to retrace the evolution of the four rules of arithmetic in the early books. MASTER GYÖRGY lists 9 operations (species): counting, addition, subtraction, doubling, halving, multiplication, division, arithmetic progression, extraction of square root.

The *Arithmetics of Debrecen* and *of Kolozsvár* include 6: counting, addition, subtraction, multiplication, division, progression.

TOLVAJ and ONADI list 5 species: counting, summation, subtraction, multiplication, division.

MARÓTHI has 4: addition, subtraction, multiplication, division.

What accounts for the initial plethora of arithmetic operations is the fact that the operational rules and computing methods that characterize each of the four species and are well known today took a long time to evolve. That explains the infinite number of "regulae" or rules in which our old arithmetics gave recipes for modes of computation, instead of summarizing the operations under more general rules which applied to all the specific cases they discussed separately.

The so-called "nine-test" for the verification of an operation can be found in all the arithmetics of the age in more or less detail. This popular process of the Middle Ages had already been known to the ancient Hindus but it fell into oblivion for a long time. The test is based on divisibility by nine and applies to all four species (PETRUS DE DACIA even extended it to the extraction of square roots): e.g., the sum of the digits of the summands being congruent *mod 9* with the sum of the digits of the result is a *necessary* but *not sufficient* condition for the correctness of the addition.

---

[3] The contemporary Hungarian word *drága* 'dear, expensive' derives from this word.

[4] Naturally enough, in MASTER GYÖRGY's book the Hungarian equivalents of the Latin thousand times thousand, etc. (milies milia) are used. Our 16th—17th century arithmetics and multiplication tables adopted these phrases, translated as 'thousand times thousand', 'a thousand thousands times thousand', etc. But, e.g. the author of the *Arithmetic of Kolozsvár* denotes million by the word *töményezer*, lit. 'dense thousand'.

When performing the test, we "throw away" (cross out) the digits that add up to 9. If the remainder is the same for the summands and the final outcome, and the addition was carried out by ourselves, the result is likely to be correct. If, however, the summation is performed by someone else, the risk of purposeful deceit is too high. Our arithmetics warn that the "nine-test" is not of universal validity.[5]

When introducing the multiplication tables, our first mathematicians were careful to mention the "law of the lazy" (regula pigrorum). This rule dating from the 12th century (and spreading extensively through the works of PETRUS DE DACIA) requires one to memorize the products up to 5 times 5 only. If one or both of the factors are 6, 7, 8 or 9, we may resort to one of the identitives below:

$$a \cdot b = 10a - a(10-b) \qquad (A)$$

$$a \cdot b = 10 \cdot [a - (10-b)] + (10-a)(10-b) \qquad (B)$$

$$a \cdot b = (10-a)(10-b) + 10(a+b) - 100 \qquad (C)$$

If we apply (B), for example, the process is as follows. We are to calculate the product of 7 times 8. We write the two factors one below the other and beside them the numbers that complement them to ten (this explains the other name — *complementary multiplication* — of the process), in this way:

The product of multiplying the right-hand numbers (less than five) gives the units digit, while the difference between the lower left and upper right (or upper left and lower right numbers give the tens digit). (If the product of the right-hand factors is over 10, its units digit will be the units of the result, while the difference computed as above must be enlarged by one.) The process must have evolved from finger numbers, just as the origin of the multiplication sign × can be retraced to them.

Our early arithmetics all make mention of the *abacus*, or calculating with calculi (= pebbles), or — as MARÓTHI called it — the "peasant numbers". This method of computing, which had originated from Babylonia and had only been known in Europe from the 10th century, was very popular in mediaeval Hungary, its vestiges surviving until the end of the 18th century in the mathematical literature. The abacus is a rudimentary form of early mechanical computation that is very easy to learn. All you need for it is a few beans, corns or pebbles, or at a higher stage a few disc-shaped wooden pieces (Hung. "fabatka", surviving in the idiom: it's not worth a *fabatka* 'brass farthing'), or any coins out of use. You have to draw several equidistant parallel lines on the ground, or a metal or marble tablet. The lowest line represents the units, the next line the tens, etc. A counter placed in the space between two lines is worth 5 times a counter on the line below it. All four operations can be performed on the

---

[5] "There can be 'Fallacia' (= deceit) here; when the quantity that is the result of the calculation is increased or reduced by nine" ([13], p. 20). "When you do not add the numbers yourself but have someone else do it for you, and you rest content with only this Nine-Test, you can be easily cheated." ([12], p. 18)

abacus, but multiplication and even more so, division, is lengthy and cumbersome. The steps of addition are as follows:

1. Represent all the addable terms on the frame (*Figure 10*, column *a*).
2. Add up the counters on each line (column *b*).
3. Now "clear up": remove every five counters from a line and place a single one instead in the space above the line; if you remove ten counters from a line, you have to replace it by one on the next line above. After rearrangement, the result can be read off easily (column *c*).

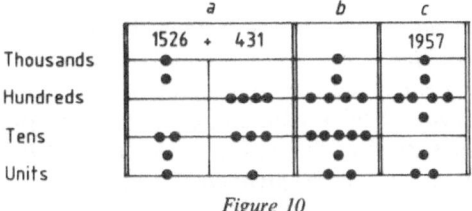

Figure 10

What accounts for the extraordinary popularity of the abacus is that illiterates can also easily learn to use it.

Our early mathematicians presented the methods of inference and the standard problems related to profit sharing among the partners in a joint business venture under the heading "social rule" (Regula societatis). Probably in view of the backwardness of mercantile activities in Hungary, the author of the *Arithmetic of Debrecen* remarks that "in Hungary this rule has not much practical use as the Hungarians are hardnecks and reluctant to pay" ([5], p. 130), but all the books of a later date devote ample space to explaining this rule.

Regula falsi, or "the rule of false position" as mediaeval Hungarians called it, is a root approximation method. The authors of the arithmetics, however, shunned all algebraic tools, and in this way, *regula falsi* meant for them solving equations by trials.[6]

The exercises and tables in our early arithmetics are a treasure-trove for the researcher of standard measures and currencies in Hungary. Let me pick out a few typical examples: in the *Arithmetic of Kolozsvár* only two quintals are used:

1 Hungarian centner = 120 pound, à 8 ferton, à 96 nehezék ('ballast')

1 German centner = 100 pound, à 32 lot

---

[6] Let us quote two examples from MARÓTHI:

"The suitors of a girl ask her how old she is. She answers: My mother is two and a half times as old as I am, my father is three times as old. Together, we are 117 years old. The question is: How old was she? Let us suppose that she was 14." Etc. (p. 236.)

"Someone who wanted to have a suit made found two kinds of cloth. One cost 9 forints a cubit (= c. 60 cm), the other 10. He wanted to get some of the latter cloth but his money was 8 forints short. If, on the other hand, he bought some of the cheaper cloth, he would have 3 forints left. The question is: How many cubits did he want to buy? and how much money did he have? I presume that he wanted to buy 7 cubits." Etc. (p. 240.)

In JULIUS CAESAR OF PADUA, five different centners are listed:

1 Vienna centner	= 100 pound
1 Leipzig centner	= 110 pound
1 Breslau centner	= 132 pound
1 Krakow centner	= 132 pound
1 upper Hungarian centner	= 120 pound

It is also from his work that the following weights can be gleaned: *kő* (lit. 'stone'), *font* (pound), *lat* (half an ounce), *gíra*, *nehezék* (lit. 'ballast'), *scrupulus*, *gran*.

MARÓTHI mentions the following cubic measures: *cseber* (lit. 'bucket'), *véka* (bushel), *kanta* (can), *itce*, *akó*, *fertály* (quarter), *veder* (bucket), *köböl* (vat), *ejtel*. Or an even more revealing list is the one culled from the table at the end of MARÓTHI's book containing the following currency denominations: *körmöci arany* (gold coin from Körmöcbánya), *belga arany* (Belgian gold coin), *császár tallér* (imperial thaler), *kurta tallér* (short thaler), *német forint* (German forint), *harminc krajcáros* (30-penny coin), *húsz krajcáros* (20-penny coin), *máriás* (silver coin), *peták, suszták, garas, poltura, krajcár, gresli* (small coins), *német vagy körmöci kis pénz* (German or Körmöc small coin), *fél krajcár* (half penny). If we also consider that several of these measures had *old* and *new*, *lower Hungarian* and *upper Hungarian*, *town* and *village* values, and that most of the conversions had to be performed with fractions (in MARÓTHI's conversion table for the above coinages 98 vulgar fractions can be found!), then the hard plight of the Hungarian citizen, often illiterate, can be vividly envisioned. As a matter of curiosity, the officially given conversion rates for various measures were often incorrect. That is why MARÓTHI, for example, checked each measure of weight, length and capacity experimentally and defined the ratios on the basis of his own measurements. One can find the rectified ratios at several places of his book.[7]

The early arithmetics are illuminating readings for other reasons as well. Their examples unravel a portrait of Hungary under Austrian, Turkish and feudal oppression, with genre pictures of the common people burdened with usury, tithe, innumerable levies, military raids, etc.

MARÓTHI offers the following exercises:

"38 soldiers, out on a raid, earned 23 thousand forints. If they divide it equally, how much will one get?" (p. 78). "30 soldiers on a raid with a captain, two sergeants and two corporals, seized 950 forints, which they wanted to share so that the captain might get 6 times as much, both sergeants 4 times as much, both corporals twice as much as an ordinary soldier; besides, 3 of the common soldiers should take 3 times as much as the rest for their valiance. The question is: How much does each of them get?" (p. 202). "In all of Hungary there are $5405\frac{1}{2}$ households; 46 of these are in Debrecen; 30 in Buda; 13 in Pest; 40 in Pozsony; 9 in Eger; 200 in the county of Bihar; 78 in the county of Szabolcs; etc. If, for example, the whole country wished to give the King 200 thousand forints as a present, the question is how much would be the share of Debrecen, Buda and Pest each?" (p. 215). An exercise in MENYŐI: "The Magyars settled in Pannonia in the year 380 and made ATTILA their prince in 401." ([13], p. 251). The question is how many years passed from ATTILA's election to the current date.

[7] Relying on the 171 arithmetics published in Hungary between 1499 and 1876, MRS. IMRE GÖDÉNY has analyzed the material and compiled a dictionary of 556 weights and other measures in use in Hungary: *Old Hungarian measures of weight, length and capacity*. Vols 1—2. Debrecen, 1984 (Manuscript at the library of the Kossuth Lajos University of Sciences, Debrecen). Another rich source is: BOGDÁN, ISTVÁN: *Linear and land measures in Hungary up to the end of the 16th century* (Budapest, 1978). (In Hung.)

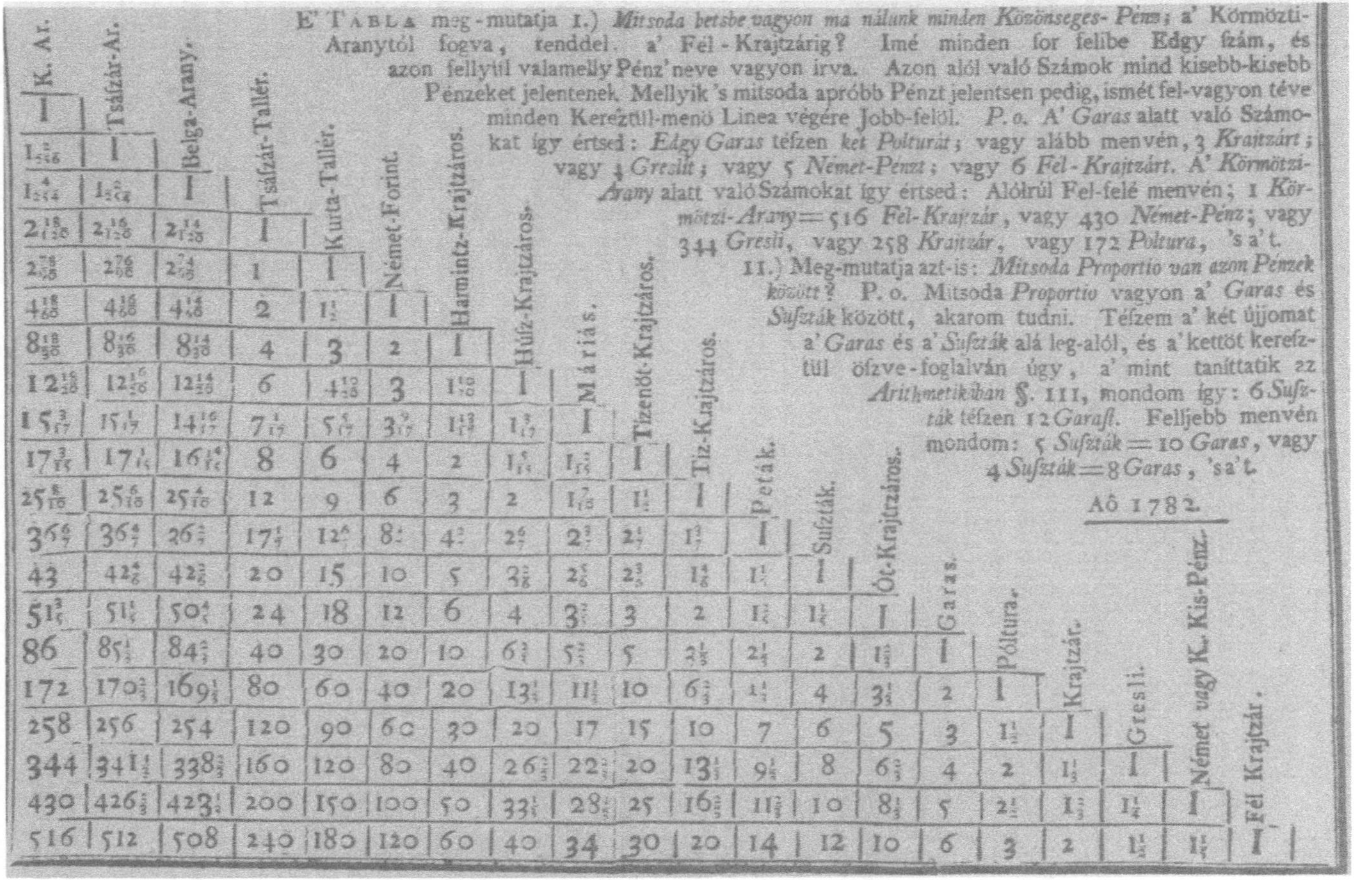

E' TABLA meg-mutatja I.) *Mitsoda betsbe vagyon ma nálunk minden Közönseges-Pénz*; a' Körmözti-Aranytól fogva, renddel. a' Fél-Krajtzárig? Imé minden for felibe Edgy fzám, és azon fellyűl valamelly Pénz' neve vagyon irva. Azon alól való Számok mind kisebb-kisebb Pénzeket jelentenek. Mellyik 's mitsoda apróbb Pénzt jelentsen pedig, ismét fel-vagyon téve minden Keréztül-menö Linea végére Jobb-felöl. *P. o.* A' *Garas* alatt való Számokat így értsed: *Edgy Garas* téfzen *két Polturát*; vagy alább menvén, 3 *Krajtzárt*; vagy 4 *Greslit*; vagy 5 *Német-Pénzt*; vagy 6 *Fél-Krajtzárt*. A' *Körmötzi-Arany* alatt való Számokat így értsed: Alólrúl Fel-felé menvén; 1 *Körmötzi-Arany* = 516 *Fél-Krajtzár*, vagy 430 *Német-Pénz*; vagy 344 *Gresli*, vagy 258 *Krajtzár*, vagy 172 *Poltura*, 's a't.

II.) Meg-mutatja azt-is: *Mitsoda Proportio van azon Pénzek között?* P. o. Mitsoda *Proportio* vagyon a' *Garas* és *Sufzták* között, akarom tudni. Téfzem a' két újjomat a' *Garas* és a' *Sufzták* alá leg-alól, és a' kettöt kerefztül öfzve-foglalván úgy, a' mint taníttatik az *Arithmetikában* §. III, mondom így: 6 *Sufzták* téfzen 12 *Garaſt*. Felljebb menvén mondom: 5 *Sufzták* = 10 *Garas*, vagy 4 *Sufzták* = 8 *Garas*, 's a't.

Aô 1782.

K. Ar.	Tsászár-Ar.	Belga-Arany.	Tsászár-Tallér.	Kurta-Tallér.	Német-Forint.	Harmintz-Krajtzáros.	Húsz-Krajtzáros.	Máriás.	Tizenöt-Krajtzáros.	Tiz-Krajtzáros.	Peták.	Sufzták.	Öt-Krajtzáros.	Garas.	Póltura.	Krajtzár.	Gresli.	Német vagy K. Kis-Pénz.	Fél Krajtzár.
1																			
1 2/256	1																		
1 4/254	1 2/254	1																	
2 18/120	2 16/120	2 14/120	1																
2 78/90	2 76/90	2 74/90	1	1															
4 18/60	4 16/60	4 14/60	2	1 1/2	1														
8 18/30	8 16/30	8 14/30	4	3	2	1													
12 18/20	12 16/20	12 14/20	6	4 10/20	3	1 10/20	1												
15 3/17	15 1/17	14 16/17	7 1/17	5 5/17	3 9/17	1 13/17	1 3/17	1											
17 3/15	17 1/15	16 14/15	8	6	4	2	1 5/15	1 2/15	1										
25 8/10	25 6/10	25 4/10	12	9	6	3	2	1 7/10	1 1/2	1									
36 6/7	36 4/7	26 2/7	17 1/7	12 6/7	8 4/7	4 2/7	2 6/7	2 3/7	2 1/7	1 3/7	1								
43	42 4/6	42 2/6	20	15	10	5	3 2/6	2 5/6	2 3/6	1 4/6	1 1/6	1							
51 3/5	51 1/5	50 4/5	24	18	12	6	4	3 2/5	3	2	1 2/5	1 1/5	1						
86	85 1/3	84 2/3	40	30	20	10	6 2/3	5 2/3	5	3 1/3	2 1/3	2	1 2/3	1					
172	170 2/3	169 1/3	80	60	40	20	13 1/3	11 1/3	10	6 2/3	[illegible]	4	3 1/3	2	1				
258	256	254	120	90	60	30	20	17	15	10	7	6	5	3	1 1/2	1			
344	341 1/3	338 2/3	160	120	80	40	26 2/3	22 2/3	20	13 1/3	9 1/3	8	6 2/3	4	2	1 1/3	1		
430	426 2/3	423 1/3	200	150	100	50	33 1/3	28 1/3	25	16 2/3	11 2/3	10	8 1/3	5	2 1/2	1 2/3	1 1/4	1	
516	512	508	240	180	120	60	40	34	30	20	14	12	10	6	3	2	1 1/2	1 1/5	1

6. Money conversion table from the 3rd edition of Maróthi's *Arithmetic*

For instance, ONADI summarizes the elements of commercial arithmetic in an appendix to his book. In the introductory lines to the section he notes that "... in our miserable country usury is almost the most common way of business ..." and that this was true is borne out by the wording of the exercises.

For practical usefulness, the exercises of our first arithmetics were taken from everyday life. Their wording is often unnecessarily redundant, their loquacity being not in proportion with their contents. Now and again there are examples from history, while ONADI, especially in the first part of his book, seems to develop a fondness for Biblical events. This elicited some criticism from MARÓTHI: "The aspiration is laudable, but the Holy Writ is not to be learnt from Arithmetic" ([12], p. 3). Some other examples are anecdotal, testifying to the healthy mentality and sense of humour the Hungarians had retained in the teeth of all their afflictions.

In MASTER GYÖRGY: "A dying man, whose wife was expecting a baby, willed that his wealth of 1000 golden forints be distributed in the following way: If my wife gives birth to a boy, he shall have two parts and my wife one; if she has a baby girl, my wife shall have two parts and the girl one. That is what he willed and he passed away. But the time came and the dead man's wife gave birth to twins, a boy and a girl. How much shall each of the three get now, according to the provisions of the testament?" ([4], pp. 27—28). In the *Arithmetic of Kolozsvár:* "There are 12 friars, each has 12 shops, in each shop there are 12 bags, in each bag 12 loaves of bread, in each loaf 12 holes and in each hole 12 mice, and each mouse has 12 sons. How many mice are there in all?" One can read the following well-known riddle in ONADI: "If you want to tell who shall have odd or even number in which hand, do like this. Tell him ...", etc.

Both the professional and methodological aspects of MARÓTHI's arithmetic compare well with his day's European textbooks of the highest standards.[8] It is not without reason that it was published three times (in a total of 9,200 copies!), and it is not difficult to explain why it was still in use in some schools as late as the beginning of the last century. While all the books that had come out before his, simply listed the rules of arithmetic in a dry and vague language, what MARÓTHI did was real *teaching*. His work can be held up as an example to be followed even today for its lucidity. He often gives methodological clues as well. For instance, he warns the reader of the only possible way of learning mathematics: "In calculation, one had better put everything to paper; if possible, nothing should be left to *memory*, for it will deceive one before long" (p. 15). Although he tries to find the easiest solutions, he does not regard his procedure as the only possible one and asks the reader: "If any one of you should know a better rule, and can show a better way, I would be only too glad to learn about it" (p. 82). Some statements of his book reveal that out of consideration for the reader he did not go into a detailed discussion of certain questions lest he should be boring to read.

The lively Hungarian language of the book must have also contributed to its success. MARÓTHI's style is effortless and albeit it has an archaic flavour, its vocabulary nearly coincides with contemporary usage. Some mathematical terms he coined proved most practicable. He was systematic in Magyarizing the technical terms: his aim was to find new expressions that faithfully reflected what they denoted and "even the womenfolk could understand them" (p. 3). This suggests that the Latin or German

[8] For a detailed discussion, see: [18].

PRACTICA

ARITHMETICA,

az-az :

SZÁMVETŐ

TÁBLA,

Mellyben mindenféle

Adásról és Vételiről,

akárminémű Kereskedés-
ben-is , bizonyos Számoknak
Summája kéfzen és könnyen
fel - találtathatik:

PADVAI

JULIUS CAESAR

által írattatott;

*Moſt pedig újobban meg-
bővíttetett, és e' kisded fór-
mában kibotsáttatott.*

KASSÁN,

Ellinger János' betűivei.

1803.

7. The title-page of Julius Caesar of Padua's "Calculator table"

language was even harder for the women to decipher than for men who picked up some foreign languages in the public offices. "... if only many followed me," he writes, "the Hungarian language would not be so narrow and poor" (p. 4). Some of the Hungarian mathematical terms he reviewed or coined fell into oblivion, but some others are still in use today, such as: *számlálás* (counting), *összeadás* (addition), *kivonás* (subtraction), *osztás* (division), *maradék* (remainder), *kerület* (perimeter); the word *szorzás* (multiplication) evolved from his *sokszorozás* (lit. taking many times).

*

As has been mentioned, quite a few multiplication tables, or Pythagorean tables as they were then called, appeared besides the books on arithmetic. They contained the products of integers; what distinguished some from the others was that they might include conversion tables for currencies, weights and other measures. All in all, they do not offer much information about our mathematical culture.[9]

*

The geometry material is next to nothing in these arithmetics. Some computations of area and volume (area of a triangle, square or rectangle; cubage of rectangular prisms) was all they contained, mostly as examples to demonstrate the arithmetic operations. It is rather a sign of laziness that they used comparatively rough approximating procedures to compute certain geometric values where they could have resorted to far more accurate formulae as well. For example, MARÓTHI rests content[10] with $\pi = 3$ for mechanical problems, although in the third edition[11] of his work he states: $\pi = \dfrac{22}{7}$. Exercises asking the height of landmarks reveal that the theorems on similar triangles were well known.

There is, however, a geometry by a Hungarian that by far exceeds the rest of contemporary Hungarian works and is among the best in the world. KRISTÓF PÜHLER (Puehler)'s *Ein kurtze und grundliche anlaytung zu dem rechten verstand Geometriae* (Dillingen, 1563) is indeed one of the oldest and most remarkable records of our mathematical literature (completed on 9 February 1561, as the author notes).

We are told in the preface that PÜHLER, who was born at Siklósd (SIGLESS, A.) and studied at Ingolstadt, was driven to a study of geometry by his protracted ill health.

---

[9] They have been thoroughly investigated in a noteworthy thesis by PÉTER SÁRDY on the history of Hungarian mathematics. (Manuscript. Budapest, 1966. Centre of Library Science and Methodology, the National Széchenyi Library.) (In Hung.)

[10] "In Geometry it is said that any circle or wheel whose diameter is 7 feet (or inches, or cubits) has a circumference (or periphery) that equals 22 feet (or inches, or cubits). And conversely, when the circumference is 22 feet, the diameter is always 7 feet." (p. 292.)

[11] That is what we get for π from the following recipe for computing the area of a "full-vaulted" ceiling (with semi-cylindrical nappe): "... when the arc is exactly semi-circular (though the masons call it a full circle), measure the width and length of the vaulted house at the pediment, and multiply the two numbers with each other ... Then add to the product its half. So many (units) will be the vault. That is, one and a half times as much as a straight wall." (p. 289.)

Ein kurtze vnd grund=
liche anlaytung zu dem rechten
verstand Geometriae.

Durch Christoffen Puehler von Syclas
in Vngern/ gemacht vnd von
newem beschriben.

Was nun ordenlich hierinn begriffen/ wirdt in
dem nechsten blat angezeigt.

Mit Röm. Kay. May. Freyheit.

Getruckt zu Dilingen/ durch
Sebaldum Mayer.

Anno Dni M. D. LXIII.

8. The title page of Puehler's *Geometry*

ist : also halten sich auch die vergleichten meßstäb / die zwischen den zweyen rechten ständen des messers gefunden werden / vñ seind 8. meßstäb vñ ⅟₄ eines meßstab / zů dem basim des grossen triangels in dem erstē stand gemacht / weñ ich dz nach der regel detri : mach / so finde ich / dz der basis des grossen triangels in dem ersten stand 22. meßstäb ist haben : vnd der Cathetus, das ist / der Thurn wirdt in der höhe / 33. meßstäb haben.

22

Das

9. A page from Puehler's *Geometry*

The beautifully finished book with ornately illuminated letters and exemplary figures is in fact *applied geometry*. In the well-organized 72 chapters (on 122 pages) he discusses in theory and demonstrates on exercises the problems that occur in the practice of land-survey and astronomy. They include: computation of depth and cubic content of ditches; height of trees, towers and buildings; division and area of land; geographical longitude and latitude, etc. Deep geometrical knowledge is necessary for the solution of such problems. It is to PÜHLER's credit that he relies not only on Greek but also on more recent achievements of geometry including the writings of REGIOMONTANUS.

An additional asset of the book is the precise definition of various weights and measures used in Europe as well as the description of the most important geodetical and astronomical instruments.

Had the book been more widely circulated in Hungary, it would have largely increased the standards of our specialists' work as, being written in German, it was accessible to many. But most regrettably, we have no knowledge of such a beneficial impact of the book. This work is just as "salient a peak" in our geometrical literature as is APÁCZAI's *Encyclopaedia*, which did not find its match for a long time.[12]

Bibliography

[1] Arithmetica, or the science of numbers, translated (for the use of those taking delight in this science and the advancement of their minds) into the Hungarian language from the Arithmetic of the scholarly Gemma Frisius. Debrecen, 1577, 1582. In short: *Arithmetic of Debrecen*. Full text in [5].

[2] *Arithmeticae summa tripartita Magistri Georgii de Hungaria*. 1499. Probably published in Utrecht. Hungarian translation in [4].

[3] DÁVID, LAJOS: *Mathematicians from Debrecen in the old times*. Debrecen, 1927. (In Hung.)

[4] HÁRS, JÁNOS: *How did Master György of Hungary count in 1499?* Budapest, 1936. (In Hung.)

[5] — The Arithmetic of Debrecen. *Publications from the Mathematical Seminar of the University of Debrecen*, Vol. XIV, Sárospatak, 1938. (In Hung.)

[6] HELTAI, GÁSPÁR JR.: Hungarian Arithmetic, or the science of numbers. Kolozsvár, 1591. In short: *Arithmetic of Kolozsvár*. (In Hung.)

[7] HORÁNYI, ALEXIUS: *Memoria Hungarorum et Provincialium scriptis editis notorum*. I—II. Vienna, 1775—1776.

[8] JAUSZ, BÉLA: György Maróthi, a pioneer of education in Hungary in the 18th century. *Acta Univ. Debrecen*, III/1, 1956, pp. 31—62. (In Hung.)

[9] KOPP, LAJOS: Old Hungarian Arithmetics. In the *Yearbook of the central school of science in the 8th district of Budapest, for 1892—93*, pp. 3—21. (In Hung.)

[10] LIGETI, BÉLA: The Bicentenary of György Maróthi's Arithmetic. *A Cselekvés Iskolája*, Vol. 12, 1943—44, pp. 11—27. (In Hung.)

[11] — *A history of Hungarian mathematics up to the late 18th century*. Budapest, 1953. (In Hung.)

[12] MARÓTHI, GYÖRGY: *Arithmetic, or the science of numbers*. Debrecen, 1743, 1763, 1782. (In Hung.) The references here are to the pages of the 3rd edition.

[13] MENYŐI TOLVAJ, FERENC: *The science of arithmetic, or the five species of computing summarized in short Hungarian rules*. Debrecen, 1674; Kolozsvár, 1694, 1698, 1703; Lőcse, 1693, 1701, 1729; Pozsony, 1727. (In Hung.)

[14] ONADI, JÁNOS: *Practici Algorithmi Erotemata Methodica*. Kassa, 1693.

[12] It is a major contribution to the history of science in Hungary that PÜHLER's book was published in careful Hungarian translation with remarkable introductory chapters and annotation. See: PORONYI, ZOLTÁN L.—FLECK, ALAJOS: *Pühler's Geometria Practica from 1563*. Pécs, 1974. (In Hung.)

[15] Julius Caesar of Padua: *Practica Arithmetica.* Lőcse, 1668, 1696; Kolozsvár, 1671; Szeben, 1678; Brassó, 1702.

[16] PÜHLER (PUEHLER), KRISTÓF: *Ein kurtze und grundliche anlaytung zu dem rechten verstand Geometriae.* Dillingen, 1563.

[17] SÁRKÖZY, PÁL: *Mathematicians of Nagyszombat in the old times.* Pannonhalma, 1933. (In Hung.)

[18] SZÉNÁSSY, BARNA: György Maróthi. *Építünk,* 1952, Vol. 2, pp. 52—60. (In Hung.)

[19] SZILY, KÁLMÁN SR.: Hungarian arithmetics in the 16th century. In: *Some data on the history of the Hungarian language and literature.* Budapest, 1898, pp. 161—163. (In Hung.)

[20] — The oldest Hungarian arithmetic. *Ibid.,* pp. 164—168. (In Hung.)

[21] — The arithmetic of Master György from 1499. *Ibid.,* pp. 169—175. (In Hung.)

5. János Apáczai Csere's contribution to mathematics

In respect of mathematics, the most significant college in 17th century Transylvania was that of Gyulafehérvár. A patron of arts and science, GÁBOR BETHLEN, the prince of Transylvania, promoted the school to the rank of "academicum collegium" in 1622 so that the Transylvania youths might acquire higher education at home without costly study trips abroad. Actually, the standards of the school did not rise to college level until three renowned German professors, ALSTED, BISTERFELD and PISCATOR arrived in Gyulafehérvár around 1630.

The atmosphere of scholarship at the college and especially the efforts of ALSTED and BISTERFELD made a deep impression on a young and enthusiastic would-be teacher of the college, JÁNOS APÁCZAI CSERE.

JOHANN HEINRICH ALSTED was born in 1588 near Herborn, Germany. At first he taught philosophy and theology in his native town, then came to Gyulafehérvár in 1629 as guest professor. He taught here until his death in 1638. The bulk of his profuse literary work is on philosophy. His aim was to reconcile the modern philosophical systems with the teachings of Aristotle. His best-known work is his *Encyclopaedia* (Herborn, 1630) which enjoyed several editions still in the life of the author. The impressive work of 2404 paginated pages and nearly 100 pages of subject and author indices, an outstanding feat of printing as well, is an almost complete repository of contemporary scientific knowledge. ALSTED devoted some 150 pages to mathematics or, more precisely, Volume 14 (pp. 803—874) deals with *arithmetic*, volume 15 (pp. 875—954) with *geometry*. — JOHANN HEINRICH BISTERFELD (Nassau, 1605—1655, Gyulafehérvár) was a philosopher and scientist. He studied at Heidelberg and came to teach at the college of Gyulafehérvár in 1629. In his books, some of which appeared in Gyulafehérvár, he took a stance against Aristotelian scholasticism.

The over riding principle of pedagogical activity of APÁCZAI was to present at school as wide a range of up-to-date scientific achievements as possible, trying to avoid superficialness at the same time. He found that especially the teaching of natural sciences and mathematics needed urgent reforms in order to bring up a stratum of intellectuals in Transylvania who could help the ordinary people with everyday arithmetical, commercial and geodetical problems. His writings on both the subject-matter and the teaching of mathematics are significant in the history of this science in Hungary. His inaugural address as college instructor includes the following noteworthy sentences: "... if the time during which we now make efforts to ram their heads almost to surfeit with grammar and in some cases with rhetoric or logic were used for lecturing on the interesting subjects of mathematics and physics, we would give them a source of unspeakable joy for all their lives ..." On another occasion he stated:

"Without mathematics there is no true science mere by futile and brain-exhausting mental exercise."

Just like his teacher ALSTED, APÁCZAI sought to raise the cultural level of the country by summarizing the sciences in an encyclopaedic manner. He devoted ample space in his large and also typographically attractive *Encyclopaedia* to arithmetic and geometry.[1] The Latin preface to the Hungarian book describes the purpose of publication in the above vein: "My aim was", APÁCZAI writes, "to make up for the alarming lack of books in my native language to the best of my power and provide the students with a single volume from which they could unravel all the threads of science in their mother tongue."[2] Thus his motive was patriotism, while his aim was the promotion of the nation's cultural level by making the work of the young people easier.

The *Encyclopaedia*'s sections on mathematics do not contain original results as that lay outside APÁCZAI's purview. Also, the genre and purpose of an encyclopaedia by themselves exclude all attempts at originality or even at presenting the peaks of knowledge attained at a given period of time. What we have here is the verbatim translation of foreign books, the sources for the mathematical parts being PETRUS RAMUS, WILLEBRORD SNELLIUS and LAZARUS SCHONERUS, as the preface says.

PETRUS RAMUS (Pierre de la Ramée, 1515—1572) was a professor at Paris University. Constant persecution for his anti-Aristotelian philosophical views drove him to Germany in 1568. — WILLEBRORD SNELLIUS OF HOLLAND (1581—1626) was one of the discoverers of the law of the refraction of light. In the history of mathematics he is best known for his posthumous work *Doctrina triangulorum canonica* (Leyden, 1627). — LAZARUS SCHONERUS is only mentioned as an editor and commentator of one of RAMUS's works (Petri Rami Arithmetices libri duo et algebrae totidem a Lazaro Schonero emendati et explicati. Eiusdem Schoneri libri duo: alter de Numeris figuratis; alter de Logistica sexagenaria. Frankfurt, 1592).

Comparing the *Encyclopaedia* with the works of the authors just mentioned, one finds that most of the *arithmetic* part (covering: the basic species, the concepts of prime and composite numbers, the greatest common divisor and the least common multiple, operations with vulgar fractions, simple and complex inference, progressions) comes from RAMUS, and to a much lesser degree from SNELLIUS and SCHONERUS. The main source of the *geometric* part (notions and components of the line, angle, triangle, rectangle; inscribed and circumscribed triangles, regular polygons, area and circumference of the circle, solids with flat and curved faces) was RAMUS: *Geometriae libri septem et viginti* (Basel, 1569). Obviously, his master ALSTED's *Encyclopaedia* also influenced APÁCZAI. He borrowed from these sources the theorems only, omitting the explanatory notes and examples. Consequently, his work is very sketchy, rather like a compendium of mathematical theorems difficult to follow even for those who know the material.[3] Evidently, APÁCZAI meant the mathematical part to be a guideline for higher education, with additions and detailed explanation left to the teachers.

[1] *Part Four:* On counting things, pp. 27—47. *Part Five:* On measuring quantities, pp. 47—83.
[2] BÁN, IMRE: *Apáczai Csere János.* Budapest, 1958, p. 163. Those interested in the mathematics material of the *Encyclopaedia* in detail will find useful guidance in this thorough work.
[3] That is exactly why it was a great contribution by the Hungarian Academy of Sciences to publish the mathematical chapters of the *Encyclopaedia* with textual criticism and annotations. (Budapest, 1961.)

XVIII. 1. Mikor valamely beſzéllőnek tzéllya nem
az hogy tſak ſzinte tanitſon, hanem hogy gyönyörköd-
teſſen, fel inditſon, meg tſallyon, akkor elkel ennek a'
rendnek rejtetni, némellyeket egymáſon általtévén, e-
gyéb dolgokra ki kapván, imit amot mulatozván, és
mindeneket okoſon alkalmaztatván a' beſzéllő tanatſára
és a' halgatoknak haſznokra.

Ekkédig, a' dolgoknak közönséges tekintetekről
mind magokon s' mind a' mondáſokban, okoskodáſok-
ban, és el rendeléſekben, következnek már magok a'
dolgok.

NEGYEDIK RESZ.

A' dolgoknak meg ſzámlálájáról.

1.1. A 'Miket még a' közönséges tekenteteken kívül az
oda fel megtalált dolgok körül tudhatunk, az ere-
detet véſen avagy tſak a' termeſzetből avagy az Iſten je-
lentéſéből is. 2. A' termeſzetből a' mit meg tudhatunk
az a' teſtes dolgoknak nézi avagy tſak menniségeket, a'
vagy a' menyiſeggel egyéb tulajdon tekinteteket. 3.
A' menyiſég avagy annyiban gondoltatik a' menyiben
ſok avagy a' menyiben nagy, ki mez bőnnak mondatik,
e' penig nagyſágnak. 4. A' a ſummaza' mely által akar
miis meg ſzámláltatik. Annakokaért, a' ſum avagy egy-
ſég avagy ſokſag, s' lehet penig leg kűſebb mint az egy-
ſeg, leg nagyob peniglen melynél nagyob ne adattat-
hatnék, nem lehet. 5. Ez avagy magán gondoltatik vagy
maſokkal egybe vettetvén. 6. A' magában nem való ſzám-
ban előſſör nézettetik a' lejegyzés, az után a' ſzámlálás.
7. A' táblara le jegyzendő és le irando ſzámnak jegyei ti-
zen vannak, mellyek ezek, 1, 2, 3, 4, 5, 6, 7, 8, 9, 0. Kik kö-
zül az elſő jelent egyet, a' máſodik kettőt, a' harmadik
harmat, a' negyedik negyet, az ötödik ötöt, a' hatodik,
hatot, a' heſedik hetet, a' nyoltzadik nyoltzat, a' kilen-
tzedik kilentzet, 8. A' kerület (cyphra) mely az utol-
ſo jegy, magán ſemmit nem jelent: jób kezfeli tettetven
peniglen, a' tőb jegyeknek jelentéſeket meg bővíti. 9.
Innek a' bővitéſnek harom graditſi (rendi) vannak, mel-

B 2   leſz

It is now easy to sum up the fundamental differences between APÁCZAI's *Encyclopaedia* and the first arithmetics: the arithmetics were, methodologically more or less successful, *textbooks* with patterns for computing, complementary text and exercises, while the *Encyclopaedia is a collection of mathematical theorems;* the arithmetics use a lot of international terms, APÁCZAI Magyarized all the terminology; the arithmetics were written with a view to contemporary demand in Hungary thus remaining confined to elementary arithmetic and geometric knowledge while the *Encyclopaedia* relying on foreign books of a higher standard encompasses a far broader area of mathematics.

This shows that APÁCZAI had grave difficulties to overcome. The *Encyclopaedia* discussed and popularized, in Hungarian, something which was unknown in Hungary at the time and for which the conditions of teaching were not ripe. Apparently, APÁCZAI himself was not quite clear about the material at all points. This is no wonder, however: even the masters of numbers in his time were not free from confusion and, in a strict sense, APÁCZAI was not a mathematician.

The language of the *Encyclopaedia* must have put the reader to the test as well, just like today. The Hungarian language, and mathematical terminology in particular, were rather poor, and APÁCZAI even coined Hungarian equivalents for many mathematical notions that were well beyond the level of arithmetic and geometric knowledge in Hungary, some being completely unknown in content. His attempts at Magyarizing the mathematical language were significant, even though he could only put some notions into Hungarian by sophisticated circumscription. Several of his technical terms are used unchanged today (*azonosság* 'identity', *egynemű* 'homogenous', *osztó* 'divisor', *egyenlőség* 'equality', *hasonlóság* 'similarity', *különközepű* 'eccentric', *tompa-, hegyes szegelet* 'obtuse and acute angle', *sík* 'plane', etc.) and a much greater number of them are used now in a modified form due to the evolution of the language and our mathematical knowledge. Many other of his terms fell into oblivion though they are not worse than the respective terms used today. "If we used *elbontás* for ≪*elemzés*≫ (analysis), *metsző* for ≪*szelő*≫ (secant), *sokasítás* for ≪*szorzás*≫ (multiplication), our terminology of mathematics would not be less Hungarian or expressive." ([3], p. 377)

As far as we know, APÁCZAI's endeavours in mathematics did not exert any measurable impact. Of the many causes behind this failure let us pick out but two. One is that although there were some outstanding masters of numbers in mid-17th century Hungary, teachers in general were trained inadequately in mathematics to be able to comprehend and disseminate the material in the *Encyclopaedia*. APÁCZAI set the standard too high for Hungarian conditions, somewhere near the level at major foreign colleges, and this got the better even of the authors of the arithmetics (ONADI, MENYŐI, etc.). DUGONICS's *Mathematics* published in 1784 contains about the same material as the *Encyclopaedia* but in a better and more systematic presentation. And yet the sources tell us that the material proved too difficult even in DUGONICS's presentation, and *for university students*, to boot.

But there is a *methodological* flaw as well that accounts for the failure of the mathematical part of the *Encyclopaedia*. As we know from the history of mathematics, the Greeks did not use algebraic symbols. They dressed algebraic relations in geometrical garments, and it is not unwarranted to assume that apart from the political and social causes, the decline of Greek mathematics was precisely due to the lack of algebraic symbols. This deficiency was eventually made good one and a half millenia later in 1591 by VIÈTE who introduced several useful algebraic symbols. After initial resistance, his efforts came to be generally recognized in the West and largely contributed to the rapid progress of algebra that began in the 17th century. During the 17th century no traces of this notation can be detected in Hungary. The sources that APÁCZAI used all followed the old, conservative and ponderous trend. Let us mention but one example: APÁCZAI stated the identity

$$\left(\frac{a}{2}\right)^2 = b(a-b) + \left(\frac{a}{2} - b\right)^2,$$

easy to verify in algebraic terms, in the form of the following geometric theorem ([2], p. 140).

Let us cut a segment of length a into two equal parts, then into two different parts. (*Figure 11*; A and B denote the points of division.) We obtain the area of the square produced from the half of the segment a if we add to the area of the rectangle constructed from the different parts the area of the square made from the segment between the two points of division.

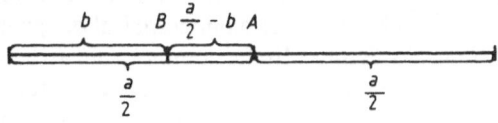

Figure 11

When the wording is so overcomplicated, the proof of the elementary algebraic identity requires inventive dissections of area. It is not likely that teachers in APÁCZAI's time stood a good chance of proving such theorems.

It is to be noted that the above-mentioned theorem corresponds nearly word for word to the fifth theorem in EUCLID's Book II, which is one of the fundamental theorems of "geometric algebra": it is another form of the identity $m^2 - n^2 = (m+n)(m-n)$.

Even though no new mathematical results can be gleaned from the *Encyclopaedia* and the *methodological* and *didactical* treatment of the material in it falls short of the highest contemporary standards abroad, posterity must appreciate the noble aspirations, untiring industry, pedagogical foresight and wide horizon of the author. From the mathematical point of view, his work is "a steeply rising, solitary peak without foothills in the middle of a barren plain and wasteland" ([4], p. 113) which deserves to be considered far more than mere "servile compilation" (KÁLMÁN SZILY).

Bibliography

[1] APÁCZAI CSERE, JÁNOS: *Hungarian Encyclopaedia, or a fine arrangement of all true and useful wisdom made available in the Hungarian language.* Utrecht, 1655. (In Hung.)
[2] — *Hungarian Encyclopaedia.* Vol. II: Mathematics. Budapest, 1961. (In Hung.)
[3] BÁN, IMRE: *Apáczai Csere János.* Budapest, 1958. (In Hung.)
[4] SZILY, KÁLMÁN: Apáczai's Encyclopaedia from the point of view of mathematics and physics. In: *Some data on the history of the Hungarian language and literature.* Budapest, 1898, pp. 112—120. (In Hung.)

6. 17th century mathematical manuscripts

The sketchy picture outlined on the basis of the early printed arithmetics can be given subtler hues by 17th century manuscripts. Systematic research has kept discovering such manuscripts, some of them having already been thoroughly analysed [3, 5]. In what follows we are going to place three manuscripts under scrutiny.

A bulky dossier preserved in the Hungarian National Archives contains a coherent geometric material written in French. The bibliographical designation of the manuscript is: *XI. Liber Regius Gabrielis Bethlen*. In the company of various Latin minutes, oaths and still blank sheets, it contains 45 pages of elementary geometry built up systematically. Somewhere among the sheets one comes across a name: GYÖRGY GÖNCZY, and a date: Gyulafehérvár, 1625. Yet these data do not give much information of the writer and date of the manuscript. Namely, the watermark of the paper (small circles like grapes) was typical of French manuscripts between 1604 and 1611. It is very likely that the geometrical theorems were taken down by a Hungarian student at Paris University in those years, and back at home the notes found their way to the court of GÁBOR BETHLEN where different official notes were added to them. This manuscript is one of the *first* Hungarian documents containing some non-trivial mathematics.

The manuscript is headed by the following title: *Principes de Géométrie*. The introductory definitions and theorems, up to a certain place, show a strong influence of EUCLID's *Elements*. But soon one comes to realize the aim of the author: an approach to the *quadrature of the circle*. Although all the theorems included are of Greek origin, this aim makes the manuscript most valuable in the Hungarian literature.

Specifically, the writer proceeds from the well-known basic constructions with straightedge and compasses (halving a distance, drawing a perpendicular, etc.), through the construction of regular polygons and some formulae for the area of plane figures, to the presentation of 14 rules, also of Greek origin concerning *equal-area* transformations by dissection of plane figures ("De Changement des Figures"): triangle into parallelogram, two congruent squares into one square, two incongruent squares into one square, etc. Finally, he includes an age-old — rather inaccurate — approximation process for squaring the circle.

Thus the manuscript contains no new results but, for the first time in Hungary, attracts attention to a group of problems which will later (in FARKAS BOLYAI's *Tentamen*) lead to remarkable discoveries and definitions (end-like equality of areas).

HARMADIK PROBA

11. A page from Géresi's manuscript

The *Teleki–Bolyai Library* at Marosvásárhely has preserved a hand-written arithmetic (registration number: 380). It was put to paper in 1626 by ISTVÁN GÉRESI, a one-time teacher at the school of Nagybánya. The arithmetic part consists of 169 sheets. Although some sheets got lost over the time, the material is easy to read and almost complete despite the gaps.

The manuscript has close kinship to the printed arithmetics of that period, first of all to the *Arithmetic of Kolozsvár*. Sometimes the author borrows certain passages word for word; the methods of computation of the moral adminishments inserted and the majority of the examples are exactly the same in both works. This, however, was not exceptional, as shown by all the Hungarian arithmetics of those days and most of the foreign ones. A conspicuous difference is that in contrast to the 6 arithmetic species of the *Arithmetic of Kolozsvár*, the later manuscript of GÉRESI enumerates 8 (counting, addition, subtraction, multiplication, division, proportional division, halving, progression), and even regards the four different kinds of the rule of three as separate arithmetic species. This also indicates that the present-day grouping of elementary arithmetic took a long time to develop and had many relapses.

The Hungarian text of the manuscript verifies that a common master of numbers in the 17th century fought a winning battle, despite the general use of Latin, for the creation of a Hungarian mathematical language. GÉRESI, for example, applies the following words for *addition:* eösve adni 'to add up', egy sommába hozni 'to sum up', hozzá adni 'to add', eösve számlálni 'to count together', eösve tenni 'to put together', ahoz számlálni 'to count to', summálni 'to sum'. For *subtraction*, GÉRESI's equivalents are: ki lopni 'to steal from', ki vinni 'to carry from', ki venni 'to take from', subtrahálni 'to subtract'.

The arduous task of deciphering the manuscript was done by SÁNDOR TÓTH, whose careful study [5] excuses us from going into detail here. Let us, however, call attention to some interesting points. Besides the 9-test of basic operations, and on the same principle, GÉRESI proposes a 3-test, 7-test and 11-test, which are missing for most arithmetics. The arithmetic operations are quite systematically discussed by GÉRESI, keeping to the following pattern: definition, problem, recipe for the solution, exercise test. He does not use the operational symbols; in the discussion of regula falsi + and − are employed but instead of indicating addition and subtraction, they denote excess and defect. Let us note here that the symbols of addition and subtraction were introduced to Hungarian literature by MARÓTHI, though he did not use them consistently either.

Another word GÉRESI uses is *czifra*, always with the meaning *zero*. In contemporary works *czifra* meant zero as well as *digit*. Practical usefulness also being a motive for GÉRESI, he listed the values and conversion rates of current coinages ("the Hungarian money") and weights (*mázsa* 'centner', *magyar font* 'Hungarian pound', *magyar ferton* 'Hungarian ferton', *német mázsa és font* 'German centner and pound', *német loth* 'German loth').

The noun *számvető* 'calculator' also occurs in the manuscript, corresponding to the more polite "Mr. Calculator" in the *Arithmetics of Debrecen* and *of Kolozsvár*. They refer to those teaching arithmetic, i.e. teachers well-versed in the arithmetic of the age.

The Kolozsvár branch of the Library of the Romanian Academy of Sciences has preserved a manuscript of about 600 pages. Its author was ANDRÁS PORCSALMI, a student of Collegium Bethlenianum, who took down in this volume what he learnt between 1638 and 1642. Accordingly, the manuscript contains material on physics, cosmology, astronomy and geometry. *Arithmetic* is given little space on a meagre 16 pages. V. MARIAN [3] has found that the notes were put down after BISTERFELD's lectures who, in turn, relied on ALSTED's book *Methodus admirandorum mathematicorum* (Herborn-Nassau, 1613) for his course in arithmetic.

The arithmetic section contains 77 "theorems", but they cannot be taken in the modern sense as they also include definitions, computation rules, detailed interpretation of terms, etc. The manuscript is an arithmetic compendium in the bad sense, with a dull and tiring list of the major arithmetic rules and concepts. There is a world between this writing and say, the *Arithmetics of Debrecen* and *of Kolozsvár* with their responsiveness to reality and amusing style. Of course, this fact alone does not justify the conclusion that the student must have been dogmatic as PORCSALMI probably only wished to record the most important points in a series of elaborate lectures.

There is, however, something revealed here which remains less apparent in the printed arithmetics: notably, that in the period at issue the structure of arithmetic was still rather rudimentary, and so was its method. There were, for instance, lots of criteria — some quite insignificant — for the classification of natural or composite numbers. It must have been a heavy burden for students to store in mind all those categories and their Latin names. The point was often lost in the sea of all the unnecessary material, as can be judged today. V. MARIAN arranged the *over 120 Latin terms* (denoting numbers, composite numbers, proportions in 4 perspicuous tables. Even the simplest table — that of the proportions — lists as many as 21 categories (arithmetic—geometric, continuous—discontinuous, simple—composite, direct—inverse, etc.).

To sum up: the mathematical manuscript surviving from the 17th century do not contain more information than the printed arithmetics. Besides, their style is laboured and deciphering them puts the researcher on his mettle on account of the bad handwriting and the mistakes. On the whole, however, they are instructive documents of the level of mathematical culture in the period studied.

Bibliography

[1] GÉRESI, ISTVÁN: *A useful way of computing* ... (Manuscript at the Teleki–Bolyai Library in Marosvásárhely). (In Hung.)
[2] *XI. Liber Regius Gabrielis Bethlen* (Manuscript at the Hungarian National Archives).
[3] MARIAN, V.: *Un manuscris ardelean de aritmetica din veacul al XVII-lea.* Cluj, 1957.
[4] ANDRÁS PORCSALMI's manuscript (at the Kolozsvár branch of the Library of the Romanian Academy of Sciences). (In Hung.)
[5] TÓTH, SÁNDOR: *István Géresi's arithmetic.* I—II. Cluj, 1962, 1963. (In Hung.)

7. The first Hungarian dissertations in mathematics

Before enlarging on the material proper of this chapter, I feel obliged to make a few remarks about the development of the meaning of the word mathematics.

"Mathematics" and "mathematician" have undergone major changes over thousands of years. The Greek equivalent of "mathematics", as is well known, originally meant "the science". In the Middle Ages and later on, the meaning of the word narrowed down, and — everywhere but especially in Hungary — "mathematician" came to denote all polymaths with an interest in natural science. The titles of books and treatises in astronomy, land survey, physics, geography and technology often included the epithet "mathematical" in this narrower sense which, however, was still very broad if compared to present-day usage. This is highly misleading as we may try to find mathematics where there is no trace of it.

Our earlier documents suggest that whenever a book or tract was focussed on a mathematical discipline in the modern sense, its title featured one or another of the words "arithmetic", "algebra", "computing", "differential and integral calculus", etc.

Let us demonstrate the above-said on a few examples culled from the literature. According to Du Cange[1] the original meaning of "mathematics" was *magic* or *science*; in the 1767 edition of Ferenc Pápai Páriz's Latin–Hungarian dictionary[2] we find this: *mathematica* = mathematics, *mathematicus* = one who has learnt mathematics well, *mathesis* = a craft consisting of certain skills.

Thus the adjective "mathematical" as well as the frequent — hence probably easily obtainable — title of "imperial and royal mathematician" do not mean at all what the unsuspecting reader might expect them to mean: a "mathematical" treatise would have any topic of natural science for its subject usually except, mathematics; on imperial and royal mathematician" could be competent in physics, astronomy, geography, land survey, calendar making, etc. but may have been hard put to perform the fundamental arithmetic operations.[3] In the 17th century Miklós Bethlen called the experimental instruments of physics "mathematical instruments"; for the "imperial and royal

---

[1] Du Cange, Charles: *Glossarium.* Vol. 4. Graz, 1954, 305 p. (Reprint of the 1883—1887 edition.)

[2] Páriz, Ferenc Pápai: *Dictionarium.* Szeben, 1767, 352 p.

[3] This probably suffices to clarify the problem raised at several points in Zemplén, Jolán M.: *A history of physics in Hungary up to 1711* (pp. 75, 94, 120, 125, 126, 138, 185, 190, 206, 237), namely that the word "mathematics" did not denote mathematics proper in each case. (In Hung.)

mathematician" DÁVID FRÖHLICH descriptive geography was "scientia mathematica"; from the 1730s SÁMUEL MIKOVINY held lectures on the improvement of riverways, map-making and architecture at the mining academy of Selmecbánya under the collective title "mathematica".

In the course of the 18th century, the expression "mathematica pura" began to denote mathematics proper (exclusive of geometry) while "mathematica applicata" was similar to today's applied mathematics (including geometry), naturally at a lower level.

In Hungary, the word "mathematica" was ultimately appropriated by the science of mathematics in the early 19th century, although the older usage cropped up at later dates as well. In 1834 the Hungarian Scholarly Society published a *Mathematikai Műszótár* containing technical terms of navigation, architecture, painting, mining, forestry, strategy and mathematics. The periodical of the Academy launched in 1861 under the title *Mathematikai és Természettudományi Közlemények* carried articles on all branches of natural science but none on mathematics.

*

These remarks had to be put forth as there are some 15 doctoral dissertations surviving from the 17th and 18th centuries whose titles might give one the impression that they contain mathematical material, let alone new discoveries. All our hopes to this effect will, however, be shattered in no time, for several reasons. First of all, the majority of these tracts treat subjects of natural science *except mathematics*. This already halves their number. The dissertations that fell through the sieve merely show that the Hungarian youths studying abroad did not waste their time, several of them returning with scientific degrees. With a measure of goodwill, we may see this as a sign of increasing interest in the natural sciences.

The dissertations that are left after the first screening pose several other problems: Are we justified to include in a history of Hungarian mathematics those scholars who, although born in Hungarian territory, were not ethnic Hungarians and lived most of their lives abroad? It is no longer possible to find an unmistakable criterion or standard to decide retrospectively who belongs where. Anyhow, the dissertations of these scholars published abroad are only known to us by title as no copies have been found yet. JÁNOS RÖSCHEL (Sopron, 1672—1712, Wittenberg), doctor of philosophy and theology, studied and lived in Wittenberg; MÁRTON LETZ (Medgyes, ?—?), though a pastor for some time at Waldhid (west of Segesvár), spent most of his life abroad; PÁL PATER (Ménhárd, south of Késmárk, 1656—1724, Danzig) taught in Torun and Gdansk in today's Poland. Judged by their titles, their dissertations (RÖSCHEL: *Exercitationem Mathematicam De Proportione* . . . Wittenberg, 1679; LETZ: *Disputatio Mathematica de Puncti Mathematici fluxu et inde resultantibus Figuris.* Wittenberg, 1701; PATER: *Disputatio decadem Miscellaneorum Mathematicarum sistens.* Danzig, 1707) were summaries of long-known mathematical and physical facts.

An analysis of the dissertations in astronomy belongs to the history of physics in Hungary.[4] In terms of philosophy, they represented a progressive attitude, while

---

[4] Concerning this, see also: M. ZEMPLÉN, JOLÁN, *op. cit.*

DISPUTATIO MATHEMATICO-PHYSICA

DE

LUMINE,

PARS PRIMA.

QVAM,

Favente Deo Optimo, Maximo,

SUB PRÆSIDIO,

Celeberrimi acutissimique Viri,

D. BURCHERI DE VOLDER, Medi-
cinæ & Philosophiæ Doctoris, hujusque in Illustr
Academia Lugduno-Batava, Professoris
ordinarii, felicissimi,

Publicè ventilandam proponit,

SAMUEL KÖLESERI, Ungarus,
AVTHOR & DEFENDENS.

Ad diem 12 Martii, loco horisque solitis, ante meridiem.

LUGDUNI BATAVORUM,
Apud Viduam & Hæredes JOANNIS ELZEVIRII,
Academiæ Typograph. 1681.

12. The title page of Köleséri's dissertation

DISPUTATIO MATHEMATICO-PHYSICA
DE
LUMINE.
PARS SECUNDA.

THESIS XII.

uantum radii objectu corporum à via recta
deflectantur, ex hisce principiis clare patebit: *Radios nimir. luminis ex diaphano medio
rariore, incidentes in medium densius refringi ad
perpendiculum*; contra verò, *Radios ex medio
densiori incidentes in medium rarius, refringi a
perpendiculo.* Ex quibus sequitur, refractio-
nem exigere diaphanum seu transpicuum
Corpus, quod poros habet rectis lineis seu radiis recipiendis ac-
commodatos; neque enim omne corpus etiam porosum radiis
transmittendis aptum esse potest, si nimir. non habeat poros in recta
linea constitutos; sed obliquos & quaquaversum deflexos ac decli-
nantes, quales sunt in spongia aliisque.

XV. Ut igitur praemissorum principiorum veritas clarius pa-
teat, inquiramus in rationem prioris. Concipiamus ex. gr. radium in Corpus pellucidum obliquo motu transferri, manifestum
est, radium in aerem translatum magis inclinare, quàm si transfer-
retur in aquam, minùs verò inclinare cùm transfertur in vitrum.
Cum enim subtilis materia per poros aëris moveatur, in istis magis
materiae subtilis actio impeditur, eo quod partes aëris uncis, hamisq;
male sibi mutuo nexis, tenuiorum funiculorum & mollium plumula-
rum instar constent, adeò ut partem sui motûs radii cum iis com-
municare debeant, & per consequens cum debiliores habeant vires,
magis inclinant: Minus verò inclinant radii cum ex aëre transferun-
tur in aquam, eamque pervadunt, quia partes aquae ramosae, laeves ac
lubricae anguillarum instar sibi mutuo nexae magis cohaerent, atque
ita magis actioni materiae subtilis resistunt, neque tantopere ab ea
commoveri possunt, quàm particulae aëris: Magis tamen adhuc

A 2   ab

in terms of astronomy they discussed topical issues. In terms of mathematics, however, they are insignificant. The more notable ones (ÉZSAIÁS PILARIK: *Dissertatio Astronomica De Eclipsibus In Genere, et Solis In Specie*. Wittenberg, 1680; TAMÁS SZIRMAY: *Eclipsin Lunae Totalem*. Greifswald, 1707; GYÖRGY BUCHHOLTZ: *De Conjunctionibus Planetarum* ... Greifswald, 1710) contain some seemingly original observation data and may also present e.g., meridian conversions based on tables by others but all they needed for that was elementary arithmetic.

*

The doctoral dissertations dating from this period are the first documents of essay-writing in Hungary. Normally, they are 30–150 pages in length with the following structure: after an ornate title-page with a long title come the names of the candidate and the opponents, and the place and exact date of defending the dissertation. The second page carries the names of the "honourable" gentlemen or maybe schools to whom the candidate dedicated his work. After the body of the dissertation we can sometimes read the answers of the candidate to the opponents' questions, and at the end a few panegyrical odes are attached. These were written by friends about the merits of the paper, apparently commissioned by the author.

*

The dissertations to be analysed here in somewhat more detail have a similar structure. Although not all are dissertations, in mathematics, they all contain references to it. Anyone who reads them will immediately see that at the beginning it was not a requirement to present original achievements in order to obtain the master's degree, which was slightly lower than today's doctorate. It was enough to choose a part of the professor's lectures and elaborate it in more detail, or to analyse a less frequently treated topic of arithmetic or geometry.

Not even the germs of new ideas can be discerned in the first "mathematical" dissertations by Hungarian authors. They appear to be belated representatives of scholasticism both in content and method. Their introductory pages would fit well in theological tracts with only a few passing remarks referring to recent trends in physics or to spreading rationalism.

JÁNOS DECSI CZIMOR's *Synopsis Philosophiae* (1591) is, as the title reveals, on philosophy. Today we would define its subject-matter as *science taxonomy*. Philosophy as the universal science also comprised "mathematical" disciplines including, besides arithmetic and geometry, music, optics, mechanics, geodesy and astronomy. The few rules for arithmetic operations and the geometric theorems are far flimsier than their counterparts in contemporary professional books. However, we cannot blame the dissertation for this as its main interest lay elsewhere.

The dissertation of SÁMUEL KÖLESÉRI, *Disputatio Mathematico-Physica de Lumine* (Leyden, 1681) published in two parts addresses a central issue of the time, the nature of light — again not a mathematical topic. Mathematics is represented by a few closing sentences in the second part. Partly in Euclidean spirit, partly in opposition to it, KÖLESÉRI makes some axiomatic statements there. For example: unit is not a number

Q. D. B. V.

DISPUTATIO GEOMETRICA 5
SOLENNIS,
Statutis PRO LOCO *nuncupata,*

EXHIBENS

PROTHEORIÆ
GEOMETRICÆ
PRINCIPIA,

Publicæ placidæq; Disquisitioni

EXCELLENTISSIMORUM

Academiæ Marburgensis

PROCERUM

submissa à

JOH. GEORGIO BRAND

Philosophiæ ac SS. Theologiæ Doctore

& Mathematum Professore,

Ad Diem 23. *Februar.* Anni M DC LXXXII.

In augusto JCtorum Auditorio;

QUAM

PRO POSSE ET DEO DUCE

defendendam suscepit

GEORGIUS POLGARI,

Etsedino Vngarus.

MARPURGI CATTORUM

Typis JOH. JODOCI KÜRSNERI, Acad. Typogr.

14. The title page of Polgári's dissertation

(Unitas non est Numerus), the movement of the point does not produce a line (linea non oritur a motu puncti), etc. Though he is not alone with his conviction, the categorical tone of his statement claiming that the quadrature of the circle is impossible (Non datur quadratura Circuli) is striking.

If there is a dissertation at all that may be called mathematical, it is GYÖRGY POLGÁRI's *Pro Theoriae Geometricae Principia* (Marburg, 1682). The treatise of 32 numbered pages contains no formulae or concrete mathematical results, but it presents a more or less acceptable definition of several geometrical figures together with examples of their occurrence in physics, astronomy and even biology. In POLGÁRI's interpretation the *point* without extent is exemplified by the moment of time, the angle by the knee-bend, the *straight line* — as in Euclid — by the shortest distance between two points, the *surface* by a geometric figure with length and width, etc. Some of his definitions are meaningless: e.g. *infinite* is what has no end, *vacuum* is where is nothing, etc. Probably upon the influence of CLAVIUS, POLGÁRI was aware of the distance line (see p. 150) and defined parallelism with its help: equidistant lines are parallel. The dissertation abounds in names and data on the history of mathematics, but this fails to improve the paper for want of an adequate elaboration of the relevant literature.

Dissertatio Philosophica De Studii Mathematici Utilitate Ejusdemque Certitudine (Franecker, 1695) by ISTVÁN KIRÁLY from Debrecen is an average dissertation whose features are shared by many others in the Hungarian mathematical literature up to the most recent times. It is a paean to the formative power of the mathematical sciences with the stress on their versatile application. Starting from "cogito ergo sum" of Cartesian philosophy, the author proceeds through arguments for truth, wisdom, logic, evidence, to discuss *rationalism* as the only possible standpoint of natural sciences.

> "The second part of the paper is a treatment of the role of mathematics in practice, heralds indeed a new era with the aspirations of urban middle classes to improve technology by scientific methods ... For, KIRÁLY says, how could navigation, military engineering or geodesy exist without mathematics and the three branches of geometry: longimetry, planimetry and stereometry? The argument ends with quotations from famous people in support of what the author is convinced of: that 'Mathematics is very useful and many of its areas are indispensable.'"[5]

These were then the first Hungarian "mathematical" dissertations. They were probably no better or worse than the *average* writings of the kind abroad — after all, they had to prove their worth in competition abroad. But they lagged far behind the foreign treatises that contained new contributions to algebra, number theory, geometry and infinitesimal calculus then published already in large numbers already. Such studies were still far away in Hungary.

[5] M. ZEMPLÉN, JOLÁN: *A history of physics in Hungary up to 1711.* Budapest, 1961, p. 207. (In Hung.)

DISSERTATIO PHILOSOPHICA
DE
STUDII MATHEMATICI
Utilitate Ejusdemque Certitudine:
QUAM
DEO VOLENTE,
SUB PRÆSIDIO
Celeberrimi ac Clarissimi Viri

D. HERMANNI ALEX: RÖELL,
S. S. Theol. & Phil. Doct. ac Professoris
Acutissimi. Solidissimi.

Publice defendet

STEPHANUS KIRALY DEBRECINUS
AUCTOR & RESPONDENS.

Ad diem   Mart. loco horisque solitis.

FRANEQUERÆ,
Apud JOHANNEM GYZELAAR, Illustr. Frisiæ Ordd.
atque Eorundem Academiæ Typograp. Ordinar. MDCXCV.

15. The title page of Király's dissertation

Bibliography

[1] DECSI CZIMOR, JÁNOS: *Synopsis Philosophiae.* Strassburg, 1591; Wittenberg, 1595.
[2] KIRÁLY, ISTVÁN: *Dissertatio Philosophica de Studii Mathematici Utilitate Ejusdemque Certitudine.* Franecker, 1695.
[3] KÖLESÉRI, SÁMUEL: *Disputatio Mathematico-Physica de Lumine.* I—II. Leyden, 1681.
[4] POLGÁRI, GYÖRGY: *Pro Theoriae Geometricae Principia.* Marburg, 1682.
[5] SZABÓ, KÁROLY—HELLEBRANT, ÁRPÁD: *Old Hungarian Library.* Vol. 3. Budapest, 1896, 1898. (In Hung.)

III. The beginnings of mathematical research in Hungary

8. The deepening of scientific thought

What characterized the 18th century in the evolution of mathematics is not so much the new discoveries as a widening and deepening of the application of the great discoveries dating from earlier times in physics, technology and astronomy. The rapidly advancing mathematical disciplines (differential equations, calculus of variations, probability theory, etc.) concentrated first of all on practical applicability. In most cases one can reveal the practical problems behind the various original mathematical investigations. In Western Europe economic development was making rapid progress toward capitalism and large-scale manufacture. The scientists of the age were also eager to submit the weapon of mathematics to a social order that had a growing demand for scientific performance. Despite the lack of exact foundations, the infinitesimal calculus proved to be a most efficient aid to natural science.

Mathematics in Hungary failed to come abreast of Western standards even after the overthrow of the Turkish rule (1699). Obviously, this had social causes. The absolutism of MARIA THERESA and even more so of JOSEPH II relied primarily on growing Austrian capital. Although the government policy of JOSEPH II included the project to extended the principles and methods of capitalism to Hungary, the ultimate aim was to subject Hungary as a whole to Austrian colonialism. The conditions for colonization had evolved earlier, and Hungary was assigned the role of a non-autonomous unit within the Monarchy. However, JOSEPH II did not block up to channels that conveyed progressive trends to Hungary, and Hungary's crises-laden feudal society was ripe to receive these fermenting ideas. Yet the entire social structure should have been transformed for a radical change in our mathematical culture. Substantial change, however, was not to come for a long time. So, mathematical knowledge that had become a must elsewhere was still useless in Hungary. In some other parts of the world mathematics was already a prerequisite of capitalistic development while in Hungary it remained the object of humble aspirations of a few individuals. Nevertheless, this period also witnessed the birth of some original mathematical results partly as an outcome of the intrinsic development of mathematics and partly elicited by the demands of practical life.

It does not contradict the above discussion that also gained in popularity over the 18th century mathematics in Hungary. It became fashionable to converse and discourse on mathematical topics and to prove one's all-round *versatility* by a competence in the allegedly most difficult branch of science. The history of Hungarian culture keeps a record of several persons who are remembered first of all for their contribution to literature or history but showed an interest in mathematics as well. Although their efforts in this respect left no indelible mark on mathematics, they deserve due appreciation. The well-known scholar of versification JÓZSEF RÁJNIS (1741—1812), the pioneer of Hungarian linguistics MIKLÓS RÉVAI (1750—1807) and the folksong collector ÁDÁM PÁLÓCZI HORVÁTH (1760—1820) embarked on discussions about the quadrature of the circle.[1] Besides being a lawyer, ÁDÁM HORVÁTH also dealt with land survey. The ardent advocate of Finnish and Lapp kinship JÁNOS SAJNOVICS (1733—1785) was a teacher of *mathematics* at the academy of Buda. In a speech the great Hungarian poet MIHÁLY CSOKONAI VITÉZ (1773—1805), the rebellious student expelled from the college of Debrecen, said of mathematics:[2]

> "The Mathesis, Gentlemen, is one and the same in every religion, every country, every system and every mode of thinking. Have you ever seen a superstitious fool or a reckless heretic who denied any branch of it? Do you know of a country with however different a system where the truths of mathematics were enacted by law and this or that paragraph was modified or rescinded? Neither climate, temperament, conversation, education, nor prejudice can spin as blurring a mist on its perfectly clear sky as that which dims the horizon — either with a thin haze or with heavy black clouds — of religion, philosophy or legislation. Thus it deserves respect by all who wish to avoid vagueness, strife and forbidden thought. That is why I became a follower of its elevated thinkers along the true path of reason that leads us along truths beyond doubt."

We also know that CSOKONAI wanted to be a land surveyor and at the end tried to mitigate the pains of his fatal illness by mathematical readings:

> "And finally he passed away under the weight of WOLFF's huge mathesis;[3] when, bed-ridden, his hands no longer had the power to hold it, he propped it against his chest to read it so as to train his mind to imagine the infinite that he was soon to enter."[4]

---

[1] Mathematically, each of the three persons' treatises is erroneous. E.g., JÓZSEF RÁJNIS (*Perfecta Quadratura Circuli.* Győr, 1793) speaks contemptuously of the results achieved previously and seeks to prove through mistaken formulae that $\pi = 3\frac{3}{23}$.

[2] *Complete Works.* Vol. 2. 1922, pp. 588—589.

[3] The German mathematician CHRISTIAN WOLFF's (1679—1754) *Anfangsgründe aller mathematischen Wissenschaften* (Halle, 1710 and other dates) and *Elementa matheseos universae* (Halle, 1713) were popular in Hungary for a long time.

[4] DOMBY, MÁRTON: *The life of Csokonai.* Budapest, 1955, p. 50. (In Hung.)

When considering that apart from their significance in other areas, ANDRÁS DU-GONICS and IGNÁC MARTINOVICS earned a place in the history of Hungarian mathematics as well (see later), we can contend that our assumption has been illustrated convincingly.

In addition to Jesuitic and Protestant schools, the *Piarist* institutions were coming to the fore in the dissemination of mathematics in these times. The Piarists were more flexible than the Jesuits, adapting more quickly to the changing requirements of the times, responding sensitively to the ideas of the Enlightenment. Besides, their system of education had always been determined by practical usefulness. They placed special weight on teaching mathematics in their secondary schools of six grades. Their reform proposal for education released in 1753 spelt out that every secondary school have a teacher capable of teaching mathematics. Further, the proposed reform also stipulated that higher mathematics be taught to those who wished to learn it. Contemporary data reveal that 5–6 students were enough to start a "course in arithmetic". These schools also provided information on commerce and agriculture, as well as the elements of architecture.

As is well known, MARIA THERESA regarded education as a political issue in accordance with the ideas of the Enlightenment and tried to confer state supervision over the ecclesiastical schools as well, without distinction as to denomination. As a consequence, the system of education had gradually changed from the middle of the century to the first *Ratio Educationis* (1777), which interrupted this process for quite some time.[5] The decree laid special emphasis on the interests of the ruling house, and its hostility to national aspirations faithfully expressed the colonial dependence of Hungary. Some of its practical measures, however, should not be underestimated. For example, a major outcome was a cutback in the ideological influence of the church, more tolerance toward *non-Catholic* schools and a reduction of the role of the church to mediation between the executive authorities and the people. The Ratio Educationis subordinated education in Hungary to the educational commission in Vienna in which, however, personages of great learning represented the cause of Hungary (e.g. MAKÓ, ÜRMÉNYI). The educational commission of the Buda-based *Gubernium* was merely the executive organ of the Vienna commission. The Protestant schools never regarded the Ratio as binding for them, and as a result, they did not adopt the mathematical textbooks used in Catholic schools. In fact, they used nearly as many textbooks as were major centres in their network of schools. That is one reason why literature on mathematics began to grow mainly in number as yet — from the last decade of the century.

According to FINÁCZY [1], the Vienna court designed the *Ratio* with a view to the following: in market-towns arithmetics should be taught with respect to housekeeping and gradually rising industry, emphasis being laid on the needs of everyday life. In urban schools the examples should be taken from trade and the crafts. In grammar

[5] MRS ERDÉLYI OLÁH, MÁRIA: *Teaching mathematics in Hungary at the time of the two Ratios, 1777—1848* (Dissertation for candidate's degree. Manuscript at the Library of the Hungarian Academy of Sciences. (In Hung.) discusses the teaching of mathematics during the studied period on the basis of rich archives material.

schools the currencies, weights and measures should be taught. Academies should concentrate on those areas of mathematics which are useful for agriculture, land-survey, hydraulics and architecture — the rudiments of engineer training can be detected in the latter. Basically the same principles applied to teaching mathematics at university as well.

After the suppression of the Jesuitic order (1773) and the release of the *Ratio Educationis*, MARIA THERESA transferred the university from Nagyszombat to Buda in 1777, and JOSEPH II shifted it to Pest in 1784. As has been mentioned, by the teaching of mathematics had become part of the Jesuitic curriculum as well, and although the old departments of the university resumed work practically unchanged after the transfers, there were substantial modifications in the mathematical department: the number of chairs was increased in an effort to adjust teaching to practical demands and to raise the standards through a more precise definition of their scope. To attain these goals, within a short time three mathematical departments were set up:

1. in 1773, the department of "Higher mathesis" (later called "Higher mathematical analysis"),
2. in 1774, the department of "Elementary pure and applied mathesis" ("Mathesis elementaris, pura, applicata"),
3. in 1777, the department of "Applied higher mathesis" ("Practical geometry").

The first department has been in existence continuously ever since; the second stopped in 1922 (when MANÓ BEKE retired) and resumed work somewhat later under another name.

The name of the third department is misleading as the central duty of its head was not to deliver mathematical lectures but to run the "Institute of Engineer Training" founded in 1782. When the Institute separated from the university of art and science in 1850, the department was also dissolved or partly merged with the technical university then in the making. Consequently, its professors (FERENC RAUSCH, 1777—1800; JÁNOS GYÖRGY SCHMIDT, 1800—1837; JÓZSEF PETZELT, 1841—1850) have failed to attract the attention of the historians of mathematics, as they were mostly experts of technology.

Most regrettably, the heads of the other two chairs do not have very much to their credit professionally, either. Our only university of art and science at that time, in its structure a poor copy of the Vienna institution, had mediocre or downright poorly trained professors of mathematics during the Enlightenment and the following half a century. Thus the university failed to become a source of mathematics and got stuck at the level of a certificate-issuing office. The faculty members were underpaid. Let us mention a few of the professors. Higher mathesis: JÓZSEF MITTERPACHER 1774—1788, JÁNOS PASQUICH 1788—1797, XAVÉR FERENC BRUNA 1798—1817, JÓZSEF WOLFSTEIN 1821—1832, JÓZSEF PETZVAL 1835—1836, OTTÓ PETZVAL 1839—1883. Elementary mathematics: ANDRÁS DUGONICS 1774—1808, KÁROLY HADALY 1809—1831, JÓZSEF WOLFSTEIN (who left the development of higher mathematics for better paid elementary mathematics) 1832—1848, and ANTAL VÁLLAS 1848—1849.

Of all the persons listed, only JÓZSEF PETZVAL had internationally acknowledged scientific merits (particularly in ordinary differential equations and optics), but his professorship of a mere two years in Buda could not exert significant influence on the university. It is instructive to ponder the question why there was actually no university-level teaching of mathematics in Hungary until the second half of the last century. Most of the professors had no ambition to reap scientific laurels and if they had talents, they capitalized on them in other areas. It is mere delusion that several of them were members of foreign scientific societies (MITTERPACHER, PASQUICH, HADALY and others), firstly because Saint Petersburg, Bologna, Berlin or Göttingen meant something different than today, and secondly, a single well-written textbook or the highly esteemed versatility could open the doors of scientific societies abroad.

The description of two of the above-listed professors given by ANTAL VÁLLAS ([12], p. 148) applies to several of them:

"Hadaly and Rausch were ignorant of the more advanced theories, and in particular Hadaly was incompetent even in the elements of science, doing no little harm to the university of Pest."

Some other professors, however, deserve our respect for their endeavours in writing textbooks and in other scientific areas, or in teaching. The mathematical writings of the director of the faculty of arts, PÁL MAKÓ KEREKGEDEI will be discussed in detail later. The faculty recommended JÁNOS PASQUICH to the chair of higher mathematics vacant after the death of MITTERPACHER with the following qualification (originally in German):

"... according to the statement of the director of the faculty, he is so competent in the sublime science of higher mathematics that we may fully trust his thorough knowledge, ability and diligence proven by many an examination." ([3], p. 6).

It is rewarding to have a look at the examination requirements to be met by someone aspiring to the post of mathematics professor at Pest University those days. PASQUICH was given the following questions for the *written* examination:
The theory of computing extreme values and demonstrating it on a concrete example.
Computation of area with integral, illustrated on a given curve.
Calculation of the centre of gravity.
For the *oral* examination, he had to lecture on some marked passages in the books of EULER and MAKÓ on analysis and SCHERFFER on mechanics.
A well-known mathematician of Vienna, VEGA (1756—1802) reviewed one of PASQUICH's books in these words:

"This treatise by far surpasses all previous works on the subject that I know by virtue of the extraordinary richness ... of the most important material of higher mathematics, the arrangement and treatment of this material, the originality of thought and the succinct, pithy and lucid style of discussion." ([3], p. 7)

However, PASQUICH secured a place in the history of Hungarian science not by his books on mathematics but by his efforts to boost astronomy. His correspondence with GAUSS and ENCKE is a highly stimulating document of the history of Hungarian astronomy, affording us an insight into the small-minded intrigues going on behind the scenes at the imperial court in Vienna and in the Hungarian scientific community.

PASQUICH's successor XAVÉR FERENC BRUNA devised the first programme for teaching higher mathematics at the university under the title *Planum Matheseos Sublimioris Docendae* (1798). It reveals that the most profound parts of the two-year course on mathematics highlighted: NEWTON's binomial theorem, the concept of function, the theory of linear and higher-degree algebraic equations and curves, and some introductory chapters of calculus.

Teaching technical subjects dates back to the reign of CHARLES III: he founded the two-year mining school at *Selmec* in 1735 which raised gradually to the rank of academy (between 1763 and 1770). The foundation of the school was an outcome of the semi-colonial position of Hungary: for the best possible exploitation of natural resources in Hungary, the Vienna court needed trained experts, and it had to see to the necessary level of technical equipment for mining. Later, in 1763, MARIA THERESA founded a trade school, the *Collegium oeconomicum* of Szenc (where double-entry bookkeeping was also taught), and a mining school at the same place, becoming an academy in 1770. Similar schools were set up in the south of the country as well, but they soon closed down when a central institution of engineer training was organized. These institutions did not distinguish themselves in mathematics in any particular way.

It has been a landmark in the development of Hungarian technical culture that in 1782 JOSEPH II enlarged the faculty of art at Pest University with an engineer training institute called "Institutum Geometrico-Hidrotechnico-Practicum" (the name was to change often). When it was founded, the Gubernium decreed that only those be allowed to pursue engineering in Hungary who had taken a certificate from its courses. It is due to this ordinance that Hungary even preceded the Western countries in the training of engineers. The first technical school of France, the "École des Ponts et Chaussées" was not a high school, and the famous "École Polytechnique", the centre of French mathematics in the early 19th century, was only founded in 1795 (by MONGE).

The Institutum was headed by the professor of "applied mathematics" and its lectures were included among the extraordinary subjects of the faculty of arts up to 1827. Although later the ties between the university and the engineering institution slackened, the Institutum retained its significance up to the middle of the last century. It has been to the credit of this technical course that the mathematical mastery and professional expertise of our land-surveyers kept rising and our engineers could greatly contribute to the implementation of the projects in the Age of Reforms (1820—1849).

The question arises: Why did JOSEPH II give consent to the foundation of an engineering institution? The emperor realized that he could not achieve his absolutistic aims unless he had a team of well-trained Hungarian engineers to support him. He also initiated the institution of county engineers. The enforcement order attached to the founding document of the Institutum also specified some engineering problems: frequent floods would wash away the borders of estates, so the boundaries had to be staked out nearly every year. Expert work was needed to raise embankments along the rivers, build mills, design military fortifications, make the just starting coal mining economical, etc.

One may add to this list that in 1784 an ordinance provided for a national census with a view to introducing new tax rates, and two years later the mapping of the country began for the sake of the new land tax. Engineering tasks like these remained just as pressing after the death of JOSEPH II, and no matter in which direction the relationship between the Vienna court and Hungary changed, their significance kept rising. In 1795, for instance, the *First Hungarian Shipping Company* was founded which set itself the goal of designing and constructing the *Francis Canal* (Great Canal of Bácska, Yugoslavia). The canal was opened to traffic in 1802. The same company made plans to make several rivers including the Danube and the Tisza navigable, which entailed the charting of the rivers.

*

Even though the university did not abound in well-trained professors, some country schools already had very enthusiastic and proficient mathematics teachers at that time. Their endeavours are attested by many textbooks on mathematics written in Hungarian, German or Latin, and by far more manuscripts remaining unpublished for financial reasons.

This was the time when the emigration of our mathematicians began to gravely afflict our scientific life especially after the leave of JÁNOS ANDRÁS SEGNER, LÁSZLÓ CSERNÁK and JÓZSEF PETZVAL. It also hindered development that the young people who travelled abroad and tried to keep abreast of the times by acquiring versatile knowledge, dissipated their energies by studying too many subjects, while back at home they only pursued those which proved financially lucrative. So, the less developed circumstances in Hungary made them slowly forget their knowledge of higher mathematics redundant.

The setting up of the *Hungarian Scholarly Society* (original name of the Hungarian Academy of Sciences) did not change the position of mathematics at a stroke. What really happened was that, taking over from the university the task of organizing the scientific scene, the Academy tried to gather the best forces of science and direct the mathematical activities more systematically.

Bibliography

[1] FINÁCZY, ERNŐ: *A history of public education in Hungary in the age of Maria Theresa.* Vol. 2. Budapest, 1902. (In Hung.)

[2] FRIDREICH, ENDRE: The Piarist school in the 16th century. In: *Magyar Középiskola,* Vol. 3, 1910, pp. 327—334, 414—425. (In Hung.)

[3] JELITAI (WOYCIECHOWSKY), JÓZSEF: Archives data on the history of astronomy in Hungary. *Csillagászati Lapok,* Vol. 1, 1938, 3—4, and Vol. 2, 1939, 1—2. (In Hung.)

[4] LÓSY-SCHMIDT, EDE: *First traces of engineering regulations in Hungary from the end of the 18th century.* Budapest, 1925. (In Hung.)

[5] MIHALOVITS, JÁNOS: *Die Gründung der ersten Lehranstalt für technischen Bergbeamte in Ungarn.* Sopron, 1938.

[6] PASQUICH, JÁNOS: *Unterricht in der Mathematischen Analysis und Maschinen Lehre.* I—III. Leipzig, 1790, 1791, 1798. For the use of engineers, based on the posthumous manuscripts of JÓZSEF MITTERPACHER.

[7] — *Elementa analyseos et geometriae sublimioris.* I—II. Leipzig, 1799.

[8] — *Anfangsgründe der gesamten theoretischen Mathematik.* I—II. Vienna, 1812—1813.

[9] — *Abgekürzte logarithmisch-trigonometrische Tafeln.* Leipzig, 1817.

[10] SZENTPÉTERY, IMRE: *A history of the faculty of arts.* Budapest, 1935. (In Hung.)

[11] SZÖGI, LÁSZLÓ: *The Engineer Training Institute at the faculty of arts 1782—1850.* Budapest, 1980. (In Hung.)

[12] VÁLLAS, ANTAL: Recent Hungarian literature on mathematics, etc. *Tudománytár,* 1837, pp. 143—172. (In Hung.)

9. The mathematical ideas
of István Hatvani

The posthumous papers of the well-known professor of Debrecen, ISTVÁN HATVANI, contained two tracts that deserve mention for their merits in applied mathematics. One is his inaugural address [1] from 1749, the other is a chapter of his book on philosophy, *Introductio* [2] published in 1757.

HATVANI started his inaugural address by exposing the backwardness of mathematical culture in Hungary and went on to prove how important a role this science played in a whole range of exact sciences. He went so far as to declare that mathematics was the only science whose findings could be surely relied on and whose conclusions were beyond doubt. Later on, his argumentation reveals that his main concept was not factual knowledge but the methods of mathematics, the logical reasoning. He stressed that young Hungarians who had travelled abroad often became the importers of scientific thought. But HATVANI stood with one leg in rationalism and with the other in religion, for — as befits a good theologian — he endowed mathematics with significance for theology as well, trying to make such concepts as the *infinite* more palpable through notions like the asymptotes of the hyperbole, continued fractions or infinite series. Even more important, he discussed some mathematical problems of planning mills, buildings, roads, digging ditches, entrenchments, etc., although without giving concrete examples.

For the history of mathematics, however, more significant is the chapter "De probabilitate" on pages 259—296 of his book *Introductio*.

As is well known, problems of computing probabilities arose as early as the Renaissance through the popular games of chance. "At the time of early capitalism with developing trade, when risk became an important factor for bankers and merchants, the notion of chance understandably attracted much attention and the laws of chance were passionately investigated, first of all in areas where they could be observed almost in laboratory conditions: in gambling" ([5], p. 674). Various games of dice or cards provided problems in which the players wished to know the probability of winning under a certain set of conditions. The solution of some problems raised during the Renaissance earned credit to the Italian mathematicians LUCA PACIOLI (1445—1514), NICCOLO TARTAGLIA (1500?—1557) and GERONIMO CARDANO (1501—1576), whom we number among the pioneers of the theory of probability, later followed by FERMAT (1601—1665), PASCAL (1623—1662), HUYGENS (1629—1695), MOIVRE (1667—1754) and others.

INTRODVCTIO
AD
PRINCIPIA
PHILOSOPHIAE
SOLIDIORIS.

CONSCRIPTA

A

STEPHANO HATVANI,

MED. DOCT. ET PHILOS. PROF.

CVI

ACCEDIT

OBSERVATIO

ELEVATIONIS POLI

DEBRECINENSIS,

IN VSVS AVDITORVM.

Numquam aliud NATVRA, (aliud) aliquid SAPIENTIA
dicit. Iuuenal. Satyr. XIV.

DEBRECINI,
Per GREGORIVM KALLAI, Typogr.
Annô P. C. N. cɔ Ɔcc lvii.

16. The *Introductio* of Hatvani

infantibus, tres ex Epilepſia deceſſiſſe;
quartum vero morbo alio. Anno 1751
Infantes intra annum mortui fuere 304.
ex his in Epilepſia 210. obiere: Ergo
$\frac{210}{304} = \frac{105}{151} = \frac{21}{30} =$ fere $\frac{4}{6}$ ſeu $\frac{2}{4}$. Anno 1752.
Annotini mortui ſunt 260. quorum 214.
deceſſere ex Epilepſia: Ergo $\frac{214}{260}$ fere $\frac{4}{5}$.
Anno 1753, ex 312. mortui ſunt
ex Epilepſia 236. Ergo ratio erit
$\frac{236}{312} = \frac{118}{156} = \frac{3}{4}$ vel quod proximum eſt
$\frac{60}{80} = \frac{6}{8} = \frac{3}{4}$ Anno 1754. ex 250. mortui
ſunt Epileptici 210. Ergo $\frac{210}{250} = \frac{21}{20}$ fere $\frac{4}{5}$.
Tenendum vero eſt Annis 1752. 1754.
Variolas & Morbillos, graſſatas epidemi-
ce non fuiſſe: vti reliquis annis, in quibus etiam multi ſemeſtres his morbis ſub-
lati decedebant. Hinc liquet, inter An-
notinos apud nos decedentes, hunc vel
illum morbo Epileptico obiiſſe, probabi-
litas eſt, cum variolae graſſantur $\frac{3}{4}$ morbo
alio autem $\frac{1}{5}$ cum variolae non graſſantur
autem $\frac{4}{5}$ morbo alio autem $\frac{1}{5}$. &c.

S C H O L I O N.

Cur tam ingens numerus infantum apud
nos moriatur, non fine graui cauſſa
quaerere quis poſſet; cum primis autem,
cur tanti Epilepſia, e medio tollantur?
Iuxta Tabulas enim Halley ex numero
viuen-

17. A page from Hatvani's textbook

At any rate, there was no consistent theory of probability up to the 18th century, but the great number of successfully solved concrete problems indicated that the mathematicians when faced with phenomena of chance were not ignorant of the laws governing those, either. These were the antecedents to the publication in 1713 of JACOB BERNOULLI's (1654—1705) posthumous work *Ars coniectandi* (The Art of Conjecture), today considered the first scholarly summary of the theory of probability. BERNOULLI defined some central notions of probability on the basis of concrete problems of gambling, presented the elements of combinatorics and proved the (weak) law of large numbers in an exact way. The work gave impetus to further research in which LAPLACE, BUFFON, EULER, CONDORCET and others joined over the 18th century.

HATVANI was familiar with *Ars coniectandi* which, together with lectures of JOHANN and DANIEL BERNOULLI, was the basis for the chapter on probability in his *Introductio*. This is evident from the whole structure of the short tract. Incidentally, HATVANI often makes explicit reference to JACOB BERNOULLI as well as to JOHANN and DANIEL BERNOULLI, both of whom were his masters listened to in Basel.

HATVANI did not probe deeply into the problem of probability as it was not his express aim to enrich it with new findings. He merely wanted to acquaint the reader with a few basic concepts and their relevance to the Hungarian context. Anyway, it is noteworthy that a synthesizing summary of probability in Hungary — even if written in Latin — lagged only four decades behind the first such summary in the world literature.

In the introduction to the 40-page chapter HATVANI notes that one who is sufficiently well-versed in arithmetic will not understand the succeeding discussion. At first glance this remark may seem redundant as the treatment practically does not require any preliminary mathematical knowledge, and relies on the basic arithmetic operations. HATVANI's remark might suggest that citizen educated better than the average of the age, namely one who could read Latin, was not sure to be good at these basic arithmetic operations.

After the introductory lines, HATVANI defines the probability of the occurrence of an event or its complement, as well as the probability of the alternative or joint occurrence of events; then he defines the certainty and the impossibility of an event. He clarifies some concepts of actuarial mathematics like mortality and life expectancy. The definitions are illuminated by a few examples. These are related to casting the dice as well as to data in tables of mortality. HATVANI attaches three mortality tables to his study based on compilations by HALLEY, KEERSEBOOM and DÉPARCIEUX, which were well known and approved in those days. The data of these tables enable him to explain the *law of large numbers*. Also indicated in the chapters is the detailed solution of a problem on deserters.[1]

The last part of the chapter contains interesting medical data which reveal the state of health care in Hungary. Arranging these data in a table, we get:

---

[1] An exercise for the probability of the sum of mutually exclusive events: "Let us suppose that a commander despatches 100 troops to defend a camp; 12 of them are German, 4 Hungarian and 84 Croatian. One of them escapes. Now, I wish to know the probability of the deserter being Hungarian or Croatian." ([2], pp. 268—269.)

Year	1750	1751	1752	1753
Number of births in Debrecen	1022	890	832	936
Deaths in first year of life	235	304	260	250

There was a year when 34.2% of newborn babies in Debrecen died before their first birthday, while the corresponding figure abroad was 19.2%. Being a medical doctor, HATVANI put the question as to what may have caused this sad plight in Hungary. He found the answer in parental negligence, adverse climatic conditions, unhealthy drinking water and air pollution by the marshes. Fired by these findings, he called on the authorities and though them the physicians to devote nore care to pediatry and the training of midwives.[2]

The last part of *Introductio* exemplifies that mathematical tools — not yet profound, though — were also used in geography. HATVANI recorded his evaluation of the geographical latitude of Debrecen by an old method. As we know, the Sun is in the Tropic of Cancer at 12 noon on 21st June: if we measure the angle α between the sunbeams and the horizontal plane in Debrecen at that instant and subtract this value from 90°, we get the difference between the latitudes of the Tropic of Cancer and Debrecen *(Figure 12)*. Adding the well-known value for the latitude of the Tropic of Cancer ($\beta = 23°\ 28'$), we obtain the latitude of Debrecen.

HATVANI computed the value of α from the minimal shadow of a high pole and obtained 66° 3′, which gave 47° 25′ for the latitude of Debrecen. Comparing this with the figure accepted today (47° 33′), we may rightly marvel at its precision, especially in view of the possible sources of error inherent in the method.

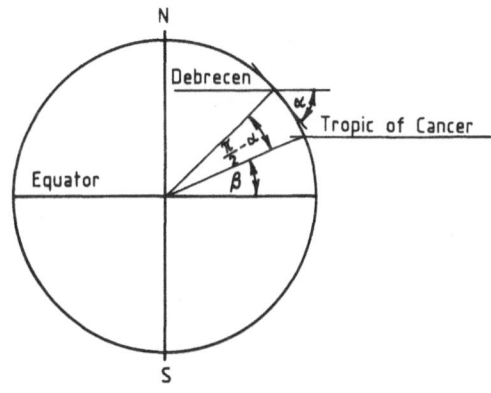

Figure 12

[2] By publishing and evaluating the above-mentioned data and mortality tables, HATVANI has become a pioneer of statistical research in Hungary. This aspect of his activity has been severally analysed by R. HORVÁTH. See, e.g. HORVÁTH, RÓBERT: *Professor István Hatvani (1718—1786) and the beginnings of statistics in Hungary.* Budapest, 1963. (In Hung.)

Bibliography

[1] HATVANI, ISTVÁN: Oratio inauguralis de matheseos utilitate . . ., etc. *Museum Helveticum*, Vol. 20, 1751, pp. 531—557.

[2] — *Introductio ad principia philosophiae solidioris.* Debrecen, 1757.

[3] LÓSY-SCHMIDT, EDE: *The life and works of István Hatvani.* Debrecen, 1931. (In Hung.)

[4] LUDÁNYI, VALÉRIA—SZÉNÁSSY, BARNA: The life and scientific work of István Hatvani. *Atom*, 1956, No. 5, pp. 126—137. (In Hung.)

[5] RÉNYI, ALFRÉD: *The theory of probability.* Budapest, 1954. (In Hung.)

[6] SZÉNÁSSY, BARNA: The mathematical work of István Hatvani. *Alföld*, 1955, Vol. 6, No. 5, pp. 76—79. (In Hung.)

[7] — István Hatvani and his contemporaries in Debrecen. *Természettudományi Közlöny*, 1964, Vol. 95, pp. 123—126. (In Hung.)

10. János András Segner

JÁNOS ANDRÁS SEGNER is the first mathematician of Hungarian origin who is on record in the universal history of mathematics. Judged by strict criteria, however, he cannot be counted among the greatest scholars of his age, as his contemporaries included EULER, D'ALEMBERT, CLAIRAUT, LAMBERT, LAGRANGE — to mention but a few. His output of textbooks is nevertheless quite remarkable, and some of his original findings also attract attention.

What characterized the majority of textbooks published before SEGNER's time was a sketchy introduction of the rules of computation, occasionally followed by one or two more recent results, but without much care for proofs or precision. These are only few exceptions such as the works of TACQUET (1656), STURM (1707), KÄSTNER (1758) and KARSTEN (1767—1778).

SEGNER had a subtle sense to discover long-forgotten values in the heritage of the past and an ability to elaborate the achievements of his age systematically so as to be understood by a wide readership. These features of his textbooks made them highly popular. The proofs of several theorems of algebra and geometry have been adopted by subsequent textbooks and some of his Latin and German technical terms (Faktor, echte-, unechte Brüche, zehnteilige Brüche, äusseres und inneres Glied einer Proportion, etc.) are still used in most languages in verbatim translation. Although his books did not score quite so much success in Germany as those of PÁL MAKÓ (see later) in the Monarchy — since SEGNER had to share the laurels with KÄSTNER and KARSTEN — his role in the development of German mathematical thought is undeniable.

His independent achievements include a proof of DESCARTES's rule of signs. In today's wording, the rule says the following: the number of positive roots — a root of multiplicity m being counted as m roots — of a polynomial equation with real coefficients is either equal to the number of changes of sign in the sequence of coefficients, or less by a positive even number than this number of changes (coefficients equal to zero are excluded).

DESCARTES published this theorem in his *La Géométrie* in 1637 without proof. The probable reason was that he discovered the rule in the course of empirical computations but did not find an acceptable verification. Several scholars have tried to prove the theorem without convincing results. It is also well known that the DESCARTES rule can be easily derived from the later BUDAN–FOURIER theorem.

Deutliche und vollständige

Vorlesungen

über die

Rechenkunst

und

Geometrie

Zum Gebrauche derjenigen, welche sich in diesen Wissenschaften durch eigenen Fleiß üben wollen,
ausgefertiget

von

Joh. Andreas von Segner

Sr. Königl. Preußischen Maj. Geh. Rath, ersten Lehrer der Mathematik und
Naturlehre bey der Königl. Friedrichs Universität zu Halle, Mitgliede der Kaiser-
lichen Academie zu Petersburg, der Königl. Societät zu London und
der Königl. Academie der Wissenschaften zu Berlin.

Zweyte verbesserte Auflage.

LEMGO
in der Meyerschen Buchhandlung
1767.

18. The title page of one of Segner's textbooks

SEGNER addressed this problem in two of his tracts [3, 6]. To understand the title of the first study [3] which was believed lost for a long time,[1] we have to know that following a mistake by J. WALLIS and LEIBNIZ, DESCARTES's rule of signs was first named after the English mathematician THOMAS HARRIOT (1560—1621) and later after SEGNER. FARKAS BOLYAI also calls it SEGNER's theorem in the *Tentamen* (2nd edition, Vol. 1, pp. 404—405).

SEGNER based his proof on the following lemma: if in the polynomial

$$g(x) = a_0 x^n + a_1 x^{n-1} + \ldots + a_n \qquad\qquad (a_0 > 0)$$

with real coefficients arranged by descending powers of x, the number of changes of sign is r, then in the polynomial

$$G(x) = g(x)(x-a) \qquad\qquad (a > 0)$$

the number of changes of sign is at least $r+1$, or greater by a positive even number than this number.

SEGNER's lemma is relatively easy to verify and helps us to arrive at DESCARTES's rule in the following way: let the positive roots of the polynomial equation $f(x) = 0$ with real coefficients be $x_1, x_2, \ldots, x_k$.
Then

$$f(x) = (x-x_1)(x-x_2) \ldots (x-x_k)\varphi(x),$$

where $\varphi(x)$ is a polynomial with real coefficients that has no positive real roots. It follows that the first and last non-zero coefficients of the polynomial $\varphi(x)$ have the same sign, that is, the number of changes of sign for the sequence of coefficients of this polynomial is even $(2p)$. Applying SEGNER's lemma consecutively to the functions

$$\varphi(x), (x-x_1)\varphi(x), \ldots,(x-x_1)(x-x_2) \ldots (x-x_k)\varphi(x),$$

we find that the number of changes of sign in the sequence of coefficients always increases by an odd number, i.e. by $(1+2m_i)$. Therefore, the number of changes of sign in the sequence of coefficients of the polynomial $f(x)$ is

$$2p + (2m_1 + 1) + (2m_2 + 1) + \ldots + (2m_k + 1) =$$
$$= k + 2(p + m_1 + m_2 + \ldots + m_k) = k + 2m.$$

Thus DESCARTES's rule is proved.

*

To determine *graphically* the value $f(x_0)$ at the point $x = x_0$ of the polynomial

$$f(x) = a_0 x^n + a_1 x^{n-1} + \ldots + a_n$$

with real coefficients there are several methods, LILL's so-called "right-angle method" being the most common (*Nouv. Ann.*, 2, 1867, VI, p. 359). Well before this date, how-

---

[1] I owe a word of thanks to Professor WERNER JENTSCH for the xeroxed copy of [3] found a few years ago. For details see W. JENTSCH's remarks on pp. 152—154 of the collection of papers [2].

ever, SEGNER offered a witty and even simpler procedure [10] which is set forth below
SEGNER's original restrictions being imposed for easier graphing only.[2]

Given the cubic polynomial

$$f(x) = a_0 x^3 + a_1 x^2 + a_2 x + a_3$$

$$(a_0, a_1, a_2, a_3 > 0)$$

with real coefficients, $f(x_0)$ $(0 < x_0 < 1)$ graphically.

SEGNER proposed the following procedure. Take a segment $\overline{AB} = 1$ *(Figure 13)* and
measure onto it the given value $x_0 = \overline{AE}$ in point A. Let us erect perpendiculars to
$\overline{AB}$ at A, E and B, and measure the values a_3, a_2, a_1 and a_0 on the perpendicular
emanating from A so that the end point of one coincide with the beginning of the next.
In this way we obtain points K, L, M and C. The line from C parallel to $\overline{AB}$ intersects
the perpendiculars erected at E and B in the points F and D, respectively.

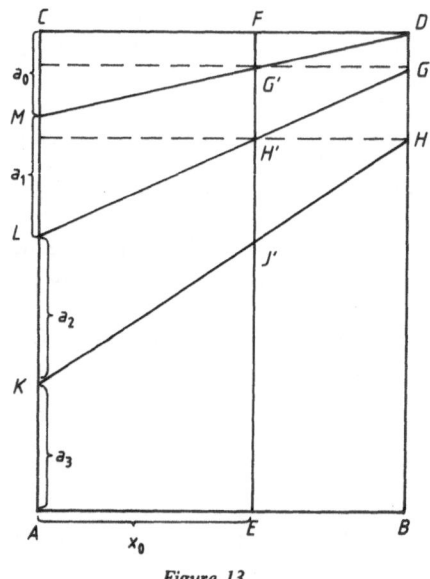

Figure 13

If we draw the segment $\overline{DM}$, it intersects $\overline{FE}$ in point G'. Constructing a parallel
line to $\overline{AB}$ through G', we obtain the point G on the segment $\overline{BD}$. Repeating this
step for $\overline{GL}$ and then for $\overline{HK}$, the distance $\overline{EI'}$ equals $f(x_0)$.

The correctness of the construction can be easily proved by theorems for similar
triangles. Further, it is easy to see that the SEGNER procedure can also be applied
when:

---

[2] SEGNER's method is also discussed by MEHMKE in his encyclopaedia entry (I, 2, pp. 1011—1012) and
by F. CAJORI in CANTOR's history of mathematics (IV, p. 141). It was used for geometric solution of interpolation problems by GYŐZŐ ZEMPLÉN (see Biographies): On graphic interpolation. *Math. és Phys. Lapok*,
Vol. 13, 1904, pp. 96—110.

90

a) the degree of the polynomial is any natural number,

b) x_0 is any non-zero real number,

c) the coefficients a_i are arbitrary real numbers.

When x varies continuously, the aggregate of the points I' obviously produces the curve $f(x)$, whose points of intersection with the segment $\overline{AB}$ provide the real roots of the corresponding polynomial equation. SEGNER planned to construct a jointed device that might draw the curve $f(x)$, but he was discouraged by the difficulty of realization. Not much later JOHN ROWING published the specifications of such a device (*Philos. Trans.*, Vol. 60, 1770, pp. 240—256).

A comparison of SEGNER's and LILL's methods reveals that both rest on the same mathematical principles and their verifications correspond step by step. SEGNER's method is more convenient when the values of a *polynomial* are needed for it provides for a better spacing of the drawing, it is more compact, it "remains on the paper" and indicates the values of the polynomial directly as ordinates belonging to the appropriate abscissas. Therefore, we have to give priority to SEGNER's method, even if LILL's procedure is somewhat more suitable for *determining the roots*.

*

Since the Greek thinkers (ANTIPHON, BRYSON, ARCHIMEDES), the classical method of computing π is by approximation with regular inscribed and circumscribed polygons. SEGNER refined this method further ([7/a], p. 281, [9], 2nd ed., p. 661).

Suppose we have a circle of unit radius and in it a segment AHD bounded by a chord of length $2h$ and the corresponding arc AD (*Figure 14*). It can be proved that the area t_1 of the circle segment is not less than two-thirds of the area t_2 of the rectangle $ABCD$, and that

$$\lim_{\alpha \to 0} \frac{t_1}{t_2} = \frac{2}{3}.$$

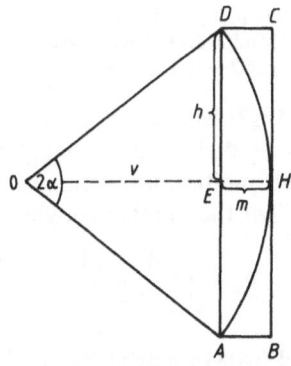

Figure 14

Taking this theorems for granted, the area of the circle is approximated by the formula below better than by the area of the inscribed regular n-gon:

$$T \approx n\left(vh + \frac{4}{3}mh \right) = nh\left(v + \frac{4}{3}m \right) = nh\left(1 + \frac{m}{3} \right).$$

SEGNER calculated the value of h and the corresponding m for $n=96$ and obtained the value of π to six decimal places.

Using the notations of *Figure 14*,

$$\frac{t_1}{t_2} = \frac{\alpha - vh}{2hm} = \frac{\alpha - \cos\alpha\sin\alpha}{2\sin\alpha(1-\cos\alpha)} = \frac{1}{2}\left[1 + \frac{\alpha - \sin\alpha}{\sin\alpha - \sin\alpha\cos\alpha} \right].$$

We can prove now, e.g., by series expansion, that for non-zero α

$$\frac{\alpha - \sin\alpha}{\sin\alpha - \sin\alpha\cos\alpha} > \frac{1}{3}. \qquad\qquad (+)$$

On the other hand,

$$\lim_{\alpha\to 0} \frac{\alpha - \sin\alpha}{\sin\alpha - \sin\alpha\cos\alpha} = \frac{1}{3}.$$

We owe historical truth the remark that certain trigonometric inequalities were also used by W. SNELLIUS in 1621 to compute π, while, in 1654, HUYGENS used the relation $\dfrac{4-\cos\alpha}{3} < \dfrac{\alpha}{\sin\alpha}$ from which $(+)$ can be obtained directly. For the lemma used by SEGNER see, e.g., JÓZSEF KÜRSCHÁK's proof in *Mathematikai és Physikai Lapok*, Vol. 1, 1892, pp. 136—139.

*

We know of a treatise by SEGNER [8] on a geometrical problem in which he offers an elegant solution to an elementary problem raised by EULER.

A sketchy discussion can be found in M. CANTOR's volume III_2, pp. 605—607. It should be noted that RODRIGUES connected the above problem with the solution of the following problem of CATALAN: in how many ways can we write the product of n different natural numbers as the sum of two-factor products formed from the same numbers? (*Journal de Math.*, Vol. 3, 1838). In addition to EULER's original formulation, we may ask how many of all possible solutions are triangulations in which every triangle has at least one side in common with the polygon. The number of such triangulations is $n \cdot 2^{n-5} (n \geq 4)$. EULER's problem was included in the "Miklós Schweitzer mathematical competition" of 1956 with this addition (*Matematikai Lapok*, Vol. 8, pp. 292—294).

In a letter to GOLDBACH in 1751, EULER raised the question: in how many ways could a convex (planar) n-gon be divided into triangles by diagonals that do not intersect inside the polygon? The question was soon answered by EULER himself who proved by complete induction that the number sought was

$$P_n = \prod_{k=3}^{n} \frac{4k-10}{k-1}.$$

In his treatise just mentioned SEGNER solved this problem, giving a recursion formula on the basis of the following argument: let us choose one of the sides of the

n-gon, say, side $1n$ *(Figure 15)* and regard it as the base of a triangle arising from the decomposition. Let the third vertex of this triangle be the r-th corner of the polygon. The triangle $r1n$ cuts two polygons off the original polygon, one with r sides, the other with $n+1-r$ sides. These can be dissected in P_r and P_{n+1-r} ways, respec-

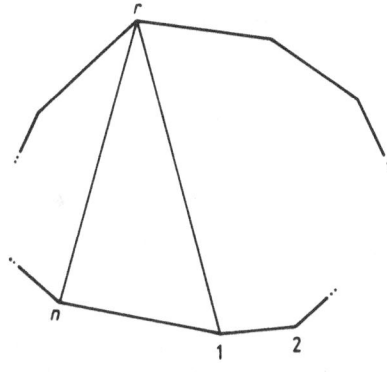

Figure 15

tively. As all the dissections of the one polygon correspond to all the dissections of the other, $P_r \cdot P_{n+1-r}$ dissections are possible when the r-th vertex and the base $1n$ are fixed. We get the *total* number of possible dissections if r runs through the values $2, 3, \ldots, n-1$. Thus

$$P_n = \sum_{r=2}^{n-1} P_r P_{n+1-r} = P_2 P_{n-1} + P_3 P_{n-2} + \cdots + P_{n-1} P_2,$$

where $P_2 = P_3 = 1$.

SEGNER computed the value of P_n for $n = 3, 4, \ldots, 20$. A few errors of calculation were corrected by GOLDBACH in the same volume.

It is worth mentioning that one of the solutions at the Schweitzer competition of 1956 was the formula

$$P_n = \frac{1}{n-1} \binom{2n-4}{n-2} \qquad (n \geq 3)$$

obtained (by ALADÁR HEPPES in his paper) via transformation of the EULER formula. (For further literature see [2], pp. 152—158.)

Let us add a few more remarks concerning SEGNER's mathematical activity. For instance, in one of his textbooks ([7/a], pp. 235—236), SEGNER began the chapter on solid geometry by presenting the long-forgotten CAVALIERI principle (1626) (without mentioning the name). For this reason, this important principle was known by SEGNER's name for a long time. SEGNER was the first to emphasize in a textbook that the criterie of the congruence of planar triangles cannot be extended to spherical triangles. Two planar triangles whose sides are pairwise equal can always be attained to cover each other perfectly, but when the orientation of the triangles is different, one has to be removed from plane and turned over. By contrast, there is no way to make a *general* spherical triangle and its diametric triangle do the same. Consequently, con-

93

gruence as defined for planar triangles will not be assured for spherical triangles unless, in addition to the coincidence of three properly chosen data, equal orientation is required.

The sections devoted to analysis in SEGNER's textbooks are poorer. He worked at a time when a wide variety of geometrical and physical applications had proved the legitimacy and usefulness of the infinitesimal calculus but the firm axiomatic foundations were still missing. SEGNER did not contribute to this area but adopted the most widely advocated theories of the age, first of all EULER's idea. Finally, if deserves mention that he was the first to suggest the application of mathematical tools for the evaluation of meteorological observation data.[3]

Bibliography

[1] DÖRRIE, H.: *Triumph der Mathematik.* Breslau, 1933.

[2] *Johann Andreas Segner (1704—1777) und seine Zeit.* Halle-Wittenberg, 1977. (Collection of studies).

[3] SEGNER, JOHANN ANDREAS: *Dissertatio epistolica, qua regulam Harriotti, de modo ex aequationum signis numerum radicum ... cognoscendi demonstrare conatur.* Jena, 1728.

[4] — *Elementa Arithmeticae ac Geometriae.* Göttingen, 1739.

[5] — *Deutliche und vollständige Vorlesungen über die Rechenkunst und Geometrie.* Lemgo. 1st edition 1747, 2nd edition 1767.

[6] — *Démonstration de la règle de Descartes, etc. Histoire de l'Académie de Berlin.* 1756, pp. 292—299.

[7] — *Cursus Mathematici.* In three parts:

a) *Elementa arithmeticae, geometriae et calculi geometrici.* Halle, 1757.

b) *Elementa analyseos finitorum.* Halle, 1758.

c) *Elementa analyseos infinitorum. I—II.* Halle, 1761, 1763. The second edition was published in six volumes in 1767—1768.

[8] — *Enumeratio modorum, quibus figurae planae rectilineae per diagonales dividuntur in triangula. Novi Comm. Ac. Petr.,* Vol. 7, 1761, pp. 203—209. A brief summary of the paper and some corrections to the computations, *ibid.,* pp. 13—15.

[9] — *Anfangsgründe der Arithmetik, Geometrie und der geometrischen Berechnungen.* Halle. 1st edition 1764, 2nd edition 1773. The German version of 7/a.

[10] — Methodus simplex et universalis, omnes omnium aequationum radices detegendi. *Novi Comm. Ac. Petr.,* Vol. 7, 1758—1759, pp. 211—226.

[11] SZÉNÁSSY, BARNA: The mathematical activity of András Segner. *Acta Universitatis Debrecen.,* VI/2, 1960, pp. 37—42. (In Hung.)

[12] WESZPRÉMI, ISTVÁN: *Succincta medicorum Hungariae et Transilvaniae Biographia.* Vol. 1, Leipzig, 1774.

---

[3] Concerning this, see GÜNTHER: Note sur Jean-André de Segner, fondateur de la météorologie mathématique. *Boncompagni Bull.,* 9, pp. 217—229.

11. Some noteworthy textbooks and tables

From the second half of the 18th century onward, Hungarian printing offices largely increased their output of mathematical books, most of them in Latin, German, Romanian, Slovakian and Hungarian, on more and more diverse subjects. No reliable data can be gleaned from the bibliographies, but it can be established that up to 1867 at least 40 arithmetics, a similar number of elementary algebras and geometries, 8 books devoted exclusively to algebraic equations and some 15 on analysis had appeared. Financial mathematics also began to have its own textbooks: about 5 works appeared in this field.

These figures might suggest that mathematics was flourishing. But we must never forget that bare figures are always misleading: the bulk of the 18th century publications is worthless, full of professional flaws and printer's errors. Their way of presentation is primitive, their style laboured, many of them having no value whatever except for bibliographies.

For this chapter those works have been singled out which had measurable influence on our mathematical culture.

The works of PÁL KEREKGEDEI MAKÓ written in Latin [13—18] were in use not only in the provinces of the Habsburg Empire but also in Germany and Italy, Austrian historians of science hold MAKÓ played a significant role by considerably raising the level of mathematics in Austria at the end of the 18th century. It is regrettable that it was Hungary where many of his works exerted the least influence; some of his textbooks, meeting the requirements of Western Europe, proved too difficult for Hungarians, and satisfied the scientific curiosity of a few individuals only.

His first work entitled *Compendiaria matheseos institutio* [13] is a textbook containing practically the same algebraic and geometric material that is required of secondary school pupils today. Its high popularity and widespread use must have been due to its systematic and light-handed presentation, lucid argumentation and inventive proofs. Reprinted several times, the book became the basis for mathematical courses at universities and academies in the provinces of the Monarchy.

His most celebrated book discusses the elements of differential and integral calculus [14]. It could not become a textbook in Hungary for the simple reason that analysis was not even included in the mathematical curriculum of universities at that time. When writing this work, MAKÓ must have kept in mind the demands of education abroad and the promotion of the level of mathematical culture at home. The

CALCVLI

DIFFERENTIALIS ET INTEGRALIS

INSTITVTIO,

QVAM

IN TIRONVM VSVM

ELVCVBRATVS EST

P. MAKO E. S. I.

LABORE ET FAVORE.

VINDOBONAE,

Typis IOANNIS THOMAE Nob. de TRATTNERN,

Sac. Caes. Reg. Aulae Typ. et Bibl.

ANNO M DCC LX VIII.

19. The title page of Makó's *Differential and integral calculus*

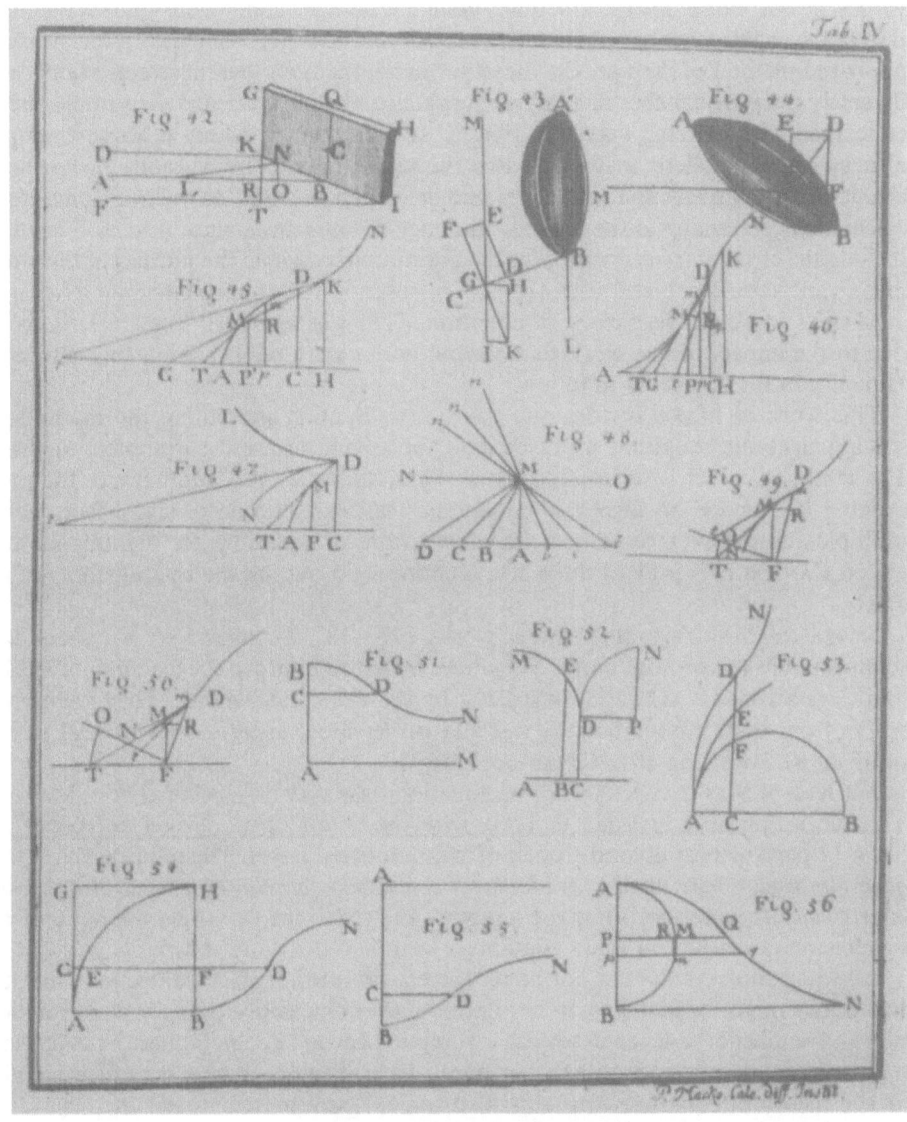

20. A figure page from Makó's textbook

book has no pretext to originality and makes no attempt at defining the basic concepts of analysis; neither does it try to put in order the various definitions of the differential quotient so typical of the "naive" age of the infinitesimal calculus, to use BOHLMANN's label. Following LEIBNIZ, MAKÓ computes with infinitesimal quantities just as freely as with finite ones, and his whole system is based on the *differential*. From today's viewpoint of he keeps being is superficial further on: e.g., looking for extreme values, he only considers the first differential quotient; he completely ignores the question of the convergence of infinite series, etc. All this, however, belonged to the generally approved method of the age. On the other hand, the book demonstrates vividly and diversely the possible uses of analysis in practice, the most attractive examples being concerned with extreme values in physics and mechanics. Many of these examples also appear in modern textbooks, like the well-known extreme value problems of elastic impact, and of the refraction and reflection of light, as well as some angle calculations pertaining to the optimal efficiency of some mechanical devices. The latter include the elementary exposition of the question what angle the normal of the windmill's vane should subtend with the axis of rotation to produce maximal efficiency.

MAKÓ's work on the theory of equations [15] was also well known. It is partly due to the impact of this work that several books were published on this subject in Hungary in the few years to follow.

This work of MAKÓ divides into two parts, the first presenting the methods of solving algebraic equations up to quartic, the second discussing graphical methods. The textbook, which also inspired IGNÁC MARTINOVICS, aptly summarizes the then-existing knowledge on algebraic equations, DESCARTES's rule of signs, facts about multiple and complex roots, and transformations of equations for making solution easier. CANTOR also praised the work, mentioning it among the best algebras of the age.[1]

Several of MAKÓ's mathematical works [16—18] are textbooks for secondary schools published without name. They were meant to improve the teaching of mathematics in secondary schooling regulated by the *Ratio Educationis*. They were good works from both the methodological and professional aspects, with several re-editions in use even long after the author's death.

The level of MARTINOVICS's work on equations falls very little below that of MAKÓ's.[2] This work appeared in Buda in 1780, with the letters of the "royal university". It has a 12-page dedication and a body of 140 numbered pages. The full title is: *Theoria generalis aequationum omnium graduum novis illustrata formulis ac iuxta principia sublimioris calculis finitorum.* VILMOS FRAKNÓI [11] mentions that the catalogue of the Franciscan monastery in Buda contains a MARTINOVICS work, *Mathesis pura* (Buda, 1780) which, however, could not be recovered. FRAKNÓI himself conjectures that the two books might be identical; in my opinion, the coincidence of the year of publication and number of pages confirms this assumption beyond doubt. Besides, MARTINOVICS had never made reference to another book. The difference in title is understandable as the theory of equations was classified under "pure mathematics" in those days, so the words entered in the catalogue may have indicated the nature of the work in order to spare the long title. There are data suggesting that a jury read an autographic study by MARTINOVICS dealing with theorems on the circle, in connection with his application to the university of Buda (see [28]).

MARTINOVICS dedicated the above book to PÁL KEREKGEDEI MAKÓ. The verbose and high-flawn lines stress the need for cooperation among those engaged in sci-

---

[1] CANTOR, M.: *Vorlesungen über Geschichte der Mathematik.* Vol. 4, p. 104.
[2] For a detailed presentation and evaluation of the contents, see [27].

entific questions and go on to eulogize about the great exploits of the "excellent" gentleman (MAKÓ).

The first pages of the work divided into six parts and 25 smaller chapters list the basic concepts necessary to discuss equations, put forth definitions and enumerate questions to be looked at. The first chapter explains theorems on equalities and inequalities called axioms here, followed by rules governing the transformation of equations. These introductory pages already reveal the merits of MARTINOVICS's writing — as well as its flaws. The great asset of the book is *gradualness* it starts from simple facts progressing toward complex ones so that in many cases the general theorems arise almost automatically from the previously expounded partial statements. On the other side, proofs are often defective and inaccurate, sometimes also missing. Mathematical rigour and precision were not MARTINOVICS's strongest points. In the first chapters he seems to have had but one aim: to get done with much of the copious material and reach the mathematically more attractive and intriguing parts that he apparently had a penchant for.

The greater part of the book is devoted to the solution of linear and higher-degree algebraic equations and systems of linear equations. To solve linear systems, he recommends the method of substitution and that of comparison but fails to specify the conditions of solubility. Next he discusses what the roots of algebraic equations of the n-th degree with real coefficients can be like. Compared to the rest of the book, these parts are more detailed and suggestive of the practical common sense of MARTINOVICS. The exercises he presents — mostly from practical life — have various (positive-negative, rational-irrational, real-complex) roots, but the author makes sure to explain these qualitative differences between the roots.

A lengthy discussion highlights the methods by which we can find *lower* and *upper* *bounds* for the roots of algebraic equations when all roots are real. According to one such method, the equation

$$f(x) = a_0 x^n + a_1 x^{n-1} + \ldots + a_n = 0 \qquad (a_0 > 0)$$

must be differentiated $n-1$ times. If all these derivatives are positive when some positive number a is substituted, then a is an upper bound for the positive roots; if, on the other hand, we get positive even derivates and negative odd ones by substituting a negative b, then b is a lower bound of the negative roots.[3]

Without going into detail, let me mention a little known root approximation method discussed by MARTINOVICS, which was first applied by NEWTON (1669). Let us assume that we have somehow found a number k that differs from some root x of the equation by a value less than 1. Thus, in the equality $x = k + d$ we have $|d| < 1$. If we apply the transformation $x = k + d$ to the equation of the n-th degree, then we obtain an n-th degree equation for the so-far unknown d. However, from the latter we can omit some terms of higher degree in d (as $|d| < 1$); the resulting lower-degree equation gives us the first approximation for d. Iterating these steps, we can approximate the root x as accurately as we wish.

[3] The method, which MARTINOVICS presented without a proof, comes from NEWTON (*Arithmetica universalis*. Vol. 2, Chapter 4).

Even this sketchy summary of MARTINOVICS's book suggests that the material he discussed is considerably large. Remembering the level of mathematical culture at those times we may conclude that he had erudition and intellectual horizon well above the average. Of course, one cannot obtain a precise picture of MARTINOVICS's personality from a single book, but a few features can be clearly drawn. This work suggests that he had a good grasp of things with his acumen, and his ability for synthesis placed every bit of knowledge in system easily. At the same time, his restless spirit may have prevented him from deep absorption in mathematics, which alone helps to meet the requirements of rigour and achieve new results. It is also obvious that he never put down anything he did not understand clearly; he never copied another author in a servile way, which cannot be said of some contemporary Hungarian mathematicians.

What brings credit to the mathematical books of ANDRÁS DUGONICS is first and foremost their language. The year 1784 saw the publication of *The first book of Science*, in which one finds the rudiments (= algebra) and bound in the same volume *The second book of Science*, which includes land-survey (= geometry). The work had a second, much enlarged edition containing two new "books", one on triangulations (= trigonometry) and one on pointed sections (= conics). The library of the Hungarian Academy of Sciences preserves several voluminous mathematical manuscripts by DUGONICS but they do not contain anything new compared to *Science* either in language or content.

DUGONICS informs us of the purpose of *Science* in his *Notes*:

"Standing up against this effort,[4] I decided to publish algebra and geometry in the Hungarian language to prove to the whole country that German is far from being as suited to the explanation of science as Hungarian. Neither the French nor the Germans have ever been able to publish mathematics in their own language without having to apply several words borrowed from foreign languages. I have shown that, should science be taught in Hungarian, no foreign language would be necessary, as in my two scientific books I have only used pure Hungarian words." ([6], p. 16)

Indeed, DUGONICS avoided all foreign mathematical terms and undeniably this was a great contributional to the coinage of new Hungarian words. In *Science*, linguists have found some 300 Hungarian mathematical terms mostly of his own coinage; however, only a few of them have struck root. The rest either proved artificial — hence useless — from the beginning, or dropped out during subsequent development of the language, giving way to more successful words and in some cases to internationally used terms. The great Hungarian linguistic innovator FERENC KAZINCZY, though admitting that DUGONICS's efforts were worthy to follow, not have a high opinion of his Magyarizing attempts and found his words impracticable ([29], p. 279).

---

[4] I.e. the Germanizing decrees issued by Emperor JOSEPH II.

A'
TUDÁKOSSÁGNAK
ELSŐ KÖNYVE,

A'

BÖTÜ-VETÉS.
(ALGEBRA)

MELLYET

KÖZ-HASZONRA IRT

DUGONICS ANDRAS,

Kegyes Oskola-beli Szerzetes Pap. A' Józan, 's egy-
fzer 's mind a' Termélzeti Tudományoknak Okta-
tója. Ennek-elötte a' Budai Tanúlmányoknak Ki-
rályi Mindenségében ugyan azon Tudákolságnak Ki-
rályi Tanítója, 's a' Tanúltaknak egygyik tagja. Mos-
tanában pedig a' Jeles Termélzeti Karnak
Leg örebbike.

Másadik meg-bővltett Ki-adás.

POSZONYBAN és PESTEN,
FÜSKÚTI LANDERER MIHÁLY
bötüivel és költségével.
1798.

21. The title-page of Dugonics's *The first book of Science*

Some older studies praise DUGONICS's mathematical terms in superlatives. This overestimation can be attributed to two reasons: first, university students of the age received DUGONICS's Hungarian works with great enthusiasm; second, his mathematical terms were analyzed by linguists and thus the aspect of professional usefulness was disregarded.

Less praise is due to DUGONICS for the professional and didactic aspects of this work. It would be boring to enumerate the contents of the four books of *Science*; for short, they embrace what we teach today in elementary and secondary schools from algebra, planimetry, trigonometry and conic sections. Thus *Science* is a far cry from higher mathematics: the instruction of mathematics at the university went already much deeper. This fact alone would, of course, be insufficient for a strict criticism of DUGONICS's work. But *Science* belongs to the less successful attempts even within its scope. And this not because of the redundancy of unnecessarily introduced terms or the all-pervasive verbosity and bombast of style, but because of the lack of mathematical reasoning and rigour. The book is written by a well-read person fairly familiar with historical data but of superficial knowledge. *Science* does not support the statement by a biographer of DUGONICS claiming that he was the "most outstanding" mathematician of the age in Hungary ([21], p. 96). DUGONICS borders on insolence in the way he praises himself and his work in general, and his allegedly original thoughts and proofs in particular. Introductory odes already outdated in poetry and prose, and long omitted from books on mathematics are also included here (one in Latin and one in Hungarian), possibly written by friends he had commissioned.

In the section on algebra, axioms, theorems, lemmas and their "proofs" follow one another whimsically, but these terms do not denote what we generally mean by them today. Actually they signify recipe-like computing rules from algebra with defective proofs. DUGONICS makes no attempt at solving more difficult problems even in this rudimentary way. For instance, all he devotes to equations of higher degree is the sentence: "solving them requires deeper knowledge". Neither does geometry fare much better, except for the fourth book which is slightly more effective. Here we can find a few theorems verified in an acceptable way.

*

Several of the tables published in the studied period are inventively and cleverly constructed. For example, PÁL SIPOS started to compile a table that was intended to contain the logarithms of circular functions. He discussed his achievements in a 20-page printed tract entitled *Specimen* [25] and in a short study, *Berichtigung* [26] dedicated to KAZINCZY. The table was compiled for astronomical purposes, commissioned from SIPOS by F. TRIESNECKER (1745—1817), the director of the Vienna observatory, as can be gathered from the latter study.

As far as we know, SIPOS's table is unique in its structure. Firstly, he takes the *diameter* of the circle as unit, and secondly, he does not use degrees but taking 90° equal to one, he divides it into 100 parts. Accordingly, angles are given in vulgar

SPECIMEN
NOVÆ
TABULÆ TRIGONOMETRICÆ

AD

COMPENDIUM

SYSTEMATICÆ CONSTRUCTIONIS

REDUCTÆ

A

PAULO SIPOS,

IN COLLEGIO REG. S. PATAKIENSI PHYS. ET MATH. PROF. P. O.
ET ERUDITÆ SOCIETATIS VIADRINÆ ADJUNCTO.

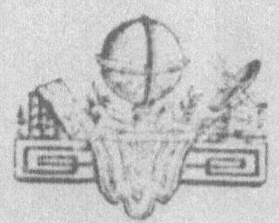

Posonii,
Typis Simonis Petri Weber,
1807.

6. Trigonometrikus táblájának címlapja

22. The *Specimen* of Sipos

fractions $\dfrac{m}{n}$. The unifinished table only contains data for angles between 0 and 0.1.

SIPOS's table contains the following columns *(Figure 16)*:

$$\left(\frac{m}{n}\right)^2; \ \log\frac{m}{n}-2; \ \log\overline{AF}; \ \log\overline{BF}; \ \log\overline{AG}; \ \log\overline{DE}; \ \log\overline{DF}.$$

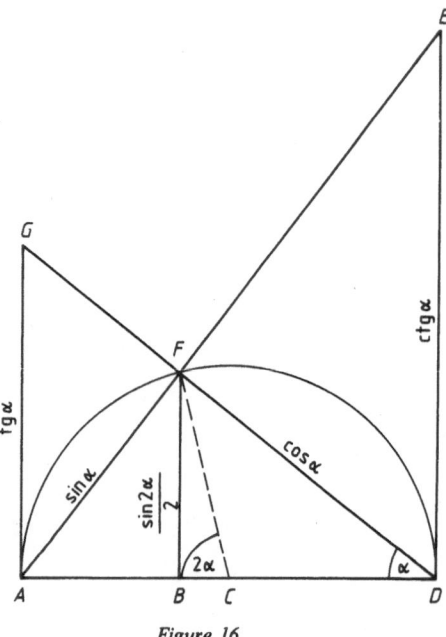

Figure 16

Or, substituting the values of the segments indicated:

$$\left(\frac{m}{n}\right)^2; \ \log\frac{m}{n}-2; \ \log\sin\alpha; \ \log\frac{\sin 2\alpha}{2}; \ \log\mathrm{tg}\,\alpha; \ \log\mathrm{ctg}\,\alpha; \ \log\cos\alpha.$$

Although the arrangement of the columns seems to be arbitrary, it follows a rigorous and purposive scheme: the inclusion of the last five columns is according to the Figure in *Specimen* (similar to *Fig. 16*). The presentation of the values of $\left(\dfrac{m}{n}\right)^2$ made the compilation of the table easier, since the following series expansions are well known:[5]

---

[5] For the calculations, see [30], pp. 92—95.

104

$$\log \sin x = \log x - Mx^2 \left(\frac{1}{6} + \frac{x^2}{180} + \frac{x^4}{2835} + \cdots \right)$$

$$\log \frac{\sin 2x}{2} = \log x - M \frac{2}{3} x^2 + x^4 (\ldots)$$

$$\log \operatorname{tg} x = \log x + M \frac{1}{3} x^2 \left(1 + \frac{7}{30} x^2 + \cdots \right)$$

$$\log \operatorname{ctg} x = -\log x - M \frac{1}{3} x^2 \left(1 + \frac{7}{30} x^2 + \cdots \right)$$

$$\log \cos x = -M \frac{1}{2} x^2 \left(1 + \frac{x^2}{6} + \cdots \right),$$

where $M = \log e$.

So the infinite series of the logarithms of trigonometry functions appearing in the table are made up from the logarithm of the angle and from *even* powers of the angle. Depending on the required degree of accuracy, for small angles it may be sufficient to evaluate the first two or three terms of the above series. SIPOS accomplished the calculations to ten places.

The peculiarly structured, unfinished table did not become known; as far as we know it has not been used anywhere. However, the mathematical roots of its structure and the procedure applied for the compilation reveal profoundness and invention.

*

The best-known table of Hungarian authorship is a prime factor table by LÁSZLÓ CSERNÁK published at his own expense in Deventer, Holland, in 1811 [2]. The *Cribrum Arithmeticum* was more frequently referred to in contemporary critical literature than later FARKAS BOLYAI's *Tentamen* (JÁNOS BOLYAI's *Appendix* remaining unnoticed for a long time). Several great masters of mathematics in those days have appreciated the work. At various places, we find remarks as:

"... a useful work for all who perform computations" (DELAMBRE); "... a valuable monument to science", "... worthy of praise and appreciation by all" (LEGENDRE); "Csernák's table must be continued without hesitation" (CRELLE); "... he did this grand, tiresome and boring work at his own expense in such a way that no one can find a single flaw either with the fine layout or the impeccable editing." (SÁRVÁRI).

The first factor table was compiled in the mid-13th century by LEONARDO PISANO. The subsequent works only differed from the original by proceeding farther and farther with the list of prime numbers and the factorization of composite numbers into primes. One of the best-known tables prior to CSERNÁK's was that of VEGA (1797) who decomposed the numbers up to 400,000 into prime factors and listed the primes up to the same number.

What immense work CSERNÁK has done by compiling his 1022-page book is also indicated by the full title of the work (originally in Latin): "An *arithmetical sieve or table,* displaying the prime numbers separately from composite numbers, carefully arranged in growing order from 1 up to one million twenty thousand (1,020,000). The table includes the prime factors of all numbers not divisible by 2, 3 and 5 — not only some, but all of them."

In the Latin foreword CSERNÁK defines some well-known concepts (prime number, composite number, odd and even number), then specifies in separate paragraphs what the aim of the table is. He says that his main aim has been to help computations: multiplication, division, extraction of roots, solution of proportions finding greatest common divisors and least common multiples. But behind this explanation one can detect the actual aim of the book — and practically of all such tables growing in number at that time —: finding proofs or counter-examples to some well-known but unsettled problems of number-theory (distribution of primes, WARING's and GOLDBACH's conjectures, the *great* FERMAT conjecture, etc.) by the aid of correct tables easy to handle.

As the title of the work reveals, the table excludes numbers divisible by 2, 3 or 5. CSERNÁK arranged the rest so that every page includes numbers between one multiple of thousand and the next. Composite numbers occur together with their factorization in primes, while primes are indicated by a conspicuous line drawn next to them. The arrangement makes the handling of the table easy; for instance, one can read off many prime twins in no time. The reliability of the table is enhanced by the strikingly few misprints. CSERNÁK himself attached a list of 35 errors to the work, and this figure did not rise above 53 after several careful checkings by A. CUNNINGHAM, either (Factor tables. Errata. *The messenger of mathematics,* 34, 1905, pp. 24—31.)

GAUSS, one of the most authentic critics of CSERNÁK's book, wrote of it at length. This is all the more surprising as he was said to be very sparing of praise:

"The full title of this significant and highly useful work precisely describes its contents: it is a table compiled with careful and arduous work of many years, containing all the prime factors of numbers from 1 to 1,020,000 except those divisible by 2, 3 and 5, in a clear and very correct printing, as several checkings have proved. Anyone who has much computing with large numbers to do will know how valuable a work like this is for arithmetic. The author deserves our gratitude for two reasons: on the one hand, for the extremely great effort, which has surely added his name to those of the unforgettable RHAETICUS, PITISCUS, BRIGGS, VLACQ, WOLFRAM, TAYLOR[6] and others, and on the other hand, for covering the presumably high costs of publication on his own without which he would probably never have found a publisher. Though usually on a smaller scale, such tables have often been compiled, yet they either remained in manuscript

---

[6] GAUSS, rather arbitrarily, mentions a selection of the over 500 tables published by then. RHAETICUS: *Canon doctrinae triangulorum.* Wittenberg, 1551; PITISCUS: *Thesaurus mathematicus.* Frankfurt, 1613; BRIGGS: *Trigonometria Britannica.* Gouda, 1633; VLACQ: *Arithmetica logarithmica.* Gouda, 1628; WOLFRAM's 48-place logarithm table was published in Schulze's collection of tables in Berlin in 1778; TAYLOR: *Tables of logaritmus.* London, 1795.

23. The title page and a page of *Cribrum Arithmeticum*

or their publication was interrupted. LAMBERT used to encourage me to continue PELL's table, which went as far as 100,000 and enjoyed editions; according to correspondence between BERNOULLI and LAMBERT, OBERREIT had continued it up to 500,000 and from him the manuscript came to the hands of Schulze. As page 223 of Volume 2 of "Monatl. Correspondenz" informs us, ANTON FELKEL reached 2 million in manuscript and intended to continue the work up to 2,460,000; the rest that appeared in Vienna on public money was only used as cartridges in the war against the Turks as no subscribers could be recruited.[7] In this way, many years of wearisome work intended for the benefit of the public has been wasted; that is why we feel obliged to make the appearance of this work widely known." ([12], pp. 181—182).

There have always been, and there will always be, practitioners of mathematics who undertake the small chores indispensable for the advancement of science. They deserve respect and appreciation, but it is doubtful if the honours of a scholar are their due. That is exactly the case with CSERNÁK. It is heartening to know that GAUSS has written at such length and in such laudatory terms about him, yet one wonders why he failed to devote more than a few laconic sentences to the *Tentamen* or the *Appendix* incomparably higher in scientific value.

During the last century several similar endeavours followed in the wake of CSERNÁK's, some embracing a richer material and typographically far outstripping the *Cribrum Arithmeticum* published over one and a half century ago. Nevertheless, it has not lost much of its usefulness. Many authors could be quoted to confirm this,[8] we content ourselves with mentioning FELIX KLEIN's appreciative criticism ([19], Vol. 1, p. 44).

Bibliography

[1] BENDA, KÁLMÁN: *Writings by Hungarian Jacobins.* I—III. Budapest, 1952—1957. (In Hung.)
[2] CSERNÁK, LÁSZLÓ: *Cribrum Arithmeticum, etc.* Deventer, 1811.
[3] DÁVID, LAJOS: *Mathematicians from Debrecen in the old times.* Debrecen, 1927. (In Hung.)
[4] DUGONICS, ANDRÁS: *Two books of knowledge.* Pest, 1784. (In Hung.)
[5] — *Four books of knowledge.* Pest, 1798. (In Hung.)
[6] — *Notes.* Budapest, 1883. Olcsó Könyvtár, Nos 401—402. (In Hung.)
[7] Eötvös Loránd: In memory of Ányos Jedlik. *Természettudományi Közlöny*, Vol. 29, 1897, pp. 387—402. (In Hung.)
[8] FARKAS, ANTAL—NAGY, SÁNDOR: *Dugonics Album.* Szeged, 1876. (In Hung.)
[9] FINKEL, LUDWIK—STARZYNSKI, STANISLAW: *Historya uniwersytetu Lwowskiego.* Lwów, 1894.
[10] FRAKNÓI, VILMOS: *The conspiracy of Martinovics and his associates.* Budapest, 1880. (In Hung.)
[11] — *The life of Martinovics.* Budapest, 1921. (In Hung.)
[12] GAUSS: *Werke.* Vol. 2.
[13] KEREKGEDEI MAKÓ, PÁL: *Compendiaria matheseos institutio.* Vienna, 1764, 1766, 1771 and other dates.

[7] The work at issue is FELKEL's *Tafel aller einfachen Faktoren der durch 2, 3, 5 nicht teilbaren Zahlen* (Wien, 1776). Volumes 1—3 proceed up to 408,000. Allegedly, Felkel compiled his table up to 2,000,000 in manuscript, but the unpublished parts have been lost.

[8] For example GLAISHER, J. W. L.: On factor tables, etc. *Proc. of Cambr.*, Vol. 3, 1878, pp. 99—108; SEELHOFF, P.: Geschichte der Faktorentafeln. *Arch. der Math. u. Phys.*, Vol. 70, 1884, pp. 413—427.

[14] — *Calculi differentialis et integralis institutio.* Vienna, 1768.
[15] — *De arithmeticis et geometricis aequationum resolutionibus libri duo.* Vienna, 1770.
[16] — *Institutiones arithmeticae.* Buda, 1777.
[17] — *Elementa matheseos purae.* Buda, 1778.
[18] — *Elementa geometriae practicae.* Buda, 1778.
[19] KLEIN, FELIX: *Elementarmathematik vom höheren Standpunkte aus.* 4th edition. Berlin, 1933.
[20] MARTINOVICS, IGNÁC: *Theoria generalis aequationum, etc.* Buda, 1780.
[21] PRÓNAI, ANTAL: *A biography of András Dugonics.* Szeged, 1903. (In Hung.)
[22] SÁRKÖZY, PÁL: Pál Kerekgedei Makó and his mathematical work. *Mat. és Fiz. Lapok*, Vol. 36, 1929, pp. 23—34. (In Hung.)
[23] — Mathematicians from Nagyszombat in the old times. *Pannonhalmi Szemle Könyvtára*, No. 6, 1933. (In Hung.)
[24] SÁRVÁRI, PÁL: A short note on the arithmetic sieve and life of the venerable László Csernák. *Tudományos Gyűjtemény*, Vol. 6, 1817, pp. 157—160. (In Hung.)
[25] SIPOS, PÁL: *Specimen novae tabulae trigonometricae.* Pozsony, 1807.
[26] — *Berichtigung über eine trigonometrische Tafel mit der Aufschrift: Specimen novae tabulae, etc.* Pozsony, 1807.
[27] SZÉNÁSSY, BARNA: The mathematical work of Ignác Martinovics. *Matematikai Lapok*, Vol. 8, 1956, pp. 277—290. (In Hung.)
[28] — On a so-far unknown mathematical treatise by Ignác Martinovics. *Ibid.*, Vol. 20, 1969, pp. 57—62. (In Hung.)
[29] *The correspondence of Ferenc Kazinczy.* Edited by VÁCZY, JÁNOS. Vol. 2. Budapest, 1891. (In Hung.)
[30] WOYCIECHOWSKY, JÓZSEF: *The life and mathematical work of Pál Sipos.* Budapest, 1932. (In Hung.)

12. Hungarian attempts at quadrature and trisection

When scanning the bibliographies of 18th—19th century Hungarian books on mathematics, we will find several treatises on the problem of squaring a circle and trisecting an angle.[1] Over the times especially the word "circle-squarer" had assumed a false tinge denoting naive dilettants, half-educated intellectuals and their ridiculous writings. Yet the question deserves more than mere criticism: our mathematical literature has preserved a few fascinating results also in this area.

For better orientation, we have divided the material on quadrature and trisection into three parts.

Naive procedures

Those who did not have sufficient mathematical knowledge tried to solve these problems in the original setting with straightedge and compasses, committing gross errors and logical skips. SÁMUEL BRASSAI passed the following ironical remark on their work:

"... many, including several Hungarians, compared to their merits in other problems of geometry, have earned a ridiculous name" ([2], p. 524).

It is of no use giving a list of those showing complete ignorance; in one of his studies ANTAL VÁLLAS [14] addressed devastating criticism at several of them. In Hungary, one of the trail-blazers in this field was the imperial and royal captain KÁROLY LAISZTNER born in Pest, who — as a "renowned mathematician" of his age [3] — discovered the great "secret" of quadrature in 1737, as he claimed in his German tract published in Vienna, and dedicated his significant discovery "zu besondern Ehren des teutschen Vaterlandes."

---

[1] In SÁNDOR SZENDREY's study *A history of quadrature in Hungary* (Debrecen, 1982. Manuscript in the library of the Kossuth Lajos University of Science in Debrecen. In Hung.) 34 earlier Hungarian writings on the subject of squaring the circle are presented and analysed mathematically, especially with a view to precision. The total number of writings on this subject is estimated at over 50.

A' Kör' négyszögíttése

(Quadratura Circuli)

lehetségesé, vagy sem; ha az, mennyiben, és mi módon lehetséges?

―――――――

A' legfoghatóbb előterjesztésben

felvilágosittani igyekezett

Somogyi Cs. Sándor.

Sz. Jakab Hava utólján

1 8 2 8.

━━━━━━━━━

Pesten,

Füskúti Landerer Lajos betuivel.

24. One of the works of "circle-squarers"

Neither does the somewhat less populous camp of naive trisectors deserve more attention. It suffices to mention the name and a brainwave of DÉNES KATONA [7]: We should triple a given angle by construction, and the procedure immediately offers us the key to trisection as well. All we need to do, Katona opines, is to perform the steps of tripling in reverse order. As if it were a film played backward.

Approximative and non-Euclidean constructions

Several people tried to figure out approximative constructions, or allowed for the use of *other tools* besides compasses and ruler. The following attempt at straightening the circular line by construction dates from 1840.

Let the angle α at the centre C of the circle be acute and denote the associated arc by $\overarc{AB}$ *(Figure 17)*. Draw the tangent to the circle at point A and bisect the arc $\overarc{AB}$. The line defined by B and the point of bisection D, intersects the tangent in E. Drawing a parallel line to CE through B, it intersects the tangent in N. The resulting $\overline{AN}$ segment is approximately equal in length to the arc $\overarc{AB}$. In order to rectify the whole circumference, we first divide it into six equal parts and rectify each part by the construction described.

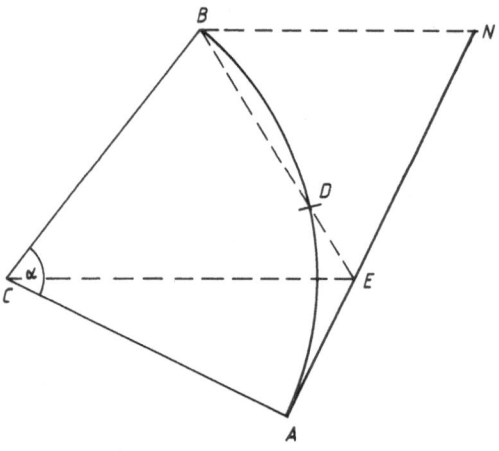

Figure 17

This construction also serves to show how controversial the situation in our mathematical life was even at the beginning of the last century: whereas the *Tentamen* and the *Appendix* were printed by students at the college of Marosvásárhely, the former — mathematically unwarranted — attempt was given forum by one of the most prestigious Hungarian journals *(Tudománytár)*.

Blind chance favoured the contriver of the above method as the construction yields quite good results for arcs belonging to certain central angles. Namely, the formula corresponding to the procedure is the following:

$$\overline{AN} = \frac{4\sin^3\frac{\alpha}{2}}{4\cos^2\frac{\alpha}{4} - 1} + \sin\alpha$$

($r=1$). For $\alpha = 60°$ from this we get π with the precision of two decimal places. If, however, we take into account that far more accurate approximative constructions had been in use well before this date, we may rightly smile at the childish joy of the author of this construction hiding behind the pseudonym ZENGEDY:

> "And that is everything. That is the treasure that as far as I know — but only as far as I know — I have been the first to find." (*Tudománytár*, Vol. 8, 1840, pp. 354—375).

There is a short note among JÁNOS BOLYAI's posthumous papers which describes how to trisect an angle with the help of an *equilateral hyperbola*: Let us place the acute angle $PO \times \angle = \alpha$ to be trisected into the rectangular system of coordinates x, y and let the equilateral hyperbola RS be given in this system *(Figure 18)*. By drawing a circle of centre P with radius $2\overline{OP} = 2r$ we obtain the point S. Let us draw the line PT parallel to the x axis. An elementary calculation shows that the angle $TPS \angle = \beta$ is one third of α.[2]

In searching for a solution to the problem of squaring a circle, PÁL SIPOS — already mentioned above (Chapter 11) — achieved some remarkable results in the field of constructions using straightedge, compasses and a ruler with a transcendental curve for edge. His main findings can be read in a study preserved in the collection of

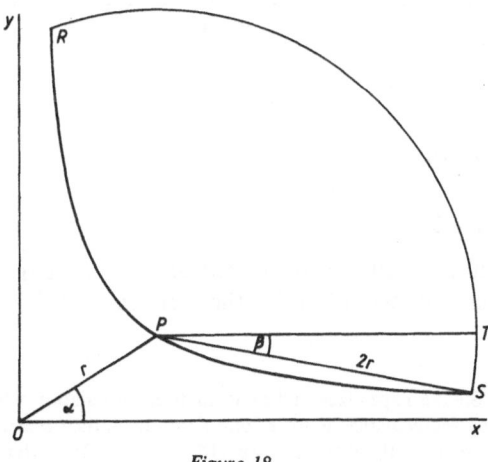

Figure 18

[2] SZŐKEFALVI-NAGY, GYULA: János Bolyai's trisection of an angle. *Matematikai Lapok*, Vol. 4, 1953, pp. 84—86. (In Hung.); TÓTH A.: *Notiuni de teoria constructiilor geometrice*. Bucureşti, 1963, p. 80. It is, however, unlikely that this construction was BOLYAI's original invention.

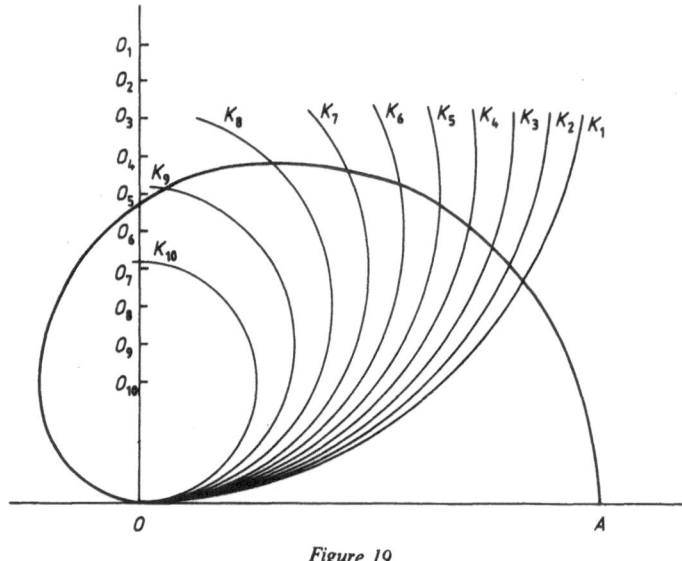

Figure 19

manuscripts of the Hungarian Academy of Sciences [11] and in a paper published in the *Acta* of the Berlin Academy [12]. In these SIPOS defined — without mathematical formulation — the *cochleoid*, already described in one of EULER's works (*Novi Comm. Ac. Petr.*, Vol. 8, 1760—1761, p. 26), in the following way. Let us regard the points of the perpendicular erected at the origin O to the polar axis as the centres of the bundle of circles passing through O *(Figure 19)*. If, starting from the origin we measure a given length k on every circles of the bundle (in the same sense obviously the radius of each circle in the sheaf must be $\geq k/2\pi$), then the endpoints determine a transcendental curve whose polar equation is:

$$\rho(\varphi) = k \frac{\sin \varphi}{\varphi}.$$

Indeed (cf. *Figure 20*),

on the one hand $\qquad\qquad\qquad\qquad\qquad\qquad\qquad\qquad\qquad k = 2r\varphi,$

and on the other hand $\qquad\qquad\qquad\qquad\qquad\qquad\qquad\qquad \rho(\varphi) = 2r \sin \varphi.$

This yields the equation of the cochleoid. EULER used this curve to rectify a quadrant, while SIPOS designed a construction for the rectification of the ellipse with its help.[3]

---

[3] Most regrettably, SIPOS's paper lacks all mathematical formulation whatever, which probably accounts for its neglect by mathematicians for a long time. Bode, then secretary of the Berlin Academy, added with an asterisk to the title of the treatise: "Mr. Sipos of Transylvania, ... returning from the University of Frankfurt, called on me during his stay in Berlin a few years ago and asked the above-mentioned tract to be presented to the Academy. That having been done, the members of the mathematical section came to the decision that the Academy should publish these ideas as the fruit of an excellent geometric thinker in its Acta, without however, regarding these mechanical constructions as geometrical theorems."

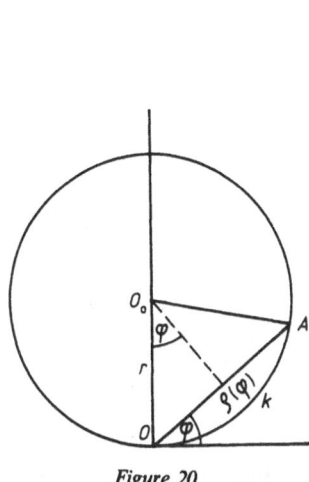

Figure 20

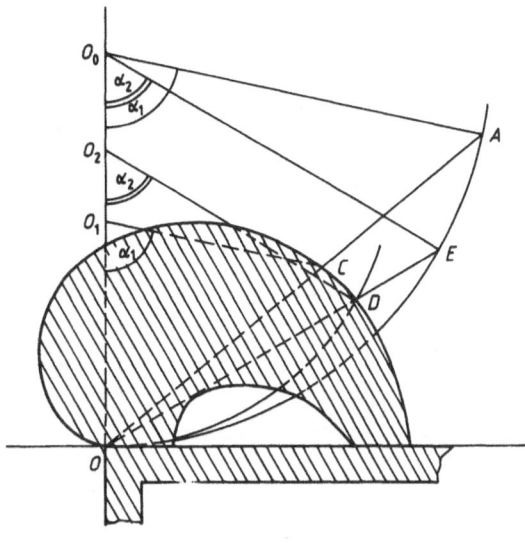

Figure 21

KÁLMÁN SZILY SR. [13] mistakenly called the cochleoid "SIPOS's spiral",[4] but — as we shall see — it does not belittle SIPOS's merits in the least that the cochleoid was not his invention.

By way of introduction, SIPOS uses the cochleoid in several constructions related to the circle. For the better understanding of subsequent material, we have to get acquainted with one of them in which a given circular arc is divided into two parts in any proportion $\frac{a}{b}$ with the help of the cochleoid.

Let us place the point O of the arc $\overset{\frown}{OA}$ to be divided in the required proportion at the origin of the polar system so that the axis be tangent to the circle at the point O. The centre of the circle containing the arc $\overset{\frown}{OA}$ is O_0. Let us place the ruler having a cochleoid for outer edge — called *isometer* by SIPOS — as in *Figure 21*. The chord $\overline{OA}$ passes through point C of the isometer; that member of the bundle of circles which passes through C has centre O_1 and radius $\overline{OO_1}$.

If we now draw that circle of the bundle which has centre O_2 and radius

$$\overline{OO_2} = \left(\frac{b}{a} + 1\right)\overline{OO_1},$$

it will intersect the cochleoid in point D. The point E obtained by extending the segment $\overline{OD}$ divides the arc $\overset{\frown}{OA}$ in the required proportional $\frac{a}{b}$.

---

[4] FALKENBERG's mistake is more serious than SZILY's; the former states in a paper (Die Cochleoide. Arch. d. Math. u. Phys., Vol. 70, 1883, pp. 259—268) that in the course of designing a kind of steering mechanism, he found an interesting new curve that he named cochleoid after its snail-like shape. E. WÖLFFING lists 29 earlier papers concerned with the cochleoid (*Loria Boll. Bibl.* Vol. 3, 1900, pp. 97—99).

The correctness of the construction can easily be verified by calculation. Let $\overline{OO_1}=r_1$ and $\overline{OO_2}=r_2$, then by the definition of the cochleoid,

$$r_1\alpha_1 = r_2\alpha_2 = \left(\frac{b}{a}+1\right)r_1\alpha_2,$$

whence $\dfrac{\alpha_2}{\alpha_1-\alpha_2} = \dfrac{a}{b}$. If we consider the angle equalities shown in *Figure 21* and easily verified, then the correctness of the construction follows. Evidently, this construction can favourably be applied to trisection: unfortunately, the rich literature on this subject does not include the name of PÁL SIPOS. Publications on attempts to divide an angle with curves related to the cochleoid mushroomed especially toward the end of the last century. For example, O. P. DEXTER (*The division of angles.* New York, 1881) used the curve

$$\rho = r\frac{\sin m\phi}{\sin (m-1)\phi}$$

and A. GRINTEN (Die $n-$ und $n+1$ — Teilung des Winkels und Kreises. *Arch. d. Math. u. Phys.*, Vol. 70, 1884, pp. 393—399) applied the curve

$$\rho = r\frac{\sin n\phi - \sin \phi}{\sin (n-1)\phi}.$$

This is the construction SIPOS uses for his method of rectifying the ellipse. In fact, let the sum of the two axes of the ellipse be $\overline{AB}$, and their difference $\overline{AC}$. Then *(Figure 22)*:

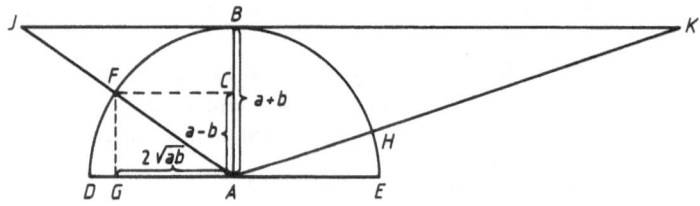

Figure 22

1. Draw the semicircle $\overset{\frown}{DBE}$ of radius $\overline{AB}$ so that $\overline{AB}$ be perpendicular to $\overline{DE}$. Drawing a parallel to $\overline{ED}$ from point C, we get the point F, whose projection on the diameter $\overline{ED}$ will be G.

2. Starting from point E, divide the quadrant $\overset{\frown}{EB}$ in the proportion $\overline{DG}:\overline{GA}$ with the isometer (in the way described above). Let the point of division be H.

3. Construct the tangent of the circle at point B; the extensions the segments $\overline{AF}$ and $\overline{AH}$ will cut the tangent at points I and K.

4. According to SIPOS, the length $\overline{AI}$ is a good approximation for the geometric mean of the distance $\overline{AK}$ and the quarter of the perimeter of the ellipse. Expressed in formula:

$$k \approx \frac{4\overline{AI}^2}{\overline{AK}}.$$

116

From the mathematical point of view, it is not the cumbersome construction which is important but the following approximation formula (not given by Sipos) which can be obtained from it by elementary tools:

$$k \approx \frac{4(a+b)^3}{(a-b)^2} \cos \frac{\pi\sqrt{ab}}{a+b}.$$

József Jelitai (Woyciechowsky) has studied this approximation formula closely [4], comparing it for accuracy with nearly fifty formulae of later dates. His finding was very favourable for Sipos's procedure: for the extreme cases of $b=0$ and $a=b$ the formula yields accurate results,[5] and compared to the values obtained from the complete elliptic integral of the second kind, we get the maximal deviation when $b = \frac{a}{10}$, but even then it is only 0.14%.

Jelitai only found two formulae that proved to be more accurate than Sipos's. One is P. Mansion's of 1882:

$$k \approx \frac{\pi}{4} \left[a+b+ \sqrt{2(a^2+b^2)} + \sqrt{a^2(2+\sqrt{2})+b^2(2-\sqrt{2})} + \right.$$

$$\left. + \sqrt{a^2(2-\sqrt{2})+b^2(2+\sqrt{2})} \right].$$

The other is Peano's (1889):

$$k \approx \frac{2\pi}{9} \left(19s - 4 \frac{2s^2+3p^2}{\sqrt{s^2+3p^2}} \right)$$

$(2s = a+b;\ p^2 = ab)$. As can be seen, both formulae are more complicated, hence their application is more difficult. As for precision, one possibility is to compare the series expansions of the Sipos, Peano and Mansion formulae with the series expansion of the complete elliptic integral of the second kind. Precision is indicated by the number of terms coinciding with those in the infinite series of the elliptic integral. It turns out that Sipos's formula can be regarded as accurate up to the term of the 8th degree, Peano's up to that of the 12th and Mansion's up to that of the 16th degree.

Approximations of π

Let us conclude the chapter with a brief mention of those Hungarian "circle-squarers" who wished to approximate the value of π as closely as possible with the help of infinite series. Chronologically, one of the first was Sámuel Mikoviny.

In his papers [9, 10] he defended a mathematician called Marinoni, who was at-

---

[5] For it is easy to prove that

$$\lim_{b \to 0} k = 4a, \quad \text{and} \quad \lim_{b \to a} k = 2a\pi.$$

tacked by LAISZTNER mentioned earlier. MARINONI[6] was among those who conjectured that π was an irrational number. Using the infinite series

$$\pi = \sqrt{12}\left[\frac{1}{3^0 \cdot 1} - \frac{1}{3^1 \cdot 3} + \frac{1}{3^2 \cdot 5} \pm \cdots \right],\tag{1}$$

$$\pi = \sqrt{768}\left[\frac{1}{3 \cdot 1 \cdot 3} + \frac{2}{3^3 \cdot 5 \cdot 7} + \frac{3}{3^5 \cdot 9 \cdot 11} + \cdots \right]\tag{2}$$

due to LAGNY (1660—1734), MIKOVINY calculated the value of π to an accuracy of 25 decimal places.[7]

ANTAL VÁLLAS represented π by several series unknown before and converging comparatively fast [15]. LAJOS BITNICZ preferred to approximate π through the area of inscribed and circumscribed regular polygons, so his approach is far from being original, yet some of his remarks deserve attention:

"... It may well be impossible to represent π even by radical quantities precisely, and thus, nothing can be done about this problem.[8] It would be good if someone cut off all future attempts by strictly proving this latter statement." ([1], p. 170).

In the opinion of FERENC KEREKES, all circle-squarers had better begin by studying analysis thoroughly because

"... with its help they have more hope to solve the problem, or to show that the solution is impossible." ([8], p. 128).

In the first half of the 19th century many mathematicians presumed that π had an extraordinary place in the realm of numbers but it was only in 1882 that LINDEMAN proved its transcendence. So BITNICZ's and KEREKES's remarks were not ahead of the times, but being personages of nationwide prestige, they helped considerably cut down the number of naive Hungarian circle-squarers. Finally, point 162 of the "Rules of Procedure" released by the Hungarian Academy of Sciences around the middle of the 19th century — one century later than those of the French Academy — spelt out that

"... papers concerned with the quadrature of the circle, the trisection of an angle, and the perpetuum mobile shall be refused unread."

[6] Mathematician of the Vienna Court, teacher of SÁMUEL MIKOVINY.

[7] (1) can be directly obtained from the series expansion of $\frac{\pi}{6} = \text{arc tg}\frac{\sqrt{3}}{3}$, while (2) follows from (1) by uniting pairs of adjacent terms.

[8] I.e. the quadrature of the circle.

Bibliography

[1] BITNICZ, LAJOS: On the quadrature of the circle. *Évkönyv*, Vol. 2, 1832—34, Part 2, pp. 151—170. (In Hung.)

[2] BRASSAI, SÁMUEL: *Euclid's Elements.* Pest, 1865. (In Hung.)

[3] HORÁNYI, ALEXIUS: *Memoria Hungarorum et Provincialium scriptis editis notorum.* I—II. Vienna, 1775—1776.

[4] JELITAI (WOYCIECHOWSKY), JÓZSEF: *The life and mathematical work of Pál Sipos.* Budapest, 1932. (In Hung.)

[5] — Pál Sipos's manuscript and the cochleoid. *Mat. és Fiz. Lapok*, Vol. 41, 1934, pp. 45—54. (In Hung.)

[6] — Sipos manuscripts in the Teleki archives at Gyömrő. *Mat. és Fiz. Lapok*, Vol. 42, 1935, pp. 134—138. (In Hung.)

[7] KATONA, DÉNES: *Tripling the acute angle.* Szeged, 1843. (In Hung.)

[8] KEREKES, FERENC: *The true fundamentals of higher geometry, etc.* (Ed. CSÁNYI, DÁNIEL) Debrecen, 1862. (In Hung.)

[9] MIKOVINY, SÁMUEL: *Epistola de quadratura circuli.* Vienna, 1730.

[10] — *Epistola ad D. Jo. Jac. Marinonium occasione questionis de quadratura circuli.* Vienna, 1739.

[11] SIPOS, PÁL: *Tractatus de conicis, etc.* Collection of manuscripts, Hungarian Academy of Sciences, Mathematics, 4—5.

[12] — Beschreibung und Anwendung eines mathematischen Instruments, usw. *Sammlung deutscher Abh.* 1790—91, Berlin, 1796, pp. 201—230.

[13] SZILY, KÁLMÁN, SR.: A Hungarian circle-squarer in the last century. *Műegyetemi Lapok*, Vol. 2, 1877, pp. 50—51. (In Hung.)

[14] VÁLLAS, ANTAL: Recent Hungarian mathematical literature, etc. *Tudománytár*, 1836, 4, pp. 143—172. (In Hung.)

[15] — Ludolph's number. *Évkönyv*, Vol. 6, 1840—1842, pp. 377—394. (In Hung.)

IV. The times of the two Bolyais

13. Some preliminary remarks

The first half of the 19th century can be regarded as an *Age of Reforms* not only in society but in mathematics as well, without trying to draw too far-fetched an analogy between Hungarian history and mathematical culture. But in mathematics the exact limits of this period cannot be defined.

The Reform Age of mathematics is predominated by the work of the two BOLYAIS. Many argue that previous to them no significant mathematical achievements had existed in Hungary. As regards *scientific creation*, this is almost perfectly true. The two BOLYAIS have brought explosive change in development — seemingly out of nothing. The foregoing chapters of this book have been intended to expound as extensively as possible the mathematical results dating from earlier times in Hungary. They show convincingly that the ambition to create something *original* had already been stimulating our mathematicians in the 18th century, although no substantial contributions ensued. But in the huge edifice of mathematics there are theorems that account to no more than a few of the infinitely many building blocks, while there are others that constitute the firm foundations for the whole. The 18th century Hungarian results all belong to the former, while the *entire* life-work of the two BOLYAIS belongs to the latter category.

Nevertheless, one must be careful not to go to subjective extremes when judging the modest results of earlier times compared with the meteoric rise coming with the two BOLYAIS. It should be realized that the *average* mathematical culture had been *increasing* step by step in Hungary from the late 18th century to the first half of the 19th. Success was often followed by failure, advancement by standstills, but on the whole the progress was measurable. The immense — almost heroic — efforts made by Hungarian intellectuals from the late 18th century onward to combat backwardness in general and in mathematics in particular were slowly bringing fruit. For instance, FARKAS BOLYAI's *Tentamen* as a textbook for higher education — not regarding here the new discoveries in it — is not much ahead in standards of PÁL KEREKGEDEI MAKÓ's or JÁNOS PASQUICH's similar efforts. F. BOLYAI did his utmost to provide his pupils with a firm knowledge for the future, but in these of his endeavours he was *one* of many: in teaching young people, the results of PÁL SÁRVÁRI and FERENC KEREKES in Debrecen, PÁL TITTEL in Eger, MÓZES KÉZY, ISTVÁN NYIRY and PÁL SIPOS in Sárospatak, LAJOS BITNICZ in Szombathely, BÉLA BRESZTYENSZKY in Győr, FERENC PETHE at the Georgicon in Keszthely, PÁL MÉHES and his son SÁMUEL in Kolozsvár, and KÁROLY SZÁSZ SR. and SÁMUEL HEGEDÜS in Nagyenyed were of the same order as those of FARKAS BOLYAI.

The spirit of reforms is evident in many contemporary technical, economic and financial problems that required mathematical tools as well. The majority of such publications, though not of higher level than before, brought a new tone into our mathematical life. Let us see a few instances of this.

Some of FERENC PETHE's technical innovations, related computations and urgent appeals to found commercial and insurance companies were aimed at accelerating middle-class development.[1] In several of his tracts, chief engineer ISTVÁN VIZER addressed the problems of geographical triangulation, and in one study [16] he discussed the elements of the *Mercator projection*. The engineer ISTVÁN GÁTY, initiator of measuring distance in an optical way — whom contemporary critics called a plagiarist — embarked in one of his studies [3] on the following question provoking much debate at that time. When a storage dam is built, the water level evidently rises in front of the sluice. The rise measured at any given point is a function of the distance from the sluice and of several data of the river and the dam. At any rate, when the level of the water rises, the previously installed paddlewheels sink deeper, which in turn modifies their efficiency and even exposes them to damage. The author of the article thoroughly examined a rather involved four-variable empirical formula for the water level and demonstrated on the basis of measurements that the formula was incorrect. The timeliness of the problem is also revealed by the author's remark that the "inopportune formula" shackled "the wings of industry and commerce". The engineer KÁROLY RAISZ used trigonometry to evaluate the force exerted when pulling a cart along a road fraught with uphill and downhill stretches [11]. An interesting consequence of his primitive considerations is that braking needs more work than pulling. The aim of his study was to give advice on how to determine the course of some roads planned in a hilly region of Hungary. Studies belonging here include ISTVÁN NYIRY's on the strength of bridges [10], and SÁNDOR GYŐRY's on how to make the Danube navigable [5]. Győry also devoted a voluminous treatise to the importance of mathematics in the industrialization of the country [6].

By far the most outstanding mathematical discussion of a technical subject was PÁL VÁSÁRHELYI's theory of the velocity of rivers. It had been a practical problem unsolved for centuries how to determine the quantity of water flowing across a section of the river per unit time. To answer this question we need the area of the river cross-section and the average velocity of the water current. The first can be well estimated by measurements, but the computation of the latter is far more difficult as velocity obviously changes from point to point from the surface downward and from the centre toward the sides. Controversy had especially centered about the relation between the depth from the surface and velocity. Most scholars thought this function was linear — or at best they suggested some quadratic function without any justification.

---

[1] For a detailed account on them, see SÜLE, SÁNDOR: *Ferenc Kisszántói Pethe* (Akadémiai Kiadó, Budapest, 1964) (In Hung.).

Auflösung

einiger

wichtigen Aufgaben

als

Beitrag

zum

geometrischen Trianguliren

durch

Paul Vásárhelyi

Donau Mappirungs-Ingenieur

Ofen,

gedruckt in der königl. ungr. Universitäts-Buchdruckerey 1827.

VÁSÁRHELYI declared that in order to formulate the relationship in question mathematically, we have to know the velocity at three points along the same vertical line: on the surface, around the middle of the depth and near the riverbed. The integral mean of the quadratic interpolational parabola can be taken as the average velocity of the current in this vertical line [15]. Measurements largely verified VÁSÁRHELYI's theory: his investigations were well ahead of the similar findings by the Americans ABBOT and HUMPHREYS in 1851—58.

VÁSÁRHELYI's works on triangulation problems also demand attention [13, 14]. In them, he described different ways to orient a surveying table, explained the basic constructions on it, and also solved several problems of constructing directions with the method of intersection, resection and side section, particularly for cases when the known points fall outside the table. Although these constructions are not original, VÁSÁRHELYI's lucid reasoning, the spotless realization of the figures attached and the stressing of the importance of a national triangulation network endow these treatises with a special value.

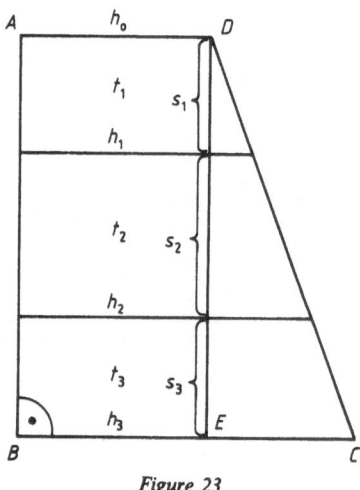

Figure 23

Finally, let us mention the NASZLUHÁCZ formulae known and used in land-survey all over the world. The problem and its solutions, published by NASZLUHÁCZ in a short paper [8] (reviewed in [12]) is briefly the following. Let $ABCD$ be a trapezoid area *(Figure 23)* in which $B \sphericalangle = 90°$. This trapezoid is to be divided by lines parallel to the base into parts (three in the Figure) of, generally speaking, different area. The following data are given: the area of the whole trapezoid and of the parts ($T = = t_1 + t_2 + t_3$), the altitude s and the two parallel sides h_0 and h_3 of the trapezoid. Define the "coefficient of the variation of length" v by the relation

$$v = \frac{h_3 - h_0}{s}$$

(that is, $v \lessgtr 0$ depending on whether $h_3 \lessgtr h_0$). Using this notation, NASZLUHÁCZ's formulae are:

$$s_1 = \frac{h_1 - h_0}{v}; \quad s_2 = \frac{h_2 - h_1}{v}; \quad s_3 = \frac{h_3 - h_2}{v},$$

and

$$h_1 = \sqrt{h_0^2 + 2vt_1}; \quad h_2 = \sqrt{h_1^2 + 2vt_2}; \quad h_3 = \sqrt{h_2^2 + 2vt_3}.$$

Obviously, the number of subareas may be arbitrary, and it is also a unnecessary restriction that the trapezoid to be dissected should contain a right angle. More recently several authors (ZOLTÁN FUTAKY, VILMOS VINCZE and others) have dealt with various generalizations of these formulae, because they are commonly used in present-day geodetic practice. It is interesting to note that the word "estimated" in the title of NASZLUHÁCZ's paper refers to weighted dissection as the subareas — due to their different estimates — must be taken with different weights given in advance.

*

Hungarian education in the Age of Reforms featured a constant, often fierce struggle with changing outcome between *science* and *humanities*, and this had an impact on mathematical life as well.

Instead of giving a detailed account of all the phases of this controversy, let us quote but a few typical facts. As mentioned earlier, the first *Ratio Educationis* — to good avail — had emphasized the utilitarian aspects of mathematics also in secondary schools. In the epoch considered, complying with the general stage of development, children of the landowner gentry usually finished secondary school, but it was also this class that insisted most strongly on the old traditions including a bias to humanities. These were the circumstances that brought along the second *Ratio* (1806) which did not modify the set up of education much, except for a tightening of the system and methods of supervision, but *reduced* the time devoted to science in secondary schools.

However, signs of an opposite tendency can also be spotted in those years: from 1791 more and more people embraced the idea of supplementing the upper grades of at least *one* elementary school per *educational district* with an industrial section. These school branches (scholae vernaculae primariae) can be regarded as the forerunners of polytechnical secondary schools. Particular attention was paid to mathematics, mechanics and drawing; the most famous two schools of this kind were Pozsony and Győr. But the polytechnical "modern schools" did not reach an equal footing with grammar-schools before the mid-19th century: in 1849 the bill "Entwurf" provided for their establishment. The first modern schools grew out of the industrial sections; initially they had 4, then 6, and finally (from 1875) 8 classes. The equal legal standing of *modern* and *grammar-schools* was, however, only declared in Act XXX of the year 1883.

A similar battle went on between the advocates of science and those of humanities in protestant schools which did not accept the *Ratios*. In Debrecen, the separation of philosophy from mathematics could not be achieved before 1798. From then onwards, mathematics was never dropped from among the regular subjects, but the time devoted to it underwent substantial changes due to the rapid succession of teaching programmes. As an outcome of the latest clash between the camps of science and arts, the curriculum introduced in 1820 abolished much of the earlier achievements. From that time Debrecen's role in our mathematical culture strongly diminished. No new upswing came before the setting up of the university of arts and sciences (1912), and more precisely, before the training of teachers of mathematics and physics began in 1924.

Conflicting trends of a similar nature characterized all our institutions of education but the demand for giving greater weight to exact sciences in the life of the country was becoming more and more resolute, as we shall see later.

In mathematics, creative work requires less organization than in most of the sciences, and almost no technical prerequisites. Nevertheless, Hungarian mathematical achievements in the 19th century had some concrete, often unnoticed, motives as well. For example, a considerable part of current mathematical problems found their way to Hungary through personal encounters abroad, and our scientists being in seclusion at home tried to intensify and extend these contacts as much as possible. The long friendship and scientifically valuable correspondence between FARKAS BOLYAI and GAUSS is well known [2]. PÁL SÁRVÁRI arrived at Göttingen in 1792 and attended the lectures of KÄSTNER (1719—1800), a contributor to the history of parallels. SÁMUEL HEGEDÜS visited Göttingen in 1807 and after his return — if his biographer can be trusted (*Évkönyv*, Vol. 7, p. 158) he maintained a long correspondence with GAUSS and another professor at Göttingen, JOHANN TOBIAS MAYER (1752—1830). What seems to confirm this statement is that HEGEDÜS often acted as a go-between for GAUSS and FARKAS BOLYAI ([2], pp. 87—88, 91, 93—95). The exchange of letters between PASQUICH and GAUSS mentioned earlier would be worthy of publication. PÁL TITTEL visited GAUSS in 1810 to improve his knowledge of astronomy. "A son's loyalty the one and a father's affection of the other kindled friendly feelings in their bosoms", as LAJOS BITNICZ put it ([1], p. 9).

But these often loose and short-lived personal contacts do not convincingly explain, even in the case of the two BOLYAIS, the immense progress of Hungarian mathematical research in the 19th century. We have to take into account also the general trends of mathematical development in the given period.

Many historians of mathematics agree that while the 17th and 18th centuries were everywhere the period of an unprecedented abundance of new mathematical results, the 19th century was rather a time of *re-examination* of this crop. This revision was inevitable in several mathematical disciplines, especially *analysis* and *geometry*, since their foundations were defective. On the other hand, some concepts of analysis (infinitely large, infinitely small, limit, etc.) and the question of the structure of *space* were closely linked to philosophical problems and as such, required the researchers to have some training in philosophy.

The philosophical approach was a typical trait of all our former scholars absorbed in mathematical research, a phenomenon rooted in deep traditions of our Calvinist schools. The philosophical ideas of the eminent masters MIHÁLY SZATHMÁRI PAP and SÁMUEL KÖTELES found fertile soil in the thoughts of FARKAS and JÁNOS BOLYAI. Although the question cannot be treated at length here, let us cite two passages. SARTORIUS VON WALTERSHAUSEN commented:

"FARKAS BOLYAI . . . is a man of extraordinary knowledge whom GAUSS allegedly said to have been the only man to understand his metaphysical thoughts related to mathematics."[2]

Undoubtedly, the word "metaphysical" refers to the foundations of mathematics and the problem of time and space. As for JÁNOS BOLYAI:

"Starting from the assumption that in the early 19th century revolution of thought philosophy was in the vanguard, and that mathematics — incapable of development according to HEGEL — was also revolutionized by ideas matured within philosophy, IMRE TÓTH examines the *Appendix* as a masterpiece of both mathematics and philosophy."[3]

In mathematics and in philosophy alike FARKAS BOLYAI advocated progressive but moderate views, while his son embraced more daring and revolutionary ideas. Thus the two trends hallmarked in society by the names of ISTVÁN SZÉCHENYI and LAJOS KOSSUTH were represented in mathematics by FARKAS BOLYAI and JÁNOS BOLYAI in that period.

The following chapters are devoted to a more detailed exposition of their scientific achievements.

Bibliography

[1] BITNICZ, LAJOS: Commemorative address over Academy member Pál Tittel. *Évkönyv*, Vol. 2, 1832—1834, Part 2. (In Hung.)

[2] *Briefwechsel zwischen Carl Friedrich Gauss und Wolfgang Bolyai.* Leipzig, 1899.

[3] GÁTY, ISTVÁN: Essays in hydraulics. *Tudományos Gyűjtemény*, Vol. 5, 1822, pp. 85—97. (In Hung.)

[4] GONDA, BÉLA: *The life and work of Pál Vásárhelyi.* Budapest, 1896. (In Hung.)

[5] GYŐRY, SÁNDOR: On the regulation of the Danube. *Évkönyv*, Vol. 2, 1832—1834, Part 2, pp. 120—151. (In Hung.)

[6] — The mathematical sciences', etc. *Évkönyv*, Vol. 2, 1832—1834, Part 2, pp. 23—43. (In Hung.)

[7] KÜRSCHÁK, JÓZSEF: The past one hundred years of the history of mathematics in Hungary. In: *Az MTA évszázada*, Budapest, 1926, pp. 451—459. (In Hung.)

[8] NASZLUHÁCZ, LAJOS: *On estimated area division.* Graz, 1856. (In Hung.)

[9] NYIRY, ISTVÁN: A history of computing life expectancy. *Tudományos Gyűjtemény*, Vol. 9, 1821, pp. 49—69. (In Hung.)

[10] — The strength of wooden and stone bridges, etc. *Évkönyv*, Vol. 1, 1831—1832, Part 2. (In Hung.)

[11] RAISZ, KÁROLY: A word about the resonable design of roads. *Tudományos Gyűjtemény*, Vol. 5, 1823, pp. 19—43. (In Hung.)

[12] SZENT-IVÁNYI, GYÖRGY: A belated book review. (A review of Naszluhácz's book). *Földméréstani Közlemények*, Vol. 5, 1953, pp. 179—181. (In Hung.)

[13] VÁSÁRHELYI, PÁL: *Introductio in praxim triangulationis.* Buda, 1827.

[14] — *Auflösung einiger wichtigen Aufgaben.* Buda, 1827.

[15] — On the velocity of our rivers. *Magyar Tudós Társaság Évkönyve*, Vol. 6, 1840—1842. (In Hung.)

[16] VIZER, ISTVÁN: On what the Earth is like, etc. *Tudományos Gyűjtemény*, Vol. 2, 1820, pp. 3—22; Vol. 4, 1821, pp. 56—75. (In Hung.)

---

[2] *Gauss zum Gedächtnis.* Leipzig, 1856, p. 17.

[3] BENKŐ, SAMU: *The confessions of János Bolyai.* Bucharest, 1968, p. 264.

14. Farkas Bolyai

His mathematical system. Basic ideas

Several written records of FARKAS BOLYAI's work have survived partly in print, partly in manuscript. As to subject-matter, these writings are most varied: they include fiction, ethnography, technical topics, mathematics, etc. Of paramount value are his literary pieces and mathematical works, as well as his letters on mathematical subjects. His much-discussed plays (*Five tragedies.* Nagyszeben, 1817; *The Trial of Paris.* Marosvásárhely, 1818) articulate valuable poetical thoughts but their artistic presentation is less successful. Contemporary critics acknowledged their importance and they had some demonstrable influence on Hungarian drama. Literary historians are, however, still at fault for a detailed analysis of his literary work.[1]

FARKAS BOLYAI's mathematical endeavours were crowned by the *Tentamen*. As the wanderer his destination, so did F. BOLYAI reach with this work the goal of long years of preoccupation: the creation of an independent mathematical system. The rest of his mathematical works [7—10] and his correspondence with GAUSS [12] touching on scientific problems as well are only stations or bypasses along his main course. *Kurzer Grundriss* [11] is a "marvellous little book" (HOÜEL), a mathematical testament summarizing the most important ideas, a greater part of which can be gleaned from other works as well. In our view, its main merit is FARKAS BOLYAI's quick-sighted comparison between the geometries of JÁNOS BOLYAI and LOBACHEVSKY.

All this eases the job of the analysts of FARKAS BOLYAI's mathematical works as they can concentrate on the *Tentamen*, drawing on his other works for the sake of completeness at certain points only.

Among his mathematical books those written in Hungarian would deserve language the linguists' special attention. Yet, however, intriguing an aspect of his life-work this may be, it has no relevance to the history of Hungarian mathematics, as BOLYAI's attempts to coin new terms proved to be a failure. The following quotation aptly describes his linguistic innovations:

· "... the overriding rationalism of his mind is well exemplified by the mathematical terms he coined. He thought that words were suitable to express the essence

---

[1] The question is treated on a high level but without an attempt at completeness in KRISTÓF, GYÖRGY: *The Two Bolyais in Hungarian literature.* (In Hung.) Kolozsvár, 1947.

of things; his main concern was to find words that expressed the nature of the denoted things or followed the formation of mental images of mathematics. No doubt
it was under the influence of the prevalent ideas of the age that he declared that
words had to stem from the nature of the language, yet in practice he treated words
as exclusive products of the mind. Whenever a word failed to suit the mathematical
notion precisely, he changed it arbitrarily, without regard to the rules of the language... Another rationalist requirement was to have short technical terms."[2]

It is exactly the abundance of these laboured technical terms that makes the reading of BOLYAI's mathematical works in Hungarian so difficult,[3] while studying the
Tentamen is aggravated by its terse and complicated Latin. Unfortunately, only a
small fragment of this work can be read in Hungarian.[4] Comprehension is further
impeded by the newly introduced — and often redundant — notations that GAUSS
also criticized in a letter ([12], p. 127), while all he acknowledged was "a respectable
attempt at thoroughness and originality" *(Ibid.)*. A reader concentrating on the mathematical aspect is frequently distracted by philosophical digressions so typical of
BOLYAI's personality, by the frequent use of metaphors, and by the profound but
far-fetched associations. It must be due to BOLYAI's modesty, and in part to the less
advanced typographic practice of the age that the results in the book do not strike
the eye; sometimes only a few easily overlooked words refer to an idea that was well
ahead of the times. Not even the second edition changed this lay-out.

In his — otherwise incompetent — paper [16] SÁMUEL BRASSAI calls the *Tentamen*
a mathematical encyclopaedia. In a strict sense this designation is incorrect. An
encyclopaedia is expected to give a possibly complete overview of the discipline concerned, summarizing the major results as fully as possible. No doubt the *Tentamen*
contains a wealth of material with various fields of mathematics being addressed
in it, yet one would consult the *Tentamen* in vain for all the diverse mathematical
knowledge accumulated by the beginning of the last century. BOLYAI's selection from
the vast material of mathematics is highly idiosyncratic: following his personal taste,
he sometimes touches on fundamentally important parts only in passing, and
sometimes looks at seemingly insignificant details at depth. Nevertheless, this arbitrary
treatment does not belittle the significance of the *Tentamen* at all: the scholar "with
little preliminary knowledge" but much intuition virtually smuggled his original thoughts
into the chapters he elaborated at length. Mathematicians working abroad under
more favourable conditions would certainly have devoted separate treatises to each
of their findings of a similar nature. FARKAS BOLYAI, who lived far from the hubs of
mathematics and was almost completely isolated from the rest of the world, and whose
self-confidence was badly undermined by a series of failures, found the only outlet
for his individual ideas in his books.

---

[2] NÉMETHY, ENDRE: *Farkas Bolyai's world view and literary works.* (In Hung.) Celldömölk, 1937, p. 7.
[3] A useful source for an endeavour of this kind is MRS.LÁSZLÓ ZOLNAI's paper *An analysis and evaluation
of Farkas Bolyai's book "Arithmetic" (1843)* (Debrecen, 1983. Manuscript in the Mathematical Library
of the University of Debrecen. In Hung.)
[4] In STÄCKEL, PAUL (Hungarian edition), Vol. 2, pp. 27—122.

AZ

ARITHMETICA ELEJE.

(az elő-szóban írt módón)

B. B. F.

Mathesist és Physicát tanító P. által.

M. VÁSÁRHELYT. 1830.

Nyomtattatott a' Reform. Kollégyom
betűivel Felső Visti Kali Jó'sef
által.

26. The title page of Farkas Bolyai's *Arithmetic*

T E N T A M E N

IUVENTUTEM STUDIOSAM

IN ELEMENTA MATHESEOS PURAE, ELEMENTARIS AC
SUBLIMIORIS, METHODO INTUITIVA, EVIDENTIA-
QUE HUIC PROPRIA, INTRODUCENDI.

CUM APPENDICE TRIPLICI.

Bolyai

Auctore Professore Matheseos et Physices Chemiaeque
Publ. Ordinario.

Tomus Primus.

Maros Vásárhelyini. 1832.
Typis Collegii Reformatorum per Josephum, et
Simeonem Kali de felső Vist.

27. The title page of the first edited *Tentamen*

Some other signs of random arrangement can also be detected in the *Tentamen*. Although the table of contents divides the material into separate fields, some distant problems are often commingled, there are also repetitions, overlappings and additions that would more organically fit into some other place.

All in all, the *Tentamen* is a valuable and rare specimen of textbooks. It contains the mathematical material of Hungarian higher education, complemented with the exacting requirements of a professor with a wide horizon and modern didactical principles who, however, overtaxed the capacity of his pupils;[5] it is a compilation of certain chapters of mathematics elaborated in different Hungarian textbooks of mathematics, interspersed with original results and ideas.

In the opinion of the eminent mathematician LAMPE of Berlin, the *Tentamen* "... is a historical document which informs today's reader of the mathematical outlook in the first half of the 19th century."[6] But that is not the whole truth, either: the *Tentamen* is more than just a document of mathematical history, as it is a repository of ideas that blazed the trail for the thinkers of the age. In the following, instead of giving a detailed survey we shall concentrate on these ideas.[7]

*

There is always an array of internal and external factors that jointly determine a scholar's interests, his choice of subject, method and style. In the case of FARKAS BOLYAI, these factors and motives form a highly intricate web. His world view is basically determined by two, seemingly contradictory features: *rationalism* stressing the rights of reason, and *romanticism*. Both his actions and writings are alternately predominated by one or the other — similarly to the case many other scientists of the period.

Quite surprisingly, both of these features can be detected in his mathematical works as well. For example, the *Tentamen* begins with a definition, not without religious elements, of concepts unusual for a mathematical book (e.g. love, faith, etc.), and goes on to describe the system of sciences. In BOLYAI's most up-to-date view, any science — including mathematics — is a collection of *theorems* which can be *verified* exactly on the basis of precise *definitions* and simple *axioms*; the true *end* of every science — as dictated by the utilitarian outlook — is *practical usefulness*. If we compare the world to a precise clockwork or a perfectly constructed church, then the duty of every science is

"... to comprehend the mechanism of the clockwork, to get to know the ground-plan, the columns, building blocks and cementing material of the church and, by

---

[5] For this, see the following studies: DAVID, LAJOS: Farkas Bolyai and the reform of the instruction of mathematics. *Magyar Pedagógia*, Vol. 30, 1921, pp. 148—156; BUJDOSÓ, ERNÖ: *The didactics of mathematics in Farkas Bolyai's works*. Szeghalom, 1934; KÖNYVES TÓTH, KÁLMÁN: Farkas Bolyai, the forerunner of the modern method of teaching mathematics. *Matematikai Lapok*, Vol. 10, 1959, pp. 12—22. (All in Hung.)

[6] *Fortschritte der Math.*, Vol. 28, 1897, p. 14.

[7] Hereafter, for convenience of the reader, page numbers will refer to the *second edition* of *Tentamen*, STÄCKEL's *German* version and to *Kurzer Grundriss* in the German STÄCKEL.

deciphering the last character, by finding the key to the code, to read all the sentences of the whole work. This is the aspiration of *physics* (in the broader sense), a branch of science using mathematics. We raise Jacob's ladder to reach the skies with the help of mathematics ..." ([6], I., p. 8.).

It was mentioned above that *Tentamen* is organized in an occasional way. This, however, does not apply to the harmonious building of mathematics as a whole designed by BOLYAI. His conception is, so to say, a bridge between the old and new approaches. As is well known, in older times geometry was not regarded of equal value with other branches of mathematics, being relegated among the applied sciences as a discipline that merely "applies" the results of "pure" mathematics. This undervaluation of geometry only come to an end with the great discoveries of the last century. FARKAS BOLYAI's viewpoint is progressive insomuch as he places geometry on equal footing with arithmetic; what is regressive in his outlook is that he strictly separates, at least in theory, these two disciplines: he claims that in arithmetic the geometric tools must be avoided, and *vice versa*. In teaching, however, "the two eternal brothers should be allowed to embrace in order to strengthen each other, instead of forcefully separating them" ([8], p. 180) because "... both need the other even if unnoticed." (*idem*, p. 178).

FARKAS BOLYAI really tried to separate the two disciplines as much as possible: the whole structure of the *Tentamen* is proof of it, and so are some explicit remarks of his. For example, BOLYAI speaks most appreciatively of GAUSS's treatise (Demonstratio nova theorematis ... resolvi posse. Helmstedt, 1799) in which the fundamental theorem of *algebra* is proved but remarks that the proof of the purely *algebraic* theorem contains some *geometrical* considerations as well ([6], I. p. 461).

According to one of BOLYAI's favorite similes, geometry and arithmetic are like two trees. The roots are the appropriately chosen axioms. The trunk of the tree of arithmetic is constituted by the knowledge that can be deduced from the axioms. As for contents, this knowledge includes elementary arithmetic and algebra. Adding the knowledge gained by the introduction of the concept of *function*, we arrive at various branches of analysis, the "crown" of the tree of arithmetic.

This felicitous simile urged BOLYAI to draw its geometric counterpart. Finally, the foliage of the trees of geometry and arithmetic are intertwined by *mechanics* which studying processes taking place in *time* and *space*.

It is at this point that we notice how important a role KANT's two basic philosophical categories: *time*, infinite in two directions, and *space*, infinite in all directions, play in BOLYAI's mathematical system. *Arithmetic is the study of time, geometry that of space.*

Before embarking on some basic ideas, let us emphasize an important fact: FARKAS BOLYAI has chosen to build up ranged both arithmetic and geometry by the *axiomatic method*. He gives the following explanation:

"... not everything can be defined, not everything can be proved ... There are things without a cause and without further, more precise words to define them." ([6], I. p. 7).

Essentially, BOLYAI, comes quite close to what modern science requires, from systems of axioms. For instance, he mentions the requirement of the mutual *independence* of the axioms at several places:

"As a matter of fact, no thing should be included among the basic causes which follows from the others." ([8], p. 199).

At another place he says that those axioms are important that "cannot be deduced from the rest" ([6], I. p. 7). The next two definitions of axiom involve the notion of *evidence:*

"Axiom is a judgment that common sense accepts without further arguments, as a matter of course." ([6], I. p. 7); "... basic truths that are beyond doubt." ([8], X).

The same is borne out by FARKAS BOLYAI's untiring efforts to replace EUCLID's parallel postulate with a more plausible one. In calling for subscriptions to the *Tentamen*, BOLYAI wrote of his method the following: "... the axioms excepted, all that is necessary for a progressive construction of the *Science* shall be strictly shown" (cited in [31], I. p. 28). The slightly archaic words express that he organized his work so that after presenting the axioms of arithmetic and geometry he proved every statement relying on these axioms and the modes of inference.

Needless to say, BOLYAI could not realize his rich programme without fail; he was unable to consistently follow the new course he had staked out in arithmetic. That was to be achieved later by PEANO and HILBERT. What is to the credit of BOLYAI in this regard — just as in various other areas — is that he foresaw the path of the future development of mathematics.

*

According to FARKAS BOLYAI's definition, "arithmetic is the science that studies quantities reduced to the form of time, and examines the results of all operations also reduced to this form" ([6], I. p. 27). If someone wonders how to interpret the word "reduced" here, he will find several clues in the *Tentamen:*

"Any surface can be reduced to a rectangle whose height is e.g. 1 fathom; any solid can be reduced to a parallelepiped whose height and width are again 1 fathom; finally, everything can be reduced so that its size be expressed by the time or the straight line." ([6], I. p. 27).

This sentence, the contents of which are analyzed in the *Tentamen* only much later, might suggest that BOLYAI here wants to introduce measurement, and derive multidimensional from one-dimensional units.

The question arises whether the involvement of *time* in the structure of arithmetic is only a figure of speech in BOLYAI's usage. In our judgment, the answer is definitely in the negative: in BOLYAI's arithmetic, time plays a more profound role than some sort of terminology. For him, arithmetic is "timology" not only by name. To see this, it is sufficient to observe that his most valuable thoughts concerning the foundations of arithmetic are all derived from the notion of time.

It is hard to decide now what made FARKAS BOLYAI include the concept of time in the arsenal of mathematics. The following two motives could be of importance:

His interest and education in philosophy must have inspired him to adopt the two basic categories of KANT's philosophy, time and space, and thereby to create a more intimate connection between philosophy, physics and mathematics.

The other, perhaps stronger, motive could be *didactical*. BOLYAI might have been influenced by a writing of WRONSKI from the year 1811, and by other attempts in the early 19th century at the introduction of real numbers in a visual manner suitable for teaching, and at replacing the almost exclusively geometrical interpretation of the continuously changing quantity with an arithmetical one.[8]

Having established, with rather strenous work a one-to-one correspondence between the points of the real line and the everyday notions related to (instant of time, past, present, future, duration of time, etc.), BOLYAI identifies time as a hole the real line.

One more fact confirms the assumption that FARKAS BOLYAI kept didactical aims in mind: the notion of time only appears in his works written for the purpose of instruction (the *Tentamen*, as well as four arithmetics in Hungarian), but *not yet* in his inaugural address in 1804, and no *longer* in *Kurzer Grundriss*, a scholarly work written towards the end of his career. Thus, in all probability, he wished to enhance the efficiency of instruction by using the concept of time. However, as will be seen, he achieved more than what he had aimed at, arriving at ideas ahead of his time with the help of this method.

We cite the axioms of arithmetic as BOLYAI conceived of them, but for easier understanding we replace of his terms by their modern equivalents.

1. Time is a continuous quantity; what is present of it is always without parts and always different . . .
2. Any finite time that has not yet been will arrive, but the totality of time will never arrive . . . (This is obviously the *Eudoxus–Archimedes* axiom).
3. What is true in a time-point p having no parts either affirms or negates B.
4. During time p, if A and B, respectively, state and negate C, and we know that A is true, then B is not true; if A is false, then B is true . . .
The latter two axioms constitute the principle of *tertium non datur*.
5. Anything is just what it is — and is perfectly equal with itself . . . ([6], I. p. 12)

As can be seen, these axioms are mostly philosophical ideas and logical principles, on which alone arithmetic cannot be based.

It is also with the involvement of time that BOLYAI defines the principle of *complete inductions*:

---

[8] F. KLEIN also speaks of it but, regrettably, he does not mention the role of FARKAS BOLYAI: "Eine sehr verbreitete Auffassung ist die, dass der Zahlbegriff eng mit *Zeitbegriff*, dem *zeitlichen Nacheinander* zusammenhängt; unter den Philosophen sei *Kant*, unter den Mathematikern *Hamilton* als ihr Vertreter genannt." (*Elementarmathematik vom höheren Standpunkte aus.* Vol. 1, Berlin, 1933, p. 11).

"Starting from a certain point, let us mark out a continuous part of time in the direction of the future and denote it by t, and let any t be followed by another t right up to infinity; let us also assume that A corresponds to the first t and A, B, C, ... correspond to the subsequent time intervals t; let us proceed from A to B, from B to C, and so on, until we reach t corresponding to Q; this t will arrive (Axiom 2), and the same is true of the corresponding Q."[9] ([6], I. p. 21).

This quotation aptly shows what problems of composition BOLYAI encountered because of the involvement of time. At the same time we should also notice that he had come very close to certain ideas that were eventually to enrich mathematics only in the second half of the 19th century (WEIERSTRASS, DEDEKIND, CANTOR, MÉRAY).

Here are, for example, the germs of the notions of least *upper bound* and DEDEKIND *section:*

"If it is certain that A is true at some instant of the continuous time interval T but A is no longer true at some instant of time t coming after T, then proceeding from the beginning of T which we let grow to infinity we can find a p which is the last among all instants of time such that A is always true between them and the beginning of T; at this p, either A is true for the last time or *non-A* is true for the first time." ([6], I. p. 20).

To understand this citation, we have to make clear that the expression "*A is true*" in contemporary terms means that certain elements of a set satisfy some prescribed condition A. If A is true at p for the last time, or *non-A* is true at p for the first time (but so that the negation of A is also true for all subsequent p), then p is the least upper bound.

FARKAS BOLYAI uses the notion of *supremum* at several places. Let us mention as typical examples: the convergence of the monotonically increasing sequence bounded from above, the definition of arc length, or the existence proof of parallel lines in the Euclidean sense by rotating one of them around a point in the common plane.

Unless viewed with absolute rigour, the definitions to be read in various chapters of *Tentamen* dealing with the foundations of arithmetic may be ascribed *set theory*. Quite understandably, for want of the appropriate technical terms, the wording is cumbersome, some idea remaining obscure until an instance of application clarifies them.[10]

Using the terminology of point sets, we come across the following notions: *set*, *intersection* of two point sets, *union* of sets, *empty set*. *Undetachable part* (pars indivellibilis) means the boundary of a closed set. *Constituent part* (pars) means proper subset ([6], I. pp. 22—24).

---

[9] I.e. the statement for A holds for it as well.

[10] According to LAJOS DÁVID, FARKAS BOLYAI's arguments presented here are the first written formulation of the so-called part theory (Teiltheorie) the foundations of which were laid around the thirties by E. FORADORI [20]. In our opinion, which tallies with STÄCKEL's, the concepts defined by FARKAS BOLYAI rather belong to set theory.

The concept of undetachable part occurs mostly in geometrical considerations (see later). Constituent part appears in other fields as well, e.g., in the surprising profound definition of the *continuum:*

"If, studying the parts, we find an A any constituent part A' of which has something in common with that B which lies outside A' but is also contained in A: then such an A is a *continuum*. Examples are space, time, line, surface, etc." ([6], I. p. 24).

In modern notation, this reads: the set A is a continuum if the condition

$$A' \cap (\overline{\overline{A} - A'}) \neq 0$$

is satisfied for every $A' \subset A$, where $\overline{A}$ denotes — as usual — the closure of A.

The exact definition of continuum was brought about in the 1870s by the activities of CANTOR, JORDAN and HAUSDORFF. The sentence quoted from BOLYAI is, in today's terminology, the definition of connected topological space.

As in the *Tentamen* the real numbers are introduced with the help of time, the notions of past and future make BOLYAI to stress the significance of "orientations". If the sign of a vector directed from an instant p to the future is ✚, then that of a vector subtending an angle 180° with the former i.e., pointing to the past, is ↦. Further, availing himself of the possibility to select the units arbitrarily, BOLYAI takes two units ✚ 1 and ↦ 1 for given, thereby warding off the difficulties which arise by the introduction of negative numbers. At the beginning, he strictly distinguishes the two symbols of *quality* ✚ and ↦ from the operational symbols + and − of addition and substraction. Indeed, today we apply the latter two symbols in two ways: to indicate the positive or negative quality of a number, and to denote addition and subtraction. When laying the foundations, it is quite reasonable to make a distinction (as is again being done in Hungarian elementary schools), but it is expedient to get rid of the extra burden of notation when it has been understood that the use of uniform symbols does not cause any confusion.

BOLYAI also contributed to the heated debate about complex numbers being in progress at the beginning of the 19th century.

FARKAS BOLYAI's theory of complex numbers developed in the *Tentamen* appears almost unchanged in a subsequent paper prepared for a competition entitled *Sigillum veri simplex*. In 1834 the *Jablonowski Society* of Leipzig invited studies for a competition to clarify the relations between imaginary numbers and geometric constructions. Before the deadline in 1838, three papers had been submitted, all three by Hungarian authors: FARKAS BOLYAI, JÁNOS BOLYAI and FERENC KEREKES. By far the best of them was J. BOLYAI's *Responsio* (see Chapter 15), yet half of the award went to KEREKES and the works of the two BOLYAIs scored no success whatever. KEREKES tried to obtain a coherent theory from his daring ideas by linguistic and philosophical arguments (*Imaginary numbers*. Debrecen, 1848). First of all, he classified the imaginary numbers into a broader category, that of "contradictory quantities". He claimed that those quantities were contradictory in whose interpretation there was an "unreconcilable" antinomy. They include the numbers expressing "existence" (positive and negative) and "nonexistence" (zero) simultaneously as well as those having positive and negative value *simultaneously*; finally, it is also contradictory to assign the same value to two *different* numbers *simultaneously*. Now, the value of $\sqrt{-1}$ must be some kind of unit, he argued, but neither $+1$ nor -1. In his opinion, there is no other choice but to realize that the imaginary unit is a contradictory quantity, its value being $+1$ and -1 *simultaneously*. This interpretation, he said, is in accordance with i being called an *absurd* number, but under this interpretation it can also be proved that $i^2 = -1$. For short, to establish the latter formula a clever handling of the two signs belonging to the imaginary unit simultaneously is sufficient. KEREKES must have been influenced in his defective theory by Hegelian dialectic which he advocated but apparently misunderstood.

Apparently, he refused the opinion widely spread abroad and predominant in Hungary that the complex numbers were inconceivable notions that could have no role in the mathematical description of the laws of the real world. In our country, this view was even reflected in terminology: complex number = "impossible", "absurd" number. FARKAS BOLYAI wanted to have the complex numbers emancipated in mathematics, quite in accordance with the view adopted by JÁNOS BOLYAI:

"... only those things, and hence only those quantities may be the subject of sober research which really exist (e.g. material things, parts of the physical or external world, or things that are at least conceivable and possible)". (*Responsio*, §11).

Unfortunately, FARKAS BOLYAI's contribution to this question was like a *modus loquendi*, not advancing the clarification of the problem, and drowright false in its conclusions. For, in his opinion, all numbers — including the imaginary ones — may be considered *real* if two completely equivalent units, $+1$ and -1, are chosen and if the number under the root sign is *freely* associated with one of the units so as to obtain a real root. $\sqrt{4}$ is real if 4 under the radical sign is combined with $+1$. In this case we say that the root is *real with respect to* $+1$. On the other hand, in the case of $\sqrt{-4}$ we have to combine -1 with -4 in order to get a real number, hence $\sqrt{-4}$ is *real with respect to* -1.

This idea of BOLYAI, however, may lead to false results, so his son rejected it. Especially sharply did he criticize his father's theory of proportions ([31], I. pp. 124—125).

All this notwithstanding, it is to be attributed to the influence of FARKAS, that JÁNOS also introduced the complex numbers in an arithmetical way and ascribed major significance to starting with appropriate *units* (see later).

*

According to the *permanence principle* generally accepted in modern mathematics, in generalizing any notion or operation it is required that the generalized concepts or operations contain the old ones as special cases; this holds even for the kind generalization usual in abstract algebra. Some authors (OHM 1822; PEACOCK 1830; BUBENDEY 1834) tried to apply the principle when developing complex algebra at the beginning of the last century, but they did not clearly state this ambition. For this reason, the permanence principle is generally believed to be of a later date and is mostly quoted in the formulation of the German mathematician HANKEL.[11] Around the turn of the century there were attempts to regard this principle as an axiom and use it *imperatively*

---

[11] "If two formulae expressed in the general symbols of universal arithmetic are equal, let them remain equal even if the symbols no longer denote simple quantities and thus the operations assume a different meaning." (*Theorie der komplexen Zahlensysteme.* Leipzig, 1867, p. 11).

(W. WUNDT). However, this rigid view is wrong, the preservation of analogies must not be prescribed with binding force.

In view of the above-said, one sentence of the *Tentamen* is particularly noteworthy. It says that it is advisable to construct the algebra of complex numbers in such a way that ". . . the operations could be continued under the sails of generality, and generality would be preserved as far as possible." ([6], I., pp. 121—122).

It would not serve any purpose to enumerate all the unusual and often unnecessary symbols FARKAS BOLYAI used.[12] Let us mention just a few of these which were introduced "with the honest intention of thoroughness". For instance, three *equality signs* can be found in Farkas BOLYAI's writings:

a) The symbol = meaning "equal to" coincides with the sign = as used today.

b) The sign ≑ of "absolute equality" means identical equality.

c) The specifically *Bolyaian* sign ⌶ of "relative equality" indicates that two things are equal in *one* property but not in *all*. For instance: two triangles are equiareal but not congruent; two arcs have the same length but different shapes. Therefore, when using this symbol, it must also be stated what the equality applies to.

Various inequality signs can also be found (they differ in the thickness of one leg[13]). Their meanings are best shown on examples.

So, for instance,

$$3 < 5, \qquad 5 > 3.$$

But: $-5 \lessdot -8$ (instead of $|-5| < |-8|$), $-8 \gtrdot -5$ (instead of $|-8| > |-5|$). Thus the symbols $\lessgtr$ denote something bigger or smaller in *absolute value*. If we consider that the sign $\|$ of absolute value was introduced by WEIERSTRASS in 1841 only, and until then the abbreviation *mod* had been used, the practicalness of the symbols introduced by BOLYAI becomes apparent.

*

Those chapters of *Tentamen* which introduce geometry also abound in remarkable ideas. The starting point is NEWTON's *space* as an objective reality which FARKAS BOLYAI assures to be a continuum, infinite in all directions and eternal, all its parts being always present and homogeneous ([6], II., p. 2). The major task is to "populate" this space with various geometrical figures. This task can be carried out by either a *descriptive* or a *genetic* definition of the figures.[14]

To give descriptive definitions one has to make use of the notions of *undetachable part* and *constituent part* mentioned previously.

Let us divide the entire space or a finite part of it into two so that the two parts have an undetachable intersection: this undetachable intersection and all its constituent parts are *surfaces*.

---

[12] A fairly detailed list can be found in CAJORI, F.: *A history of mathematical notations* (London, 1929).

[13] In the second edition of *Tentamen* these inequality signs have a different shape, probably for technical reasons.

[14] What follows here is a summary of some rather obscure arguments from [6], II., pp. 1—42 and [31], II., pp. 149—179.

The undetachable intersection dividing a finite or infinite surface into two is called a *line,* or *curve.*

The "partless" intersection cutting a non-closed curve into two is a *point.*

FARKAS BOLYAI defines curve and surface in another way as well.

A continuum is a *curve* if the point has only a finite number of ways out from each of its points.

If, however, an infinite number of ways emanate from every point of the geometrical figure and, besides, the figure is not a constituent part of space, then we speak of a *surface.*

A geometrical figure can also be defined *genetically.* For this, we have to regard the point as a primary concept, "the first-born child of space" in BOLYAI's wording, introduce the motion of "geometrically movable", and allow for some kinds of motion borrowed from physics. A figure is geometrically "movable" if it can be transferred anywhere in space without changing its shape.

In the case of geometrical motion we have to ignore a property of physical motion, namely that it takes place in time. There are *simple* and *composite motions.* Simple motions are the following:

1. *Free motion:* some point of the space moves to another point. In modern terminology this means *translation.*

2. *Rotation around a fixed point:* the geometrical figure presumed to be' rigid (the "geometrically movable") is moved keeping *one* of its points fixed.

3. *Movement around two fixed points:* two points of the rigid figure remain fixed.

The latter two kinds of motion are *rotations.*

The combinations of the above three *primitive* motions are the *composite* motions.

The question arises whether BOLYAI's "motions" can be regarded as such in the sense of present-day geometry. The question is justified since BOLYAI's only explicit restriction is that the outward form ("externa qualitas", [6], II., p. 3) of the figure must not change in the course of geometrical motion. So, considering the strict formulations of geometry, BOLYAI's usage is objectionable. Yet, in our opinion motion in BOLYAI's sense is motion in the modern sense as well. In fact, as noted above, he modelled the notion of "geometrically movable" upon the bodies of physics ("corpus") and consequently, his concept of figures "movable into each other" automatically implied congruence (the segment defined by any two points of the original figure is congruent with the segment formed by the corresponding two points of the transferred figure). Moreover, he mentions congruence and movement, as if filed, together, in one sentence: "... oritur Axioma congruentiae, et constructio mobilis geometrici, motusque geometricus." ([6], II. XII). Thus the condition that the "outward form" be unchanged requires more than *similarity.*

By introducing motion, BOLYAI has a means to *generate* geometrical figures: the line is the aggregate of all points that preserve their position when a body is rotated around *two* fixed points (they remain "unique").

If we move the line segment, while fixing one of its end points, in every possible direction then the other end point describes a sphere, "secondborn child" of space.

The sphere having been created, new possibilities arise to generate some geometrical figures including the point, the line and the plane — accepted as the basic notions today.

Let us be given two spheres lying outside each other, having fixed centres and no common part. If we let the radii increase continuously, there will be a moment when the two spheres have a "partless" intersection: the *point* of contact. Proceeding from this position by lengthening one radius and shortening the other equally, the aggregate of all points of contact will form a *straight line*. If, however, starting from the position of contact both radii increase the intersection of the spheres will be by a *circle*. Finally, if the radii of the two spheres are equal at every moment and they grow infinity, then their continuously expanding curve of intersection describes a *plane*.

FARKAS BOLYAI's further geometrical investigations also prove that the idea of motion was not secondary in his writings but he was aware of its significance and sought to exploit all the possibilities offered by physical motion for the construction of geometry. The idea itself was not new, as the ancient Greeks had already noticed how fruitfully motion could be applied in geometry and used it, for instance, in the definition of congruence.[15] Yet some two thousand years had to pass before attention was again focussed on the question; the new impulse was given by research in physics done by STOKES and HELMHOLTZ around the mid-19th century, leading eventually to FELIX KLEIN's "Erlangen programme" and LIE's theory of the groups of motion. HELMHOLTZ's oft-cited four axioms of a much later date[16] also display similarities with FARKAS BOLYAI's ideas:

1. The point is an element of space: its position is determined by three independent data or coordinates. Any motion of the point is accompanied by a continuous change of at least one of the coordinates.

2. The distance between any two points of a rigid body moving in space is unchanged during the motion; this can be expressed by an equation, independent of the motion, involving the six coordinates of the two points.

3. Two points (six coordinates) are needed to determine the position of a rigid body.

4. If two points of a moving body are fixed, it can return to its original position without changing the direction of motion.

The only thing missing from BOLYAI's theory is the interpretation of the points of space as ordered triples of numbers; in this regard, HELMHOLTZ's axioms make a step forward.

---

[15] Incidentally, FOURIER preceded FARKAS BOLYAI generating the sphere, circle, plane and line by motion (Séances de l'École Normale Supérieure. *Débats*, Vol. 1, 1795, pp. 28—33), but we have found no data whatsoever that BOLYAI would have been aware of FOURIER's completely similar ideas.

[16] Über die thatsächlichen Grundlagen der Geometrie. Heidelberg. *Verhandl. d. naturw. med. Vereins*, Vol. 4, 1868, pp. 197—202 and Vol. 5, 1869, pp. 31—32.

Farkas Bolyai's achievements in algebra and analysis

At the beginning of the last century not all branches of mathematics were separated in the way usual today. For example, many regarded algebra as a special, introductory section of analysis taken in a wide sense: the role of algebra, they said, was to find values of the independent variable that make the function equal to zero.

Algebra has a place of this kind also in BOLYAI's system. Accordingly, the point of departure for both algebra and analysis is the notion of *function*. BOLYAI gives several definitions of function:

In a narrow sense, a function is a relation which connects variables and constants through certain operations ([6], I., p. 204).

Or, in a more detailed version:

"... any variable (or variables) joined with any constant (or constants) by any operation (or operations, either in time or in space) ... may be called a function." ([8], p. 214).

But in a *broader sense* (sensu lato) any operation conceivable in time and space is a function ([6], I., p. 204).

The first two definitions are virtually the same as those given by JOHANN BERNOULLI (1718) and also accepted by EULER. These definitions require that the function can be represented by an analytic expression, and, therefore, by the early 19th century — mainly due to the discovery of the FOURIER *series* — they turned out to be far restrictive. BOLYAI's third definition, however, is rather general and reminds of the DIRICHLET definition accepted today. It makes no mention of representing the function by a formula, and at the same time fulfils the requirement of logical consistency because anything that is possible in time and space is logically consistent.

We have to add that LOBACHEVSKY, in a treatise published in 1834, also gave a very general definition of function. Among other things, he writes the following:

"The function can be given by an analytical expression, or by a condition that enables us to try out any number, or finally there may exist a relation which is unknown."

In a work published in 1810, LACROIX also defined the function in the manner usual today, but his definition was not received with the appreciation it deserved.

Thus the way to DIRICHLET's (1837) general concept of function was paved by LACROIX (1810), FARKAS BOLYAI (1832) and LOBACHEVSKY (1834).

After introducing the notion of function, a long chapter of *Tentamen* is devoted to the theory of equations, but this branch of mathematics apparently does not engage BOLYAI's attention so much as, e.g., the foundations of arithmetic and geometry. The original algebraic ideas in the book, are few, but it would be mistaken to think that BOLYAI confined his treatment to results already known: how and then he interjects witty remarks and observations.

As is well known, in 1786 the Swedish mathematician E. BRING reduced the general quintic equation to a trinomial normal form with the help of a TSCHIRNHAUS transformation. This result, as well as the rate of interest problem in the calculation of annuities — growing in importance with the advance of capitalism — directed the attention of several scholars (including GAUSS and DANDELIN) to the solution of trinomial equations. After the failure of the attempts at solving the general trinomial equations by the methods of algebra, attention was turned to approximation methods.

FARKAS BOLYAI also put forward a root approximation method for trinomial equations showing his penchant for infinite procedures. Let the equation $x^2 + ax = b$ with real coefficients be given ([6], I., p. 442). From this

$$x = \frac{b}{a+x}.$$

BOLYAI iterates here by successively substituting the whole right side for x on the right. Then he proves carefully that the continued fraction

$$\cfrac{b}{a + \cfrac{b}{a + \cfrac{b}{a + \cfrac{b}{a + \,\cdot}}}}$$

so obtained provides a root of the equation.

More remarkable is the method of root approximation he proposed both for theory and practice, is the following iterative method. Let us be given two equations ([6], I., pp. 447—449):

$$1.\ x^2 = a + x; \qquad 2.\ x^m = a + x$$

(m a natural, a a positive real number).

In both types of equations FARKAS BOLYAI substitutes an appropriate initial value x_0 for x on the right side of the equation, and expresses x from the left side so obtained, whereby he gets x_1. Iteration of this procedure gives $x_2, x_3, \ldots$.
In both cases $x_0 = 0$. Thus:

1. $x_0 = 0$
   $x_1 = \sqrt{a}$
   $x_2 = \sqrt{a + \sqrt{a}}$
   .
   .
   .
   $x_n = \sqrt{a + \sqrt{a + \sqrt{a + \ldots}}}$
   .
   .
   .

2. $x_0 = 0$
   $x_1 = \sqrt[m]{a}$
   $x_2 = \sqrt[m]{a + \sqrt[m]{a}}$
   .
   .
   .
   $x_n = \sqrt[m]{a + \sqrt[m]{a + \sqrt[m]{a + \ldots}}}$
   .
   .
   .

145

Taking everywhere the positive real value of the root, FARKAS BOLYAI shows that the sequence x_n is monotonically increasing and bounded from above, further that $\lim_{n \to \infty} x_n$ is a solution of the equation. In Case 1 the procedure is uninteresting as a method of solving equations but provides a useful way of generating real numbers.

We take a closer look at the wealth of literature concerned with the "Bolyai algorithm" in Chapter 21.

We note in our literature the first mention of solving system of linear equations with determinants appears in *Tentamen* ([6], I., pp. 416—424), however, in the clumsy and inexpedient notation of LEIBNIZ and BÉZOUT.

*

No less remarkable is the *Tentamen*'s treatment of analysis taken in the contemporary, narrow sense of the word. FARKAS BOLYAI advocated modern views in this field as well; he not only knew but also duly appreciated the investigations carried out by LAGRANGE and chiefly by CAUCHY (some of their works were in his possession) in rigorizing the notions of limit, convergence, derivative and integral. This is the least known area of his work, although it reflects his sense of criticism very clearly.[17]

In building up analysis, he hold the opinion that EULER's "infinitely small" quantities should be avoided because "then overcomplicate a simple theory . . .; common sense requires that the field be cleared of the legions of infinitesimals in order to have a more open view" ([31], II., p. 147). Accordingly, he based analysis on the notion of *limit* ("fringe-value"), using the BOLZANO—CAUCHY definition (1817, 1821):

"If p is the general name of the finite quantities that may occur under a certain condition, and ✛ or ⊢ p can become greater than any given quantity homogeneous with it, or the difference between p and K — though never zero[18] — can because smaller than any given z: then in the first case infinity, and in the second K is the limit of the sequence p." ([6], I., p. 35).

Initially, for lack of the notion of *limit*, several theories were constructed to explain the differential quotient. Finally, the MACLAURIN—D'ALEMBERT "méthode des limites" came out victorious only when BOLZANO and CAUCHY, at the beginning of the 19th century, gave a correct definition of limit. Of the Hungarians, FERENC KEREKES was involved in this one-and-a-half century long fight but his ideas are futile and confused. He claimed, for instance, that in the quotient $\frac{ds}{dt}$ both the numerator and the denominator are expressly of value zero, but only for our perception. For, he said, there are zeros of different values but we are unable to tell the difference between them. A similar position was taken around the middle of the last century by the English mathematician B. PRICE (*A Treatise on Infinitesimal Calculus*. Oxford, 1852). KEREKES's views almost automatically suggested the terms he introduced: to differentiate = *to nullify*; differential calculus = *nullifying calculation*; integration = *completion*; integral calculus = *completing calculation*, etc. (*The true basic principles of higher geometry, and a supplement on fractions and the theory of contradictory quantities*. Ed. by DÁNIEL CSÁNYI. Debrecen, 1862. In Hungarian).

[17] Only a few of FARKAS BOLYAI's ideas in analysis are presented here; for more details, see [33].
[18] A restriction usually imposed in the first times.

At the same place we can find the symbol ⌐ introduced by FARKAS BOLYAI to denote transition to the limit. Apart from its practical usefulness, it has a theoretical value as well since right up to the beginning of the 20th century the abbreviation *lim* combined with the — actually incorrect — notation $n = \infty$ had been used. The symbol → (LEATHEM, 1905) almost generally accepted today is perhaps easier to handle than BOLYAI's that's all.

A methodological novelty in the construction of analysis is that FARKAS BOLYAI does not discuss differential and integral calculi separately but developes them parallel with each other. The advantage of this approach is confirmed by the fact that several modern textbooks apply a similar method. His exposition of the rules of differentiation is otherwise quite whimsical: he just mentions how to differentiate sums, products, powers and the logarithm function while for the trigonometric functions the treatment of the same question is lengthy and cumbersome. A conspicuous defect is that BOLYAI fails to come down from general theory to concrete examples that might illuminate his procedure.[19]

FARKAS BOLYAI introduces integration via the definite integral. By itself, this fact is not very significant as it tallies with historical development; for example, LEIBNIZ also put the definite integral in the first place. Later, however, EULER started from the indefinite integral as the inverse of differentiation. Owing to his reputation, this view held sway for a long time, until BOLZANO and CAUCHY put the definite integral in the foreground again.

Question related to the definite integral and the infinite series made the study of convergence urgent as early as at the beginning of the last century. This was the point where BOLYAI joined the critical work in progress.

Infinite series were already known at the time of the discovery of differential and integral calculus, but the question of their convergence was given superficial treatment for centuries. Even outstanding mathematicians deemed a piece of work finished when they had managed to expand a function into an infinite series with the help of some formal method, regarding the examination of convergence unnecessary. In the best case, the convergence of some *special* infinite series was studied by individual methods. No general criteria were known for deciding the problems of convergence in wider classes of series. Then, at the beginning of the 19th century, parallel with the rigorization of analysis, tests of convergence began to mushroom but there was much confusion, about their being necessary or sufficient.

Though only a limited sample of FARKAS BOLYAI's results achieved in the theory of infinite series is presented here,[20] to avoid misunderstanding we should note the following: although in the first edition of *Tentamen* infinite series and remarks on their convergence occur at several places, most of the permanent ideas are to be found on pages XCI—XCVI of the *Errata*. FARKAS BOLYAI kept publishing these additions and corrections in instalments up to 1844, and attached them to the *Tentamen*. So, up

---

[19] This is typical of the *Tentamen* as a whole. FARKAS BOLYAI did not conceal this feature in his call for subscriptions: "There are only as many examples as comprehension requires; they can be got anywhere." (Quoted in [31], I., p. 29).

[20] For more details, see [33].

to that date he could — at least in principle — take into account the achievements of others. In a letter to GAUSS dated 18 January 1848 he raised some relevant questions, and received an answer three months later ([12], pp. 129—134). Obviously, he might not use the information so acquired in either the *Errata* of the *Tentamen* or the enlarged edition of the *Introduction to Arithmetic* [8] published in 1843. Therefore, in some questions of priority the year 1843 may be considered a landmark.

For example, BOLYAI proposed an interesting test for series of positive terms. Let the fraction $\dfrac{n-m}{n}$ denote the factor with which we have to multiply the $(n-1)$th term of the series to get the nth term (in general, m is a function of n). That is,

$$u_{n-1}\frac{n-m}{n}=u_n.$$

Hence:

$$m=n-n\frac{u_n}{u_{n-1}}.$$

According to FARKAS BOLYAI, if from a sufficiently large n

1. $m\geq k>1$, then the series is *convergent*,
2. $m\leq 1$, then the series is *divergent*.

In the case where $m>1$ but $\lim\limits_{n\to\infty} m=1$ nothing can be said of the behaviour of the series ([6], I., pp. 177, 588—589; [8], pp. 383—385).

BOLYAI proves this criterion by comparing certain series.

Let us slightly alter the form of BOLYAI's criterion of *convergence*:

$$m=n\left(1-\frac{u_n}{u_{n-1}}\right)\geq k>1,$$

that is,

$$\frac{u_{n-1}}{u_n}\geq\frac{n}{n-k}.$$

Hence, for $n>k$ we obtain

$$(n-1)\left(\frac{u_{n-1}}{u_n}-1\right)\geq (n-1)\left(\frac{n}{n-k}-1\right)=\frac{k-\dfrac{k}{n}}{1-\dfrac{k}{n}}\geq k,$$

which is RAABE's *criterion of convergence*. Or take BOLYAI's criterion of *divergence*:

$$m=n\left(1-\frac{u_n}{u_{n-1}}\right)\leq 1.$$

It implies that

$$\frac{u_{n-1}}{u_n}\leq\frac{n}{n-1},$$

whence

$$(n-1)\left(\frac{u_{n-1}}{u_n}-1\right)\leqq(n-1)\left(\frac{n}{n-1}-1\right)=1.$$

The latter is RAABE's *criterion of divergence*.[21]

In other words, BOLYAI's test is identical with RAABE's, only the formulation is different.

BOLYAI's test presented above was not yet included in the first edition of *Tentamen*: it first appeared in the *Errata*. However, it can also be found in one of BOLYAI's letters to GAUSS ([12], pp. 130—131). Thus, priority in print is due to RAABE but no one can deny that BOLYAI also made an independent discovery.

Ignoring here the witty counter examples he used to disprove several tests bearing the names of OLIVIER, BURG and MONTUCLA, let us point out one more remarkable fact: independently of anyone else, FARKAS BOLYAI also discovered the so-called DE MORGAN *test scale of the first kind*. The relevant passage of *Tentamen* reads, in verbatim translation and original notation, as follows (the abbreviation *log* stands for common logarithm):

"Accordingly as the series with general term $\dfrac{1}{n^a}$ converges or diverges, keeping a unchanged, the same is true for the series with general term $\dfrac{1}{n\cdot l^a}$ or $\dfrac{1}{n\cdot l\cdot l_1^a}$, or generally $\dfrac{1}{n\cdot l\cdot l_1\ldots l_t^a}$, where $l=\log n$, $l_1=\log l$ and $l_t=\log l_{t-1}$, and l_t is not less than 1." ([6], I., p. 585; [8], pp. 380—383).

The outline of the proof given by FARKAS BOLYAI for this theorem is the following. Consider the series

$$\frac{1}{10(\log 10)^a}+\frac{1}{11(\log 11)^a}+\cdots+\frac{1}{100(\log 100)^a}+\frac{1}{101(\log 101)^a}+\cdots.$$

Let the sum of the first 90 terms be α, the sum of the subsequent 900 terms β, etc. Then

$$90\,\frac{1}{10\cdot 1^a}>\alpha,\quad 900\,\frac{1}{100\cdot 2^a}>\beta,\quad 9000\,\frac{1}{1000\cdot 3^a}>\gamma;\ldots$$

Adding up these inequalities we obtain:

$$9\left(1+\frac{1}{2^a}+\frac{1}{3^a}+\ldots\right)>\alpha+\beta+\gamma+\ldots=\frac{1}{10(\log 10)^a}+$$

$$+\frac{1}{11(\log 11)^a}+\cdots.$$

---

[21] RAABE, J. L.: Note zur Theorie der Convergenz und Divergenz der Reihen. *Journal f. d. reine u. angew. Math.*, Vol. 11, 1834, p. 309.

Consequently, if the series $\sum\limits_{n=1}^{\infty} \dfrac{1}{n^a}$ in parenthese converges, then $\sum\limits_{n=10}^{\infty} \dfrac{1}{n \cdot l^a}$ on the right side is also convergent. Further, we also have the following inequalities:

$$90\,\frac{1}{100 \cdot 2^a} < \alpha, \quad 900\,\frac{1}{1000 \cdot 3^a} < \beta, \quad 9000\,\frac{1}{10,000 \cdot 4^a} < \gamma, \ldots$$

Summing them, we get

$$\frac{9}{10}\left(\frac{1}{2^a} + \frac{1}{3^a} + \frac{1}{4^a} + \ldots\right) < \alpha + \beta + \gamma + \ldots = \sum\limits_{n=10}^{\infty} \frac{1}{n \cdot l^a}.$$

Thus, if $\sum\limits_{n=2}^{\infty} \dfrac{1}{n^a}$ is divergent, $\sum\limits_{n=10}^{\infty} \dfrac{1}{n \cdot l^a}$ is also divergent. The remaining statements can be proved in a similar way.

The logarithm scale was already known to ABEL but he did not publish it; DE MORGAN published his work in 1839 (*Differential and integral calculus*. London, 1839); BERTRAND rediscovered it independently (*Journal de Math.*, Vol. 7, 1842, p. 42). Knowing the scantiness of the literature accessible to BOLYAI and the almost complete lack of verbal communication, we can take for certain that he also discovered this useful test on his own.

His achievements in geometry

In addition to the fundamental ideas discussed so far, BOLYAI's sovereign geometrical results cluster around two problems: EUCLID's *fifth postulate* (or 11th axiom: if two coplanar straight lines are intersected by a third, extending the lines they will meet on that side of the transversal where the sum of the interior angles is less than two right angles), and the *end-like equality of areas*.

His mathematical interest had been focussed on the parallel postulate and related problems from his youth. His writings on this subject are moving documents of human endeavour: they reveal the boundless self-confidence of a talented young man who saw no unsurmountable barrier the perseverance of a thoughtful and realistic man in the prime of his life no longer believing in easy success, and finally, the resignation of a man frustrated in his own efforts but taking pride in the accomplishments of his son.

At the beginning, BOLYAI was fully convinced that the parallel postulate was not independent of the rest of axioms and could be derived from them. To a letter written to GAUSS on 16 September 1804 he enclosed his paper *Theoria Parallelarium* (full German translation: [31], II., pp. 5—15), presenting a proof of EUCLID's 5th postulate in which he had not found any defect or contradiction.

CRISTOPH SCHLÜSSEL (CLAVIUS, 1537—1612), in his critical Latin edition of EUCLID's Elements (*Euclidis Elementorum libri*, XV., Rome, 1574), wanted to prove the theorem — an equivalent of the parallel postulate — that the "distance line"[22] of a straight

---

[22] The distance line of line a at distance d is the locus of points at distance d from a on one side of a.

line is also straight. FARKAS BOLYAI also embarked on proving this problem. His argument was essentially the following.

Let the segment $\overline{CH}$ be perpendicular in the plane to the segment $\overline{BD}$ (*Figure 24*). Distances $\overline{BC}$ and $\overline{CD}$ are equal. Let us denote by T the "geometrically movable" rigid figure $BCDH$ (shaped like a T upside down) and let us move T along the line AE in the directions $\pm\infty$. The question is what kind of curve the point H describes during this translation. Let be denoted the curve by g; BOLYAI asserted that g was a straight line.

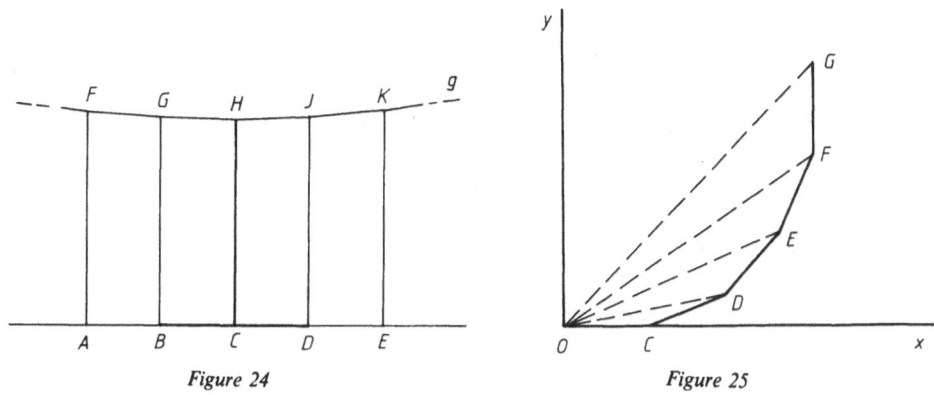

Figure 24   Figure 25

He tried to prove this statement by contradiction. GAUSS, however, spotted an error in the proof and, in a letter dated 25 November 1804 ([12], pp. 81—83), informed BOLYAI of his finding in an accurate mathematical way.

Namely, FARKAS BOLYAI argued as follows. Let us assume that g is not a straight line but some kind of curve. In that case the curve g — due to how it has been generated — would be symmetric[23] with respect to the segment $\overline{CH}$ (for example, if its right-hand branch were turning upward then its left-hand branch would wind upward in an equal measure and, moreover, the right and left branches would meet somewhere along the extension of the segment $\overline{CH}$ (they would close up, to use BOLYAI's terminology). However, the stem of the inverted T at any moment of the motion is parallel to the same at any other moment because it is always perpendicular to the base line AE. Should g "close up", these parallel lines would sooner or later intersect. That, in turn, would contradict the Euclidean definition of parallelism.

It remains to prove that the two branches of the curve g "close up". BOLYAI offered the following proof: mark out the points $C, D, E, \ldots$ on the curve g so that the arcs between them be of equal length, and draw the broken line $OCDEFG \ldots$ composed of the chords (*Figure 25*). Now BOLYAI plunges into a circuitous explanation to show

---

[23] In this reasoning, FARKAS BOLYAI also employed the following axiom given by him: "If A is equal to B, and A and B undergo the same operation then, whatever the outcome of the operation with A is, it must be equal with the outcome of the operation with B." ([8], p. 199).

151

I. Die *Arithmetik*, durch zvekmässig constru-
irte Begriffe, von eingebildeten und un-
endlich-kleinen Grössen gereinigt, an-
schaulich und logisch-streng darzustellen.

II. In der *Geometrie*, die Begriffe der geraden Linie, der Ebene, des Winkels allge-
mein, der winkellosen Formen, und der
Krummen, der verschiedenen Arten der
Gleichheit u. d. gl. nicht nur scharf zu be-
stimmen; sondern auch ihr Seyn im Raume
zu beweisen: und da die Frage, *ob zwey
von der dritten geschnittene Geraden, wenn
die summe der inneren Winkel nicht* $=2R$,
sich schneiden oder nicht? niemand auf der
Erde ohne ein Axiom (wie *Euclid* das XI)
aufzustellen, beantworten wird; die davon
unabhängige Geometrie abzusondern; und
eine auf die *Ja-* Antwort, andere auf das
Nein so zu bauen, dass die Formeln der
letzten, auf einen Wink auch in der ersten
gültig seyen.

Nach einem lateinischen Werke von
1829. M. Vásárhely, und eben daselbst ge-
druckten ungrischen.

28. A page from the *Kurzer Grundriss*

that the angles *DOC, EOD, FOE, ...* are equal as the arc lengths are equal; so, however, small the angle *DOC* may be, the curve *g* will sooner or later intersect the *y* axis.

The error in the proof is evident: only if have a *circle* for curve *g* can we assert that peripheral angles to equal arcs are equal.

In the letter mentioned above GAUSS pointed out this mistake, stressing that by taking equal chords of the curve *g*, nothing was known of the peripheral angles: they might for example, diminish as a geometrical progression with quotient between 0 and 1, and have smaller than 90° necessary for the "closing up".

No doubt GAUSS's remark disappointed FARKAS BOLYAI, yet he did not give up investigating the fifth postulate. "... I ought to have expressed myself more lucidly," "I became hesitant" — he complained in a letter to GAUSS ([12], p. 85). Then, on 27 December 1808 he sent to Göttingen a *Supplement* of his previous paper ([31], II., pp. 16—22); this, however, remained unanswered.

After this fiasco he set a humbler goal to himself and tried to simplify EUCLID's parallel postulate. He tried to find substitute axioms that would be simpler and more plausible than, but equivalent with, EUCLID's.

The question of substitute axioms engaged BOLYAI's attention for decades, although he realized more and more clearly that they would not suffice in laying the foundations of Euclidean geometry.[24] He listed eight substitute postulates in the *Tentamen* ([6], II., pp. 46—47), to which one more, published later in *Kurzer Grundriss* ([31], II., 151), can be added.

The wording of the substitute postulates proposed in the *Tentamen* is laboured, some statements are rather vague and sometimes the involvement of the notion of time (tempus) conceals the otherwise plausible geometrical contents. Their originality can also be questioned: they are much rather the re-wording of attempts made during the earlier centuries to replace the parallel postulate. What still endows them with some significance is that BOLYAI must have thought over the proof of their equivalence with the Euclidean parallel postulate.

Let us mention three substitute axioms presented by FARKAS BOLYAI. To avoid misunderstanding, it should be noted that in the works of the two BOLYAIs, in both the text and the figures the notation of the right angle, or *angulus rectus* in Latin, is chiefly *R*.

(A) If *z* does not contain *y*, then $z+v$ does not contain $y+v$, either. ("Si a *z* non capitur *y*, neque a $z+y$ capitur $y+v$") ([6], II., p. 46)

This extremely succinct substitute axiom cannot be understood without the attached figure *(Figure 26)* and the argument verifying its equivalence with the fifth postulate. The latter is given in detail by STÄCKEL ([31], I., p. 48). According to it, (A) says that if the rays AC' and BC'' intersect then the rays AC_1, and BC_2 also intersect.

FARKAS BOLYAI regarded this as the simplest of the proposed postulates and presented it to his students.

---

[24] One of his letters to JÁNOS BOLYAI shows that FARKAS himself was dissatisfied with the substitute postulates he proposed: "... none of my Axioms is like what it should be ..." ... "There are axioms among those of mine that at first glance look good, but they are not, they do not constitute a perfectly simple and sufficiently lucid foundation for a science like Geometry." (Quoted in [31], I., pp. 77—78.)

(B) No sphere may differ from another sphere in any property but its size and location ("Sphaera nulla per ullam qualitatem, praeter magnitudinem locumque ab ulla alia discerni potest"). ([6], II., p. 47).

(B) offers an easy method to find similar planar triangles — the literature also calls it a counterpart of WALLIS's substitute postulate (1663): There are similar planar triangles.

(C) Three points are on either a straight line or a circle.[25]

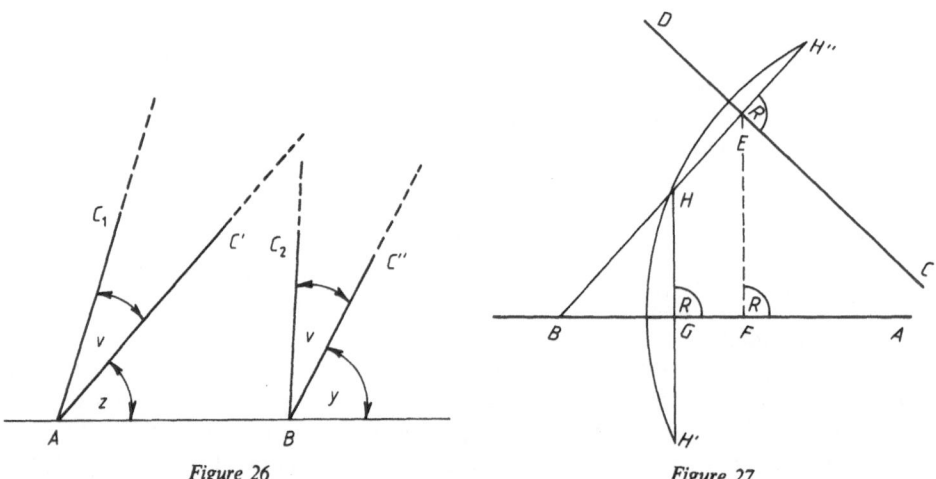

Figure 26   Figure 27

STÄCKEL quotes JÓZSEF KÜRSCHÁK's observation which makes the distinction of the case where the three points are on a straight line unnecessary. To this end, the former axiom is to be reformulated in the following way: "Given any three points in space, it is always possible to lay across two (appropriately chosen) points a sphere such that the third point is inside the sphere." ([31], I., p. 226) FARKAS BOLYAI's postulate remind us of an idea of WACHTER (1792—1817), a student of GAUSS. In 1817 WACHTER made an attempt to prove, by means other than the Euclidean parallel postulate, that through any three non-collinear points of the plane a circle could always be drawn.

FARKAS BOLYAI's latter substitute postulate is often quoted in the literature. It owes its popularity to its transparency and that the 5th postulate of EUCLID follows from it easily.

In fact, accept the BOLYAI postulate as true remaining in the plane for simplicity, and take two straight lines AB and CD (*Figure 27*).

Drop a perpendicular from the point B of the line AB to CD, and denote the foot by E. Now drop a perpendicular from E to AB; its foot F will either coincide with B

---

[25] The original wording of the axiom in *Kurzer Grundriss* is: "If any three points that do not lie on one and the same line fell always on the surface of a sphere, that would prove EUCLID's 11th axiom." ([31], II., p. 151)

154

(then AB and CD are parallel in the Euclidean sense), or not. If in the latter case we draw a perpendicular from any interior point G of the segment $\overline{BF}$ to the line AB, then — according to PASCH's axiom of order — it will intersect the line BE at some point H. Reflecting the point H in the lines AB and CD we, obtain the points H' and H''. As H', H and H'' do not lie on a straight line, a circle can be drawn through them according to BOLYAI's postulate. Then, however, the lines AB and CD which are not parallel in the Euclidean sense intersect each other in a finite point (the centre of the circle) as they are perpendiculars erected at the bisecting points of two chords of the circle which are not parallel in the Euclidean sense.

*

IMRE CSADA examined [17] the substitute postulates listed by BOLYAI, trying to answer the question how to rank them in regard to simplicity. Though in theory simplicity can be computed by a mathematical formula, such examinations do not have much use. First, the substitute postulates are today no more than historical curiosities. Further, one has to start from one or another system of axioms of Euclidean geometry and deduce from it, with the help of Euclidean axioms and theorems, the substitute postulate and, conversely, starting from the substitute postulate, one has to arrive at the 11th axiom. The axioms and theorems used in the process of verification are, however, assigned equal "weight". "Weighting" of simplicity and plausibility are, in turn, a matter of subjective judgment. For this reason we refrain from giving a detailed account of the investigations into equivalence and simplicity.

*

One of the best-known achievements of FARKAS BOLYAI is the definition of the *end-like equality of plane figures* and three related theorems. According to this definition, two plane figures are *end-like equal if they can be divided into a finite number of pairwise congruent pieces*, that is, if they can be rearranged into each other by cuts. The circle of radius r and and the square of side length $r\sqrt{\pi}$ are *not* end-like equal. BOLYAI's books reveal that he was nudged on to this question by the problem of squaring the circle — a problem the temptation of which he could not resist either. Apart from that, the definition above and the three theorems related to it are in accordance with his attempts to reduce the measures occurring in geometry to data obtainable from figures as simple as possible.

The three theorems, contained in *Tentamen*, are the following ([6], I., pp. 27, 64—66):

1. Two polygons bounded by straight lines and having equal areas are end-like equal.

2. The non-common parts of two congruent but only partially overlapping plane figures are end-like equal.

3. Taking away pairwise congruent pieces from two congruent plane figures, the remainders will be end-like equal.

Only the first of these theorems is completely proved in the *Tentamen*. The proof relies on the following lemma.[26]

If in the parallelograms *ABED* and *EFGH* we have $\overline{AB} \cdot \overline{BE} = \overline{EF} \cdot \overline{FG}$ and $A \sphericalangle = E \sphericalangle$, then the two parallelograms can be rearranged into each other by cuts ([6], II., p. 108).

FARKAS BOLYAI proves this lemma by decomposing the two parallelograms in the following way:

Let us fit the two parallelograms together, as shown in *Figure 28*, so that vertex *E* be in common. Let us measure the segment $\overline{BE}$ on the side $\overline{EH}$, and the segment $\overline{EF}$ on the side $\overline{ED}$ as many times as possible. If this can be accomplished without remainder an integer number of times then we are ready: drawing parallels through the points of division, both parallelograms can be cut up into the same number of parallelograms all of which are congruent.

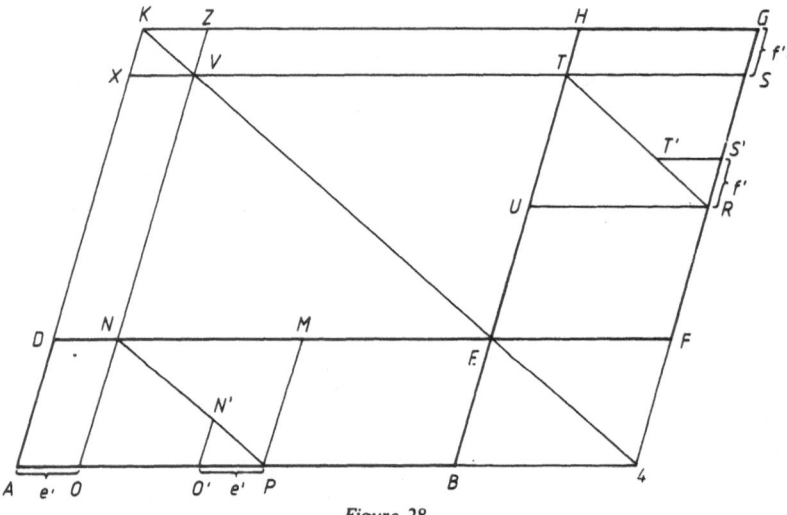

Figure 28

Now suppose that after measuring the segments on the relevant sides an integer number of times, there remains a segment f' of $\overline{EH}$ and a segment e' of $\overline{ED}$. In this case we continue the dissection as follows: we measure f' upward from point *R* and e' leftward from point *P* to obtain the points *S'* and *O'*. From these, drawing parallels up to the diagonals of the parallelograms *URST* and *PMNO* as shown in the diagram, we get the points *T'* and *N'*.

By construction, we have the following congruences:

$$FRUE \cong BEMP,$$

$$RTU \cong PNM,$$

$$T'RS' \cong PN'O'$$

---

[26] Using more appropriate notations.

156

(as each of these figures is congruent with the triangle VKX). Consequently,

$$\overline{HG} = \overline{AO'} \quad \text{and} \quad \overline{S'G} = \overline{AD}.$$

It follows that

$$S'GHTT' \cong DAO'N'N$$

(they can be superimposed by a rotation of 180° in the plane).

Thus, we have carried out the re-arrangement required ([6], II., p. 109).

It immediately follows from BOLYAI's lemma that *all plane figures divisible into triangles can by cuts be rearranged into a rectangle of unit base.*

The concept of end-like equality introduced by BOLYAI, and similar ideas raised subsequently (1833) by P. GERWIEN, assumed fundamental importance later when the basic postulates of geometry were set up. There is an abundance of comments, additions and generalizations on this subject (see Chapter 21).

FARKAS BOLYAI also touched on the question of *end-like equality of solids*, to wit:

"Whether an arbitrary three-sided pyramid can or cannot be reduced to a prism by end-like equality has not been clarified as yet." ([6], II., p. 241).

The problem has also been raised by GAUSS. At the international mathematical congress in Paris in 1900 it was presented as DAVID HILBERT's 3rd problem. Before long, M. DEHN (1900, 1903) and W. F. KAGAN published a necessary condition for rearranging polyhedra of equal volume into one another by cuts. It implies that a cube and a regular tetrahedron of equal volume are not end-like equal. This problem is still engaging the attention of mathematicians today, with the finest results being attained by H. HADWIGER and his school. For example, J. F. SYDLER proved in 1965 that DEHN's condition for dissection is not only necessary but also sufficient.

Of the many results achieved in this field, let us only mention the one which says that polyhedra end-like unequal for dissection by planes are, as a rule end-like unequal even for dissection by fairly general surfaces. On the other hand, there are examples of solids that are end-like equal for dissection by surfaces but end-like unequal for dissection by planes.

*

FARKAS BOLYAI's inaugural address held in 1804 includes the following passage:

"The knowledge that follows from certain postulates (=axioms) solely by the laws of inference belongs to another field, just as space is allowed for geometry (without our questioning whether it exists beyond its representation)." (Quoted in [34], p. 109).

The point in this clumsily formulated sentence is that we arrive at different geometrical systems if we start from different axioms. Nevertheless, it remains questionable whether the geometries constructed in this way can be applied to the real world or not.

Despite this pioneering recognition far ahead of the mathematical and philosophical knowledge of the age, FARKAS BOLYAI never planned to work out a new geometry. His main goal was to place Euclidean geometry, the only geometry known and generally accepted, on firmer foundations. Some commentators opine that he might have failed to fully understand the mathematical contents of his son's space theory and could never comprehend the significance of absolute geometry. A few sentences culled from his books, however, refute such commentaries. FARKAS BOLYAI shared the conviction of GAUSS, LOBACHEVSKY and JÁNOS BOLYAI that decision between Euclidean and hyperbolic geometry could not be made *a priori*, and only practical measurements could give an answer. The Euclidean system, FARKAS BOLYAI declared, could be accepted as true for the time being, as "... the motion of the planets corresponds to the results" that had been obtained by applying the theorems of Euclidean geometry. But he also raised the question what happened if our range of measurement would extend to the star Sirius or beyond ([6], II., p. 45).

Space theories employed successfully in modern theoretical physics and the mathematical description of the structure of the universe attest the profundity of BOLYAI's question.

That he did understand the mathematical contents of the extremely pithy and laconic *Appendix* can be proved by a few pages from the *Tentamen* ([6], II., pp. 395—398) and *Kurzer Grundriss* ([31], II., pp. 149—162). In the former he shows that the formulae of the plane trigonometry of absolute geometry are formally identical with the corresponding formulae for spheric triangles of imaginary radius, and he also gives an example (the Pythagorean theorem) that when a parameter tends to infinity, the theorems of hyperbolic geometry turn into the relevant formulae of Euclidean geometry. In *Kurzer Grundriss* he draws a parallel between the geometrical systems of JÁNOS BOLYAI and LOBACHEVSKY, and proceeds as far as to add some notes or corrections to their statements.

It is known from some old-age notes of JÁNOS BOLYAI that when in early 1825 in Marosvásárhely he presented his geometrical discovery to his father, FARKAS did not receive his son's space theory with the enthusiasm expected. This is not surprising, as the disappointment caused by his failure to prove EUCLID's parallel postulate must still have been a painful memory to FARKAS. Passages can, however, be quoted from subsequent works to show that the father was soon (at the very beginning of the 1830s) won over to his son's new geometry. Let us cite a few passages in evidence. In the first half of 1831 FARKAS BOLYAI wrote to one of his students about the then unpublished *Appendix*:

"It is an original and great work; never before has a similar mathematical work been written by a Hungarian stylus; it would deserve credit everywhere."

In the *Tentamen* and his Hungarian textbooks he praises his son's discovery in words like these:

"The author of the Appendix has solved the problem with awe-inspiring acumen, and created a geometry that is absolutely true in every case." (*Tentamen*, ed. 2, Vol. 2, p. 45.)

Or:

"The *Appendix* I mentioned is a little book worth thick tomes; it is such a great, necessary, original and colossal work for geometers committed to pure truth that one rightly expects, even demands, similar works of its author." ... "but still few are those that can see its real worth although it is written without verbosity, with marvellous lucidity." ([8], p. 185)

As we know, JÁNOS BOLYAI was commanded to Lemberg in early 1831. On his way there he visited his father at Marosvásárhely. He is alluding to that meeting in the following lines put down later:

"Had my father not happened to urge or even force me at Marosvásárhely, on my way to duty in Lemberg, to immediately put things to paper, possibly the contents of the Appendix would never have seen daylight." ([4], p. 48)

The commentaries putting the blame on FARKAS BOLYAI for the lack of recognition of the *Appendix* rely on some of the father's letters. One of them was written on 4 April 1820 to his son in Vienna. At that time JÁNOS BOLYAI, just like his father, was trying to give a proof of the parallel postulate. Here are the first lines of the letter:

"Do not try the parallels *in that way* (author's italics): I know that way all along, I have measured that bottomless night, and all the light and all the joy of my life went out there."

We know the precedents to this letter: JÁNOS BOLYAI followed in his father's foot-steps, trying to prove the substitute axiom that the distance line of a straight line is also straight. The father got to know of these endeavours from a letter; the phrase "in that way" referred to the futility of this attempt. Possibly upon his father's admonishment, JÁNOS gave up the trials to prove or substitute the Euclidean parallel postulate and turned towards a new and different direction. The outcome of this new road was absolute geometry.

*

In his commemorative address on FARKAS BOLYAI, SÁMUEL BRASSAI [16] declared that so as to earn the attributive "scientific", a work on mathematics has to meet the following requirements:

"1. Producing new tools that solve certain problems unsettled before or simplify the troublesome and lengthy solutions.
2. Settling, in any way satisfactory for all, one or more questions unsettled in science before."

But, in BRASSAI's opinion, "... the *Tentamen* did little in respect of the first and nothing of the second."

We hope to have proved on the preceding pages that this statement is erroneous. FARKAS BOLYAI's name hallmarks several new results, but more importantly, he came close to ideas of theoretical significance that are fundamental in modern mathematics as well. It is his personal tragedy that these sparks failed to enlighten the whole scientific community and thus could not save others the trouble of a lot of tiresome research. Thus they remained unknown in mathematical thought progressing at a feverish pace and spilling forth ever newer ideas in the last century, and their place was taken by the same thoughts in better and more lasting formulation by others. But, contrary to his oft-quoted saying, FARKAS BOLYAI did not "flicker out": an objective history of science records his name among the great thinkers of mankind.

Bibliography

See end of Chapter 15

15. János Bolyai

The Appendix

In the discussion of FARKAS BOLYAI's mathematical activities frequent mention was made of his results related to EUCLID's Postulate 5. Before embarking on the *Appendix*, one of the most outstanding writings in mathematical literature, let us first review the main trends in the research of the fifth postulate.

As has been seen, some tried *to simplify* the postulate. Yet, in spite of the plenitude of fascinating substitute axioms, the problems arising could not be solved in this way. Others, including amateurs as well as prominent mathematicians, sought *to prove* the fifth postulate relying on the residual axioms of Euclidean geometry. Also this line remained futile as long as *direct* proofs were being searched. By contrast, the attempt at finding *indirect* proofs by assuming that the Euclidean parallel postulate was not true brought surprising results. For although some of the resulting theorems were unfamiliar to the customary way of thinking, no inner contradiction could be detected among the results. The most profound result in these investigations was achieved by LAMBERT around 1766. Starting from "Case 3" or the "acute angle" hypothesis (that the sum of the angles of a plane triangle is less than 180° in Euclidean geometry) he arrived at such incredible results that he thought he had also proved EUCLID's Axiom XI. His study was published posthumously in 1786; as far as we know, the two BOLYAIS had no knowledge of it. Anyhow, it is true that LAMBERT "opened wide the gate, but was afraid to enter the new world of non-Euclidean geometry" (FERENC KÁRTESZI).

Starting (around 1820) in the wake of his father, JÁNOS BOLYAI also tried to give indirect proofs to substitutes of the Euclidean parallel postulate. Very soon, perhaps in the same year, however, he realized the futility of this attempt and on 3 November 1823 he discovered the basic idea of a new geometrical system, as a letter written to his father from Temesvár attests. His hypotheses rested on a definition of parallelism more general than in EUCLID's geometry. His investigations were recorded in the extremely concise and brilliantly structured work *Appendix* consisting of 43 sections.

In what follows we summarize, without aiming at completeness and omitting the proofs, the contents of the *Appendix* so as to bring out its structure while, as far as possible, are preserved BOLYAI's original methods and notions.[1]

---

[1] In presenting the material, I have freely used the profuse literature on the *Appendix*, especially [13, 19, 22, 23, 31, 32, 37, 40]. For proofs and additions, see e.g. [13].

APPENDIX.

SCIENTIAM SPATII *absolute veram exhibens:*

a veritate aut falsitate Axiomatis XI Euclidei
(a priori haud unquam decidenda) in-
dependentem; adjecta ad casum fal-
sitatis, quadratura circuli
geometrica.

———————

Auctore JOHANNE BOLYAI de eadem, Geometrarum
in Exercitu Caesareo Regio Austriaco Ca-
strensium Capitaneo.

———————

29. The title page of the first edited *Appendix*

structum sit S. Omnia, quae expresse non dicen-
tur, in Σ vel in S esse; absolute enuntiari, i. e.
illa, sive Σ sive S reipsa sit, vera asseri intel-
ligatur.

§ 16. (Fig.5). Si am sit axis alicujus L ; tum L in
Σ recta ⌐am est. Nam sit e quovis puncto b ipsius L
axis bn; erit in Σ bam+abn =2bam =2R, adeoque
bam = R. Et si c quodvis punctum in $\widetilde{ab}$ sit, at-
que cp|||am; est (per §. 13.) cp ≙ am, adeoque c
in L (§. 11.)

In S vero nulla 3 puncta a, b, c ipsius L vel F
in recta sunt. Nam aliquis axium am, bn, cp (ex. gr.
am) intra duos reliquos cadit; et tunc (per §. 14.)
tam bam quam cam < R.

§ 17. L est etiam in S linea, et F superficies.
Nam (per §. 11.) quodvis planum ad axem am (per
punctum aliquod ipsius F) ⌐re, secat ipsum F in
peripheria circuli, cuius planum (per §. 14.) ad
nullum alium axem $\widetilde{bn}$ ⌐re est. Revolvatur F
circa bn; manebit (per §. 12.) quodvis punctum ipsius F in F, et sectio ipsius F cum plano ad $\widetilde{bn}$
non ⌐ri, describet superficiem: atqui F (per §.12)
quaecunque puncta a, b fuerint in eo, ita sibi con-
gruere poterit, ut a in b cadat; est igitur F superficies uniformis. Patet hinc (per §. 11. et 12)
L esse lineam uniformem.

§18.(Fig.7). Cujusvis plani, per punctum a ipsius
F ad axem am oblique positi, sectio cum F in S peripheria circuli est. Nam sint a, b, c, 3 puncta hujus
sectionis, et bn, cp axes; facient ambn, amcp
∧lum; nam secus planum (ex §. 16.) per a, b c
determinatum ipsam am complecteretur (contra
hyp). Plana igitur, rectas ab, ac ⌐riter bissecan-
tia se mutuo secant (§. 10.) in aliquo axe $\widetilde{fs}$ (i-
psius F), atque fb = fa = fc. Sit ah ⌐ fs, et revolvatur fah circa fs; describet a peripheriam radii ha,
per b et c euntem, et simul in F et $\widetilde{abc}$ sitam

30. A page from the *Appendix*

To avoid misunderstanding we note in advance that János Bolyai, though not stating this explicitly, used Euclid's axioms which remain after deletion of the parallel postulates as well as several theorems given without proof and deducible from these residual axioms.

Let us be given a straight line g *(Figure 29)* and a point O outside g. Let the points A and B on g be located on either side of foot N of the perpendicular ON. Let us rotate OA clockwise and OB counter-clockwise around O in the plane AOB. During the rotation there will be a position when OA and OB no longer intersect the line g but "rebound"[2] from it. Let the two half-lines at the moment of "rebounding" be OA' and OB'. Two cases can be imagined in this position. First, the half-lines OA' and OB' may be extensions of each other, thus constituting a single straight line. According to Euclid's Postulate 5 this case is realized in our world.

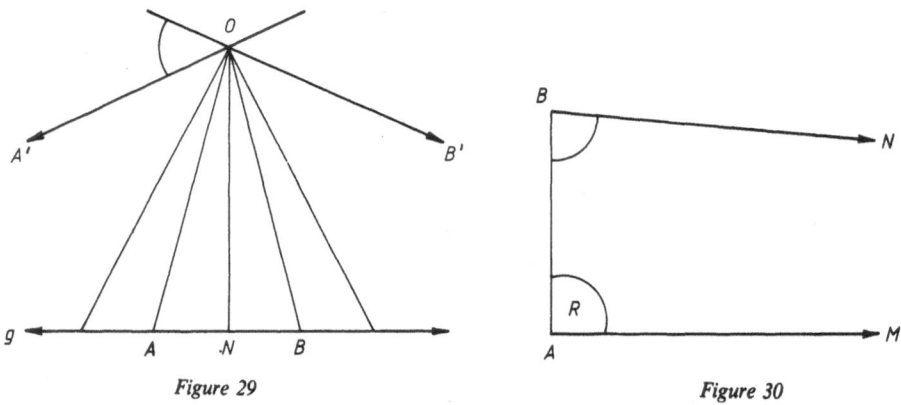

Figure 29   Figure 30

However, the second case — where the two half-lines are not one another's extension, hence after extending them beyond the point O they intersect — is also conceivable. This case is incompatible with the fifth postulate.

The synthesis of the two possibilities leads to a new definition of parallelism, more general than the Euclidean one. This definition of parallelism and the theorems deducible from it and from the residual axioms of Euclid, make up the material of the *Appendix*.

János Bolyai defines parallelism not for lines but for directed half-lines, or rays: We say that ray $\overrightarrow{BN}$ is parallel with ray $\overrightarrow{AM}$ if $\overrightarrow{BN}$ is the first of the rays obtained by rotating a ray around B that does not intersect $\overrightarrow{AM}$ *(Figure 30)*. Then, according to János Bolyai,

$$BAM \sphericalangle + ABN \sphericalangle \leq 2R, \tag{1}$$

[2] A term proposed by Károly Szász. In his notes, János Bolyai calls the "rebounding" line also the "asymptotic parallel", "nearest parallel" or "first non-secant". Earlier commentators (e.g., Dávid Kagan) proved its existence with the Dedekind cut. The modern approach makes this proof superfluous as the Dedekind axiom is also included among the axioms (see, e.g., Greenberg, Marvin Jay: *Euclidean and Non-Euclidean Geometries.* San Francisco, 1974).

which is obviously equivalent to the theorem that the angle sum of a triangle is $\leq 2R$. The notation of the BOLYAI parallelism is $\overrightarrow{AM} \parallel \overrightarrow{BN}$.

A significant feature of the definition is the directedness of the half-lines because with a change of direction we obtain another parallel half-line. If, however, the direction is fixed, there is one and only one $\overrightarrow{BN}$ parallel to $\overrightarrow{AM}$ in the BOLYAI just like in the Euclidean sense (§ 1).

Hereinafter, parallelism defined as above will be called BOLYAI parallelism, if in (1) only the sign $<$ is allowed, its is called hyperbolic parallelism. In the case of hyperbolic parallelism and fixed parameter k (see later), there are two lines parallel to AM through the point O.

In BOLYAI's definition, the points A and B seem to enjoy a distinguished role when the half-lines $\overrightarrow{AM}$ and $\overrightarrow{BN}$ are parallel. This, however, is not true: it can be proved that if we push the point B along the line BN to any point C *(Figure 31)*, the half-line $\overrightarrow{CN}$ will also be parallel to $\overrightarrow{AM}$. On the other hand, any ray $\overrightarrow{CQ}_n$ drawn in the angular domain ACN will intersect $\overrightarrow{AM}$.

Under the hyperbolic definition, there are half-lines $\overrightarrow{CP}_n$ that are neither parallel to $\overrightarrow{AM}$ nor intersect it (§ 2).

Thus, while in the Euclidean plane two straight lines either intersect or are parallel, in the hyperbolic plane there are straight lines that are neither intersecting nor parallel (they are generally called *ultra-parallel* lines).

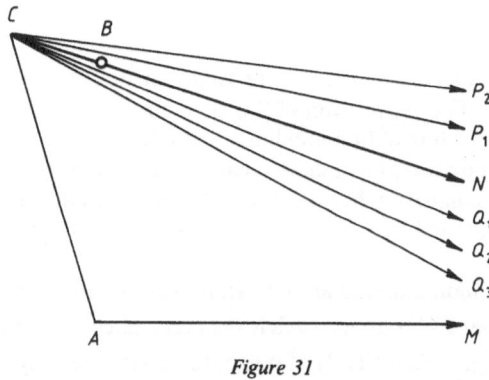

Figure 31

At this point, to make expression easier, we mention the theorem given in a later section of the *Appendix* and saying that BOLYAI parallelism is a symmetric relation: if the half-line $\overrightarrow{BN}$ is parallel to $\overrightarrow{AM}$ then, conversely, the half-line $\overrightarrow{AM}$ is parallel to $\overrightarrow{BN}$ (§ 6).

In the further phase of constructing geometry, JÁNOS BOLYAI's goal is to find the analogues of some properties of Euclidean parallelism in the case of this parallelism. Thus, as early as in § 3 he proves that if two half-lines are parallel in the BOLYAI sense to a third one then they do not intersect. In § 7 he goes on to prove that in this case all the three lines are parallel.

To prove this theorem, JÁNOS BOLYAI first introduces the concept of *isogonal* points (*corresponding* points in GAUSS's terminology). Namely, if the half-lines $\overrightarrow{AM}$ and $\overrightarrow{BN}$ are parallel then to any point A' on AM there is one and only one point B' on BN (*Figure 32*) such that

$$A'B'N \not< = B'A'M \not<.$$

Points A' and B' are said to correspond to each other, and correspondence is a symmetric relation (§ 5).

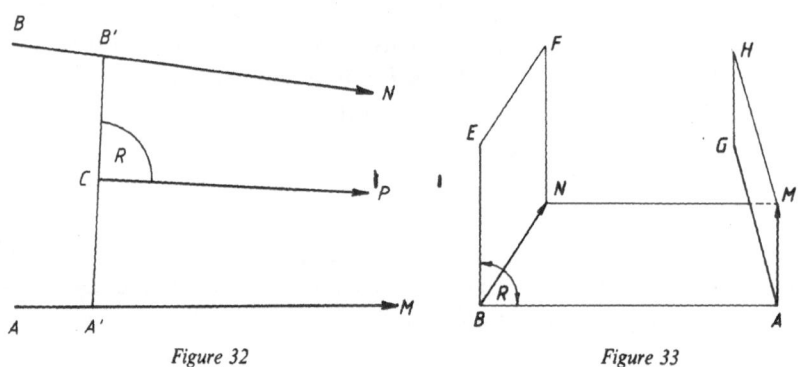

Figure 32   Figure 33

On the basis of the aforesaid, it can be proved that BOLYAI's parallelism, just like EUCLID's, is *transitive*. The verification of this theorem shows for the first time JÁNOS BOLYAI's favourite procedure of first tackling the problem in space and then specializing to the plane. As, however, he assigns great significance to transitivity, he gives a separate proof for coplanar straight lines. It also follows from this theorem that if three *planes* intersect pairwise and two *lines* of intersection parallel then the third one is also parallel to them (§ 7).

If a pair of corresponding points are known on two parallel half-lines, it is easy to find a third half-line parallel to them: such is the perpendicular bisector $\overrightarrow{CP}$ of the isogonal secant $A'B'$ (*Figure 32*). Really, if one of the given lines, say $\overrightarrow{BN}$, intersected it then, by symmetry, $\overrightarrow{AM}$ would intersect it at the same point, whereas $\overrightarrow{AM}$ and $\overrightarrow{BN}$ are parallel (§ 8).

§ 9 gives a proof of a characteristic theorem: if we lay the planes $AMGH$ and $BNEF$ through the half-lines $\overrightarrow{AM}$ and $\overrightarrow{BN}$, assumed to be parallel in the Bolyaian sense (*Figure 33*), so that one is perpendicular to the plane $AMBN$ and the other not, then these planes intersect on the side of the acute dihedral angle. Thus, if we start from BOLYAI parallelism, a spatial theorem reminding of EUCLID's fifth postulate can be proved, where lines and angles are replaced by planes and dihedral angles, respectively, JÁNOS BOLYAI proves the theorem only in the case where one dihedral angle is acute and the other is R, but states it also in the general case (where it is only required that the sum of the two dihedral angles be less than $2R$).

166

This part of the *Appendix* closes with the theorem asserting that isogonal correspondence is a *transitive* relation both in the plane and the space. Let $\overrightarrow{BN}$ and $\overrightarrow{CP}$ be parallel to $\overrightarrow{AM}$, and let the point A correspond to both B and C *(Figure 34)*: then the points B and C are isogonal $(NBC \npreceq = BCP \npreceq)$.

Here again, BOLYAI begins the proof with the non-planar case and then proceeds to the case of coplanar lines. Direct proofs for the latter case can be read at several places.

§§ 11—24 of the Appendix may be considered a structural unit where some concepts fundamentally important in the sequel are defined and relevant theorems discussed. The new concepts are the *parasphere* and the *paracycle*, or to use BOLYAI's terminology, the *F-surface* and the *L-line*. Both can be defined with the help of corresponding points.

All the points corresponding to A on the collection of those lines in space which are parallel to $\overrightarrow{AM}$ in the Bolyaian sense form a connected surface. This surface is called *parasphere*, and the half-line $\overrightarrow{AM}$ is called its *axis (Figure 35)*. Any plane through the axis intersects the parasphere in an *L-lines*, or *paracycle*.

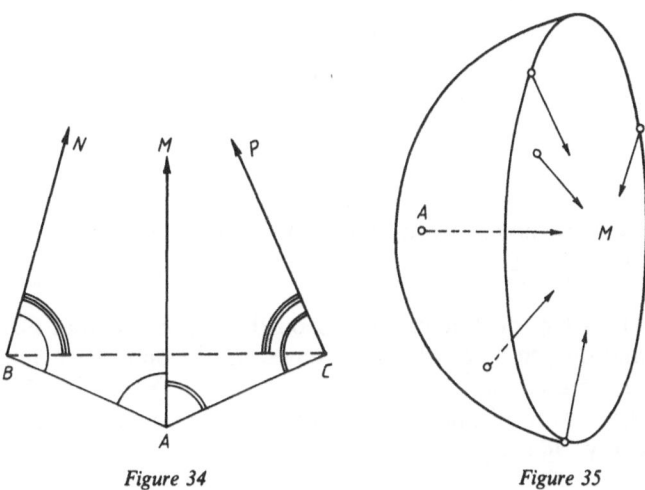

Figure 34   Figure 35

It is to be noted that the terms *paracycle* and *parasphere* were recommended to FARKAS BOLYAI by GAUSS in a letter dated 6 March 1832. LOBACHEVSKY used the terms *horocycle* and *horosphere*.

The *L-line* can also be interpreted independently of the parasphere: it is the collection of all points isogonal with A on the collection of those lines which are coplanar with, and in the Bolyaian sense, parallel to, the half-line $\overrightarrow{AM}$ *(Figure 36)*. To use a term of differential geometry, the paracycle is the orthogonal trajectory through A of the bundle of lines parallel to $\overrightarrow{AM}$ in the Bolyaian sense. Rotating the paracycle around $\overrightarrow{AM}$, we obtain the respective parasphere (§§ 11—12).

167

Now the question arises whether space with the generalized concept of parallelism is homogeneous. Is it possible that in one part of space Euclidean, while in the remaining part hyperbolic parallelism is prevalent? §§ 13 and 14 answer this questions. Really, according to the *inducing* theorems proved here, if there exist a pair of parallel lines such that the sum of the interior angles on one side of a transversal is equal to 2R, then this applies to all pairs of parallel lines. Conversely, if the sum of the angles just mentioned is less than 2R for some pair of parallels, then the same is true of any pair of parallel lines.

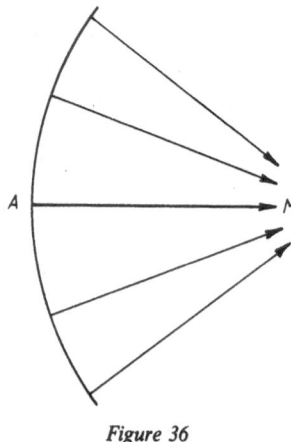

Figure 36

Having proved the homogeneity of space, JÁNOS BOLYAI deems the time ripe to distinguish between two geometrical systems:

"... denote $\sum$ the system of geometry based on the hypothesis that EUCLID's Axiom XI is true, and denote by S the system based on the opposite hypothesis. All theorems we state without expressly specifying the system $\sum$ or S in which the theorem is valid are meant to be absolute, that is, valid independently of whether $\sum$ or S is true in reality." (§ 15)

Today these are called *Euclidean, hyperbolic,* and *absolute* geometry, respectively.

This is the point where LOBACHEVSKY's and JÁNOS BOLYAI's works differ substantially: LOBACHEVSKY focussed almost all his attention on the foundation and elaboration of hyperbolic geometry, while most of the theorems in the *Appendix* are absolute and only rarely can we find a separate treatment of the Euclidean and hyperbolic cases.

The definition of F-surface and L-line easily yields a whole row of relevant theorems. BOLYAI confines himself to mentioning and proving only a few that are indispensable in the sequel. For instance, if can immediately be seen that in Euclidean geometry the parasphere is a *plane* and the paracycle is a *line*; in hyperbolic geometry they are a curved surface and a curve, respectively. Moreover, the parasphere is a "uniform" surface and the paracycle is a "uniform" curve. This means that, in System

S, F is a surface which can be turned around a suitable axis so that a prescribed point of F turns into another prescribed point of F and no point leaves the surface; on the other hand, the L-line can be shifted within itself (§§ 16—17).

§ 18 contains a theorem typical of System S. It says that any plane that passes through the point A of F and stands obliquely to the axis $\overrightarrow{AM}$ intersects F in a circle. As, due to the uniformess of the parasphere, to any point A on it there is one and only one axis $\overrightarrow{AM}$, the point A does not have a distinguished role in this theorem.[3]

In hyperbolic geometry, any line AT perpendicular to an axis AM of the paracycle is a tangent of the paracycle. Although JÁNOS BOLYAI states this theorem explicitly for System S, it is an absolute theorem in view of the fact that the tangent of a line is the line itself. It is also true that any two points of the F-surface determine one and only one L-line (§ 20).

The following theorem is a parasphere version of the results of § 9: if two paracycles of the parasphere are intersected by a third one and the interior angles on one side of the latter have sum <2R, then the first two paracycles meet each other at a point on that side. To put it in another way, *Euclidean geometry is valid on the parasphere provided that the role of straight lines is taken over by paracycles (Figure 37).*

In current terms this theorem says: the parasphere F is "model" of Euclidean geometry in hyperbolic space.

Let us introduce the notion of *equidistant* (or: *mutually parallel*) paracycles in the following way. Let the axis $\overrightarrow{AM}$ and the paracycle arc $\widehat{AB}$ be given *(Figure 38)*. If we chose an arbitrary point D on the axis $\overrightarrow{AM}$ and shift the rigid figure BAD made up of a paracycle arc and a line segment in the direction $\widehat{AB}$, then the point D describes the paracycle arc $\widehat{DE}$ parallel to $\widehat{AB}$.

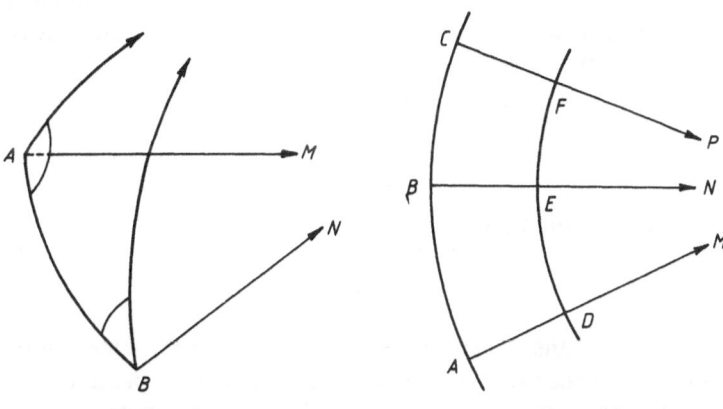

Figure 37   Figure 38

---

[3] As a start for the proof of this theorem, JÁNOS BOLYAI assumes that on the line of intersection there are two points other than A which also lie on the parasphere. The commentaries usually give separate proofs of the existence of these two points, as if making up for law of the *Appendix*. According to an observation by GYÖRGY HAJÓS, the existence of the points follows by rotating the parasphere around an axis as JÁNOS BOLYAI describes it in § 17, so no additional proof is necessary. The rotation serves the very purpose to prepare the proof of the theorem in § 18.

§§ 22—24 give proofs to theorems concerning parallel paracycles. For example, the quotient of the corresponding arcs of two parallel paracycles independent of the arc $\overset{\frown}{AB}:\overset{\frown}{DE}=\overset{\frown}{BC}:\overset{\frown}{EF}$ *(Figure 38)*. It can also be verified that the value of this quotient only depends on the distance between the two *L*-lines. Let us denote the distance between parallel paracycles by lower-case letters x, y, ...; let upper-case letters X, Y, ... denote the quotients of the corresponding arc lengths *(Figure 39)*. According what was said above $X = f(x)$; $Y = f(y)$.

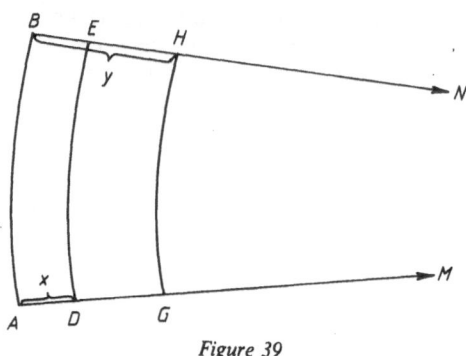

Figure 39

For any x and y it can be proved that

$$X^y = Y^x,$$

or

$$[f(x)]^y = [f(y)]^x.$$

Interrupting the order of presentation applied by JÁNOS BOLYAI and jumping forward to § 30 we can easily anticipate the important results included therein. The continuous solution of the above functional equation is

$$f(x) = X = e^{cx}; \qquad f(y) = Y = e^{cy}.$$

In Euclidean geometry $c = 0$ as in that case X, Y, ... $= 1$. In hyperbolic geometry $c > 0$.

Let us introduce the notation $c = \dfrac{1}{k}$; then

$$X = e^{\frac{x}{k}}.$$

The constant k is called the *parameter* of hyperbolic geometry. One can read off the latter equation that k is the distance of parallel paracycle arcs with quotient e. To every fixed value of k there corresponds a consistent hyperbolic geometry, the limiting case $k \to \infty$ being the Euclidean system.[4]

---

[4] In the *Appendix*, JÁNOS BOLYAI employs the letter i to denote the parameter; to avoid mixing it up with the imaginary unit, k is now the generally accepted symbol. If we were able to determine the value of the parameter k with some method in a single case, then we could choose of the infinite number of hyperbolic geometries that valid in the real world. LOBACHEVSKY tried to find k through astronomic measurements, but he obtained only a lower bound. What we now can shortly say is that the value of k in our world is very large.

And with that, the major concepts of absolute geometry have been introduced. In the remaining sections very few new concepts appear, the aim of the second part of *Appendix* being to present some applications of hyperbolic geometry to trigonometry, differential geometry and the theory of constructions, occasionally with the help of infinitesimal considerations. The preceding chapters have gradually and systematically initiated the reader into the world of the new geometry, helping him part with the Euclidean approach by and by, while in the forthcoming parts — where computations play an increasing role — some characteristic consequences of the previous theorems are given. While the first part strikes the reader with its inventiveness, the second also displays JÁNOS BOLYAI's ability in the use and coordination of the achievements of various mathematical disciplines. It becomes clear that in his youth he was not reluctant to carry out tiresome computations either. His later works mostly contain sketchy ideas without elaboration of details.

§ 25 immediately embarks on one of JÁNOS BOLYAI's most beautiful theorems much discussed by commentators, the *sine law* of hyperbolic geometry: in any rectilinear triangle, the circumferences of the circles with radii equal to the sides are to one another as are the sines of the angles opposite to them.

The theorem adds nothing new to Euclidean geometry. We fail to see its significance unless we realize that the formula $2r\pi$ of the circumference of the circle applies to Euclidean geometry only.

It is evident, however, that also in hyperbolic geometry there must be some relationship between the radius and the circumference, but at this point we do not know what (BOLYAI presents it in § 30).

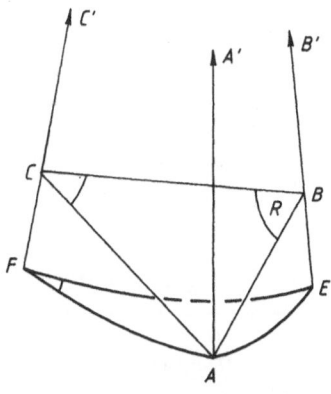

Figure 40

JÁNOS BOLYAI proves the sine law for right triangles only, which, however, immediately yields the general case. His proof rests on the fact that the relations of Euclidean trigonometry hold true on the parasphere, so it would be desirable to find a connection between the rectilinear triangle studied and some triangle on the parasphere. He achieves this aim in the following way. Let us erect a perpendicular $\overrightarrow{BB'}$ to the plane of the right triangle ABC ($B\not\!\star = R$) *(Figure 40)* and draw to rays $\overrightarrow{CC'}$ and $\overrightarrow{AA'}$ parallel to it in the Bolyai sense. The three planes thus determined cut out

171

the paraspheric right triangle AEF on the parasphere belonging to axis $\overrightarrow{AA'}$. For the latter triangle we have:

$$2\pi\widehat{AE}:2\pi\widehat{AF}=\sin F \qquad (F \sphericalangle = C \sphericalangle).$$

Let $\bigcirc r$ denote the circumference of the circle with radius r in hyperbolic geometry. Then:

$$2\pi\widehat{AE}=\bigcirc\overline{AB}; \qquad 2\pi\widehat{AF}=\bigcirc\overline{AC}.$$

That is,

$$\bigcirc\overline{AB}:\bigcirc\overline{AC}=\sin C \sphericalangle.$$

In §26 BOLYAI proves that spheric trigonometry can be constructed without EUCLID's fifth postulate. This, however, was not a new recognition; it had been known to LAGRANGE, as was pointed out by HOÜEL (*Essai critique sur les principes fondamentaux de la géométrie élémentaire.* Paris, 1883, p. 84).

In order to understand the subsequent parts of the *Appendix* better, we remind the reader of FARKAS BOLYAI's attempts at proving that the "distance line" of a straight line were straight. That would imply EUCLID's fifth postulate. Obviously, if we define parallelism in a non-Euclidean way, we cannot say that the "distance line" is a straight line and the "distance surface" is a plane. That is why JÁNOS BOLYAI now introduces the *hypercycle* and the *hypersphere*, the analogues of the distance line and the distance surface in hyperbolic geometry. The *hypercycle* of the base line a at distance d is the locus of all points at distance d from line a on one side of the latter. The *hypersphere* can be defined similarly.

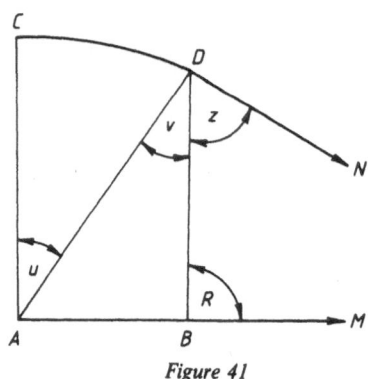

Figure 41

These terms were also recommended by GAUSS in his letter to FARKAS BOLYAI mentioned before. JÁNOS BOLYAI did not invent now names for these notions.

Let AC and BD be perpendicular to AM, and let the hypercycle arc belonging to the segment $\overline{AB}$ be $\widehat{CD}$ *(Figure 41)*. It can be proved that

$$\widehat{CD}:\overline{AB}=\sin u:\sin v, \tag{2}$$

and, if we remove AC from BD to infinity, then

$$\widehat{CD}:\overline{AB}=1:\sin z. \tag{3}$$

172

Here z is one of the most important concepts of hyperbolic geometry, the "angle of parallelism" corresponding to the distance $\overline{BD}$.

Comparing the ratios (2) and (3) we obtain finally:

$$\sin u : \sin v = 1 : \sin z \qquad (\S\ 27).$$

It can be proved without difficulty that

$$\overset{\frown}{AB} : \overset{\frown}{CD} = X = \sin u : \sin v \qquad (\S\ 28)$$

also applies to mutually parallel paracycle arcs corresponding to each other *(Figure 42)*.

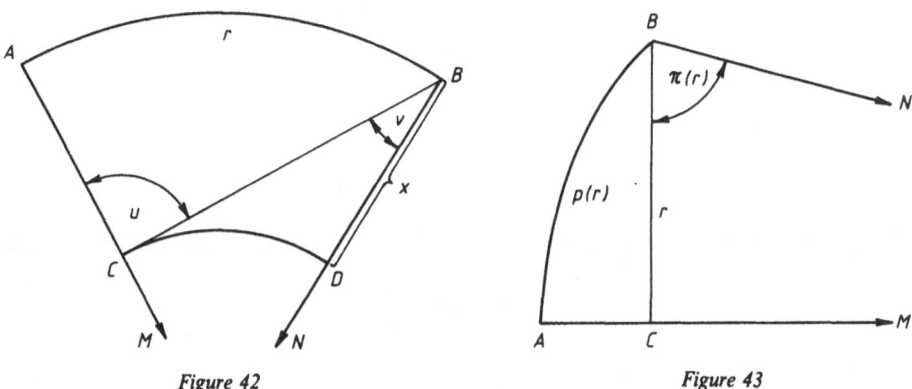

Figure 42   Figure 43

Thus X can be expressed with the help of two angles which have a geometrical meaning.

Let $\pi(x)$ denote the angle of parallelism corresponding to the distance x (LOBACHEVSKY's notation). JÁNOS BOLYAI proves wittily the following relation between X corresponding to x and the angle of parallelism:

$$X = \mathrm{ctg}\,\frac{\pi(x)}{2}.$$

In view of a former result,

$$X = e^{\frac{x}{k}},$$

therefore

$$\mathrm{ctg}\,\frac{\pi(x)}{2} = e^{\frac{x}{k}}. \qquad (\S\ 29) \qquad (4)$$

This basic, but now classic, formula gives a relation between the distance x, the corresponding angle of parallelism, and the parameter k.

Further, the following relation between the angle of parallelism $\pi(r)$ and paracycle arc $p(r)$ belonging to the altitude r can be established *(Figure 43)*:

$$p(r)\,\mathrm{tg}\,\pi(r) = k. \qquad (5)$$

173

The proof of (5) given by JÁNOS BOLYAI is incomplete because it tacitly assumes that the paracycle is rectifiable and the quotient of the paracycle arc and the altitude tends to 1 and the altitude tends to 0. In correction of this flaw, several Hungarian mathematicians (MÓR RÉTHY, JÓZSEF KÜRSCHÁK, LAJOS DÁVID, PÁL SZÁSZ, FERENC KÁRTESZI) have given a proof of (5).

Now it is no longer difficult to find the formula for the circumference of the circle in hyperbolic geometry. In fact, the Euclidean geometry of the plane being valid for the parasphere, the circumference or of the circle having the paracycle arc $p(r)$ for radius is:

$$\bigcirc \; r = 2\pi p(r).$$

Hence, using the trigonometric identity

$$\operatorname{ctg} \pi(r) = \frac{1}{2}\left(\operatorname{ctg} \frac{\pi(r)}{2} - \operatorname{tg} \frac{\pi(r)}{2} \right)$$

as well as (4) and (5):

$$\bigcirc \; r = 2\pi k \operatorname{ctg} \pi(r) = \pi k (e^{\frac{r}{k}} - e^{-\frac{r}{k}}).$$

This formula is easier to remember if written in terms of hyperbolic functions. Thus:

$$\bigcirc \; r = 2\pi k \operatorname{sh} \frac{r}{k}.$$

Finally, in view of this result the absolute sine law reads:

$$\operatorname{sh}\frac{a}{k} : \operatorname{sh}\frac{b}{k} : \operatorname{sh}\frac{c}{k} = \sin \alpha : \sin \beta : \sin \gamma. \tag{6}$$

It is worth noting that in spherical trigonometry the sine law for a triangle with sides a, b, c and angles α, β, γ cut out of the sphere with radius r is:

$$\sin\frac{a}{r} : \sin\frac{b}{r} : \sin\frac{c}{r} = \sin \alpha : \sin \beta : \sin \gamma. \tag{7}$$

Later, in the *Responsio*, JÁNOS BOLYAI stressed the formal analogy between (6) and (7) ([31], II., p. 230), but in the *Appendix* he did not yet apply the hyperbolic notation which makes the analogy striking (§ 30).

No lengthy computations are needed now to get the trigonometric relations for the right triangle (§ 31). JÁNOS BOLYAI restricts his attention to those three cases where

1. a, c, α
2. a, α, β
3. a, b, c

are given. Using again the hyperbolic notation rather than BOLYAI's, these data are connected by the following relations:

1. $\operatorname{sh}\dfrac{a}{k} = \operatorname{sh}\dfrac{c}{k} \sin \alpha$;

2. $\cos\alpha = \operatorname{ch}\dfrac{a}{k}\sin\beta; \quad \cos\beta = \operatorname{ch}\dfrac{b}{k}\sin\alpha;$

3. $\operatorname{ch}\dfrac{c}{k} = \operatorname{ch}\dfrac{a}{k}\operatorname{ch}\dfrac{b}{k}.$

The *Appendix* also contains the equation

4. $\operatorname{sh}^2\dfrac{c}{k} = \operatorname{ch}^2\dfrac{a}{k}\operatorname{sh}^2\dfrac{b}{k} + \operatorname{sh}^2\dfrac{a}{k},$

immediately obtainable from the 3rd formula, and the equation

5. $\operatorname{ctg}\alpha\operatorname{ctg}\beta = \operatorname{ch}\dfrac{c}{k}$

which can be obtained by multiplying the equations of the 2nd relation with one another and taking the 3rd relation into account.

If we use the power series of the hyperbolic functions and let $k \to \infty$, we arrive at the respective relations of Euclidean geometry. Formula 4 yields the Pythagorean theorem of Euclidean geometry, so we can regard it as one possible generalization of the Pythagorean theorem in absolute geometry. Formally, these relations coincide with the respective relations in the trigonometry of the sphere of radius $\dfrac{k}{i}$ where $i^2 = -1$.

As an indication of the thoroughness with which FARKAS BOLYAI must have studied JÁNOS's work, let us mention the *Supplement* he added to the *Tentamen* (second edition, Vol. II., pp. 295—298) which, he says, "*is the property of the author of the Appendix as a completion*". His aim was to assist the reader with understanding the *Appendix*. In it he proves some theorems concerning hyperbolic functions and the EULER relations, and shows that the 4th relation above, for $k \to \infty$, gives the Pythagorean theorem of Euclidean geometry.

The longest section of the *Appendix*, § 32, presents formulae for various measures of geometric figures. Just like in Euclidean geometry, some results of analysis are needed in these computations. Seen from today's vantage point, this chapter is not rigorous enough in this regard: we miss those considerations which have been customary in similar investigations since the appearance of the precise definitions of limit, continuity, analyticity, etc. In the manner of the old masters, JÁNOS BOLYAI treats the differential formally calculates definite integrals without strictly justifying his steps. We must admit, however, that, for instance, the concept of *limit* was accurately defined at about the same time as the *Appendix* was written and its spreading took some time; the rigorization of analysis was to be achieved in the subsequent decades. Unlike his father, JÁNOS BOLYAI cannot have known much of these deep investigations when he was writing the *Appendix*, or he must have deemed them irrelevant as the point of his work was of quite another nature.

The subjects treated in § 32 are essentially the following: slope of a curve, arc length are bounded by curves, area of a spherical cap, volume of a ball. The system of coordinates BOLYAI uses is rectangular. In it the location of a point is given by the arc length from the Y-axis to the point in question of the corresponding hypercycle of the X-axis, and by the length of the segment perpendicular to the X-axis and bounded by the point. In hyperbolic geometry the arc length of the hypercycle is not equal to the abscissa x of Euclidean geometry. Let t denote the arc length of the hypercycle belonging to the abscissa x. Then:

1. The slope of a curve at any of its points is

$$\operatorname{tg} \alpha = \lim_{\Delta x \to 0} \frac{\Delta y}{\Delta t}.$$

2. For the arc element Δz of the curve $z = f(x)$ it can be proved *(Figure 44)* that

$$\lim_{x \to 0} \frac{(\Delta z)^2}{(\Delta y)^2 + (\Delta t)^2} = 1.$$

3. The equation of the paracycle that is tangent to the Y-axis at the origin is (with hyperbolic notation)

$$\operatorname{ch} \frac{y}{k} = e^{\frac{x}{k}}.$$

4. Hence, using statement 2 and computing the integral, for the arc length $\widehat{OP}$ *(Figure 44)* we obtain:

$$\widehat{OP} = k \sqrt{e^{\frac{2x}{k}} - 1} = k \operatorname{sh} \frac{x}{k}.$$

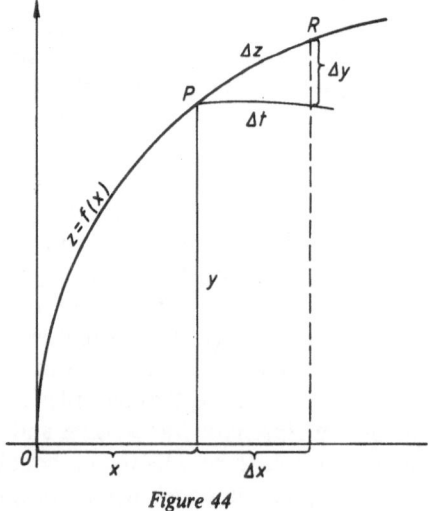

Figure 44

5. If the equation of a curve is known, the area is given by the corresponding definite integral. For example, the area of a circle of radius r is

$$4\pi k^2 \operatorname{sh}^2 \frac{r}{2k}.$$

176

In addition to a few other formulae, the *Appendix* also contains:

6. The area bounded by the parallel paracycle arcs $\widehat{AB} = r$ and $\widehat{DE}$, and by the segments $\overline{AD} = \overline{BE} = x$ (see *Figure 39*) is

$$rk(1 - e^{-\frac{x}{k}}).$$

If the respective integral formulae are known, expressions for surface area and volume can also be derived; JÁNOS BOLYAI presents only a few them.

7. The *area* of the sphere of radius r is:

$$4\pi k^2 \,\mathrm{sh}^2 \frac{r}{k}.$$

8. Its *volume* is:

$$2\pi k^3 \left(\mathrm{sh}\frac{r}{k} \,\mathrm{ch}\frac{r}{k} - \frac{r}{k} \right).$$

9. If the hypercycle arc at distance q from the straight segment $\overline{AB} = p$ is rotated around $\overline{AB}$, the surface of revolution thus obtained has *area*

$$\pi k p \,\mathrm{sh}\frac{2q}{k}.$$

10. The *volume* of the respective solid of revolution is:

$$\pi k^2 p \,\mathrm{sh}^2 \frac{q}{k}.$$

For some of the above formulae, JÁNOS BOLYAI also proves that for $k \to \infty$ they go over into well-known results of Euclidean geometry, unless what we obtain after passage to the limit is an identity.

By way of concluding this part of the *Appendix*, JÁNOS BOLYAI poses the decisive question as to which system is valid in reality: system Σ or system S. He gives no explicit answer, but his sentences suggest the belief that whether Euclidean geometry or a hyperbolic one belonging to a fixed value of k is valid, the theory outlined above leads to a geometry without intrinsic contradictions (§ 33).

With that, BOLYAI's system is completed, but not so the *Appendix*: probably under the influence of GAUSS's *Disquisitiones Arithmeticae*, which he knew well, JÁNOS BOLYAI felt obliged to extend the theory of constructions to the hyperbolic plane.

The problem itself can be briefly summarized as follows. In BOLYAI's geometry there are four "uniform" curves: the straight line, the circle, the paracycle and the hypercycle. The instruments needed to draw them might be called *rulers* with straight and paracyclic edge, and common and hypercyclic *compasses*. Some basic constructions can be set up with these four instruments just as is usual for the ruler-and-straightedge constructions in Euclidean geometry. If a construction can be carried out by combining the basic constructions with the four instruments a finite number of times, we say that the problem is solvable by a BOLYAI construction.

Only one of the construction described by JÁNOS BOLYAI is given here in detail; the rest is mentioned only in passing.

The half-line $\overrightarrow{DM}$ from the point D parallel to $\overrightarrow{AN}$ is to be constructed. Let $AN \perp BD$, $AN \perp AC$ and $AC \perp CD$ *(Figure 45)*. In view of a former result (§ 27):

$$\bigcirc CD : \bigcirc AB = 1 : \sin z.$$

If we draw a circle from point A with radius CD, it will have a point E or B in common with the segment BD. The latter is the case in Euclidean geometry ($\overset{\frown}{CD} = \overline{AB}$; $z \not\ast = R$). In the other case z at point E as seen in *Figure 45* is the angle of parallelism. Drawing a copy of z at point D, we obtain $\overrightarrow{DM} \parallel \overrightarrow{BN}$ (§ 34).

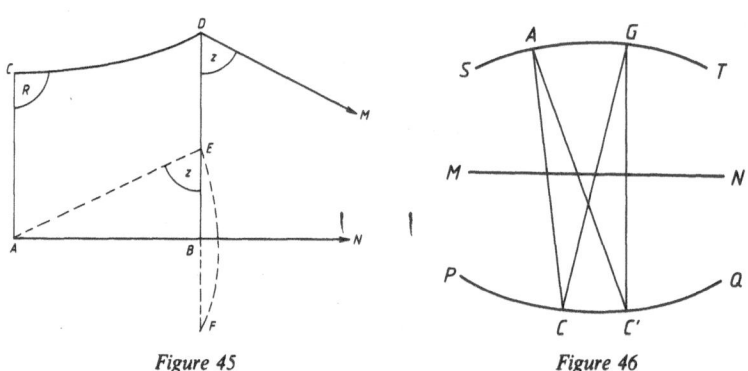

Figure 45   Figure 46

In the *Appendix*, one can also find the detailed description of how to construct: the distance of parallelism to a given angle of parallelism (§ 35), the point of intersection of a line and a plane, the line of intersection of two planes (§ 36), a pair of isogonal points (§ 37). § 38 treats two special problems: the construction of the distance x corresponding to the values $X = 2$ and $X = e$ (X denotes the quotient of mutually parallel and corresponding paracycle arcs, x is the distance between them).

The next sections discuss theorems that are needed for the most spectacular construction of hyperbolic geometry, the squaring of the circle. Let us sketch some of these theorems. Hypercycle arcs $\overset{\frown}{ST}$ and $\overset{\frown}{PQ}$ which lie symmetrically to the base line MN are given *(Figure 46)*. It can be shown that both the area and the angle sum of the triangle GAC bounded by the hypercycle arc and some straight lines remain unchanged while the point C moves along $\overset{\frown}{PQ}$ (§ 39). This result can be put in another way: two triangles with equal area and with one side in common have equal angle sums (§ 40). Moreover, the condition of having a side in common is superflous: in the *S*-system the areas of two triangles are equal if and only if their angle sums are equal (§ 41).

Let us call, in hyperbolic geometry, the value that supplements the angle sum of a triangle to 180° the *defect* of the triangle. The corresponding notion for spherical triangles is the *excess*. It follows from the previous theorems of the *Appendix* that the ratio of the areas of two triangles is equal to the ratio of the defects of the triangles.

178

In other words, the expression for the area of a triangle in hyperbolic geometry is

$$\Delta = c\delta,$$

where Δ denotes the area of the triangle, c is constant for all triangles and δ is the defect. This result was also known to GAUSS; a sketch of the proof he proposed can be found in a letter he wrote to FARKAS BOLYAI after having received the *Appendix* ([12], pp. 109—112).

In the closing section, § 43, of the *Appendix* JÁNOS BOLYAI proves that $c = k^2$. That is, in hyperbolic geometry the area of a triangle is:

$$\Delta = k^2\delta.$$

It follows directly from this formula that in hyperbolic geometry there exists a triangle of maximal area: it is the one with maximal defect, i.e., π. All the angles of this triangle are zero, hence its adjacent sides are parallel. Its area is $k^2\pi$. Comparing this result with the area formula of the circle in hyperbolic geometry (see § 32), we arrive at the following, at first glance paradoxical, conclusion: in system S, *the area of the circle does not have an upper bound while that of the triangle has.*

Let us draw a triangle of maximal area, that is, $\overrightarrow{AB}$ and $\overrightarrow{DC}$; $\overrightarrow{CD}$ and $\overrightarrow{FE}$; $\overrightarrow{EF}$ and $\overrightarrow{BA}$ are pairs of parallel half-lines *(Figure 47)*. Taking the bisecting ray $\overrightarrow{MN}$ of the segment $\overline{AB}$ for axis, we can prove that the length to the corresponding paracycle

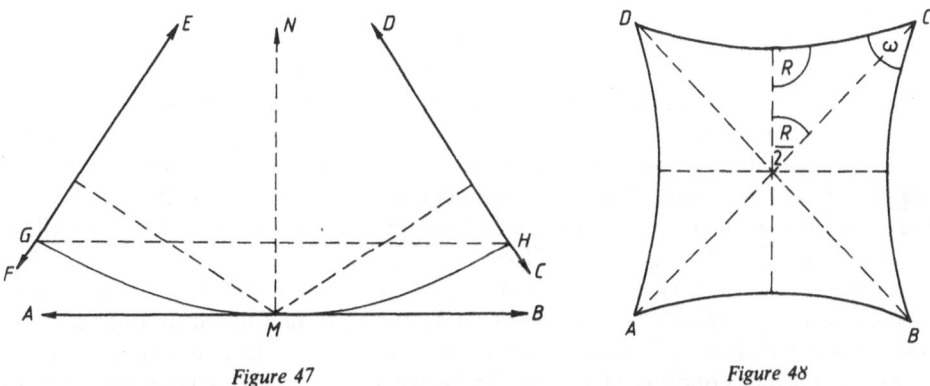

Figure 47   Figure 48

arc $\overgroup{GMH}$ is $2k$. Consequently, the area of the paraspheric circle produced by rotating this paracycle arc around the axis $\overrightarrow{MN}$ equals $k^2\pi$, the possible area of maximal triangle.

Let us define — following JÁNOS BOLYAI — the square in hyperbolic geometry as a quadrilateral all of whose sides and angles are equal. The angles, however, are smaller than R. Let us dissect the square into 8 congruent triangles with the help of the lateral bisectors and the diagonals *(Figure 48)*. The area of one triangle is

$$k^2\left(\frac{R}{2} - \frac{\omega}{2}\right).$$

179

Hence the area of the square is

$$4k^2(R-\omega).$$

If we want to have the area of the square equal to that of the paraspheric circle constructed above then the following condition must be fulfilled:

$$2k^2R = 4k^2(R-\omega).$$

Therefore, $\omega = \dfrac{R}{2}$.

We have thus found a square whose area is equal to the area of a paraspheric circle. Using the theory of geometric constructions and some theorems of the *Appendix* it can be proved that this square and circle can be constructed. In other words, we have arrived at the surprising result that *in hyperbolic geometry there is a circle which can be squared.*

Let us finally quote JÁNOS BOLYAI's words he attached to this result:

"Consequently, either Axiom XI of EUCLID holds or the geometrical quadrature of the circle is possible."

*

Leaning on the contents of the *Appendix*, it is easy to outline the "genesis" of JÁNOS BOLYAI's space theory. As a result of the scrupulous research work of several scholars in Romania and Hungary, we have now an array of scattered data that like pieces in a mosaic begin to fall into place, but further research may make this picture more complete. The documents are preserved in the Teleki—Bolyai Library at Marosvásárhely (*TK*) and the Library of the Hungarian Academy of Sciences.

As early as at the age of 17 or 18, JÁNOS BOLYAI knew much of the history of the axiom of parallelism through the teachings of his father. The problem was raised for him by FARKAS, who remained the paragon of mathematical rigour in his eye. Around 1820, in Vienna he set out along his father's path trying to prove two substitute axioms of the Euclidean postulate. However, he must have realized in the same year that the axiom of parallelism could not be proved, which might have prompted him to proceed toward the idea of a geometry different from the Euclidean. Four sketchy figures in a note-book (*TK*) attest to this, they illustrate — as STÄCKEL assumes — several important concepts of hyperbolic geometry (paracycle, hypercycle, angle of parallelism, limiting octogon of hyperbolic geometry). After some three years of deep absorption, on 3rd November 1823 he wrote a letter to his father from Temesvár, stating that he had "created a new, another world out of nothing". The letter, however, does not give accurate information on how far he had gone in his investigations, as the verb in past tense is followed a few lines later by the phrase "it's not yet invented". Our inquiry is furthered by another clue gleaned from a later note of his which says that at the end of 1823 he had already arrived at what later constituted § 29 of the *Appendix* —" ... having corrected it in winter around midnight ..." (*TK*). But all this is insufficient to answer the question when the whole material of the 43 sections of the *Appendix* got ready. A letter of 4th October 1855 (*TK*) clarifies this question. The relevant sentence reads:

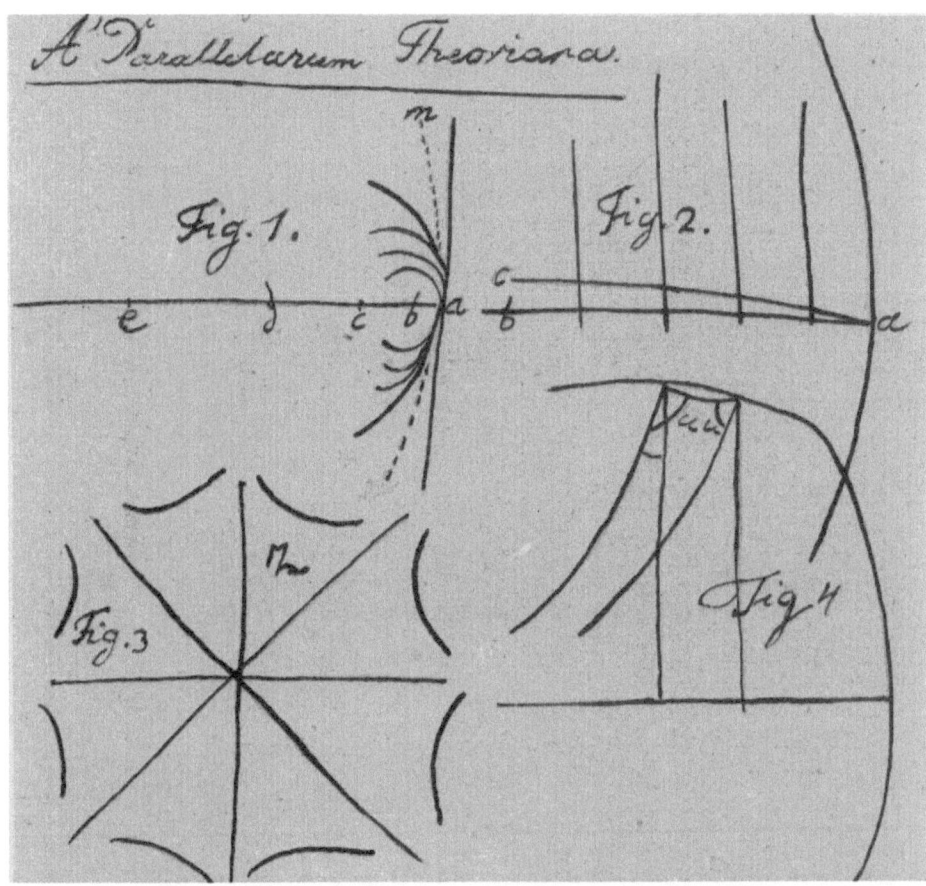

31. Four sketchy figures from one of János Bolyai's note-book in 1820

"As far as I remember I immediately told you that back in 1824 I had noticed in GAUSS's letters that he had also realized that the area of a planar Δ (= triangle), if all sides $\to \infty$, is only end-bounded (= has an upper bound) in S."

The quotation refers to "letters" of GAUSS. One of them is obviously that in which he informed FARKAS BOLYAI of the substitute axiom he had found.[5] It is dated 26 December 1799. The other one is dated 6 March 1832 and it acknowledges the receipt of the *Appendix*.[6] What has relevance here is the date "1824" as it shows that in 1824

[5] "... wenn man beweisen könnte dass ein geradlinigtes Dreieck möglich sei, dessen Inhalt grösser wäre als eine jede gegebene Fläche so bin ich im Stande die ganze Geometrie völlig streng zu beweisen. Die meisten würden nun wol jenes als ein Axiom gelten lassen; ich nicht; es wäre ja wol möglich, dass so entfernt man auch die drei Endpunkte des Δ im Raume von einander annähme, doch der Inhalt immer unter *(infra)* einer gegebenen Grenze wäre" ([12], pp. 36—37).

[6] Here GAUSS — unlike BOLYAI in the *Appendix* — uses functional equations to give a sketchy proof of the expressions for the area of the triangle and "limiting" in hyperbolic geometry. ([12], pp. 110—112).

32. "... I had created a new, another world out of nothing..."

around midnight ... the essence of the § 29 ..."

34. "... I had presented my work ... to Captain
Wolter ... still in 1826 ..."

JÁNOS BOLYAI already knew the formula of the area of the triangle in hyperbolic geometry. This latter, as has been seen, is the subject-matter of the last section of the *Appendix*. Presumably, JÁNOS BOLYAI had a comprehensive conception of the entire material of the *Appendix* in 1824.

As for the subsequent years, it has been revealed by JÁNOS BOLYAI's posthumous manuscripts that in 1826 he handed over a draft of his theory of space to his one-time Vienna teacher Captain WOLTHER VON ECKWEHR in Arad (where both were stationed at the time), probably for criticism. But from him "... God only knows through what *events* it can have drifted somewhere by some whim of fate" (*TK*).

The search after this manuscript has remained futile so far. It can, however, be documented that the treatise was in the hands of FARKAS BOLYAI before 23 June 1830, "but only in rough copy" (*TK*). We do not know as yet whether it was a copy of the first draft or a more detailed version. It was mentioned in another context that being transferred from Arad to Lemberg in the spring of 1831, JÁNOS BOLYAI was asked by his father at Marosvásárhely to give him the treatise on the theory of space for publication in the *Tentamen*. This request probably referred to the Latin *translation* of the work written in Hungarian or German. On 20 June 1831 a reprint of the *Appendix* was mailed to GAUSS ([12], p. 103).

The chronology of LOBACHEVSKY's discoveries can be reconstructed on the basis of researches by V. F. KAGAN and B. L. LAPTEV as follows. In the manuscript of his 1823 lecture entitled *Geometry*, LOBACHEVSKY made clear that all the previous attempts at proving the Euclidean 11th axiom had failed. On 11 February 1826 (Old Style) he held a lecture at a meeting of the department of physics and mathematics of Kazan university, in which he explicated the foundations of hyperbolic geometry. The text of this lecture, just like JÁNOS BOLYAI's first manuscript, was lost. In 1829—1830 he published his treatise *On the foundations of geometry* in instalments in the journal *Kazansky Vestnik*. According to KAGAN, it is "so thorough an exposition of non-Euclidean geometry that all his subsequent geometrical works are mere re-writings and improvements on that material".

What a miraculous coincidence of two great discoveries by scholars knowing nothing of one another! If we compare the data at our disposal about JÁNOS BOLYAI and LOBACHEVSKY, we must conclude that no order of priority can be established. BOLYAI envisaged the material of his treatise somewhat earlier but LOBACHEVSKY was the first to publish his work. Thus the historical facts fully justify the term BOLYAI—LOBACHEVSKY, or LOBACHEVSKY—BOLYAI, geometry.

GAUSS's ideas concerning non-Euclidean geometry have been preserved by a few notes and some letters. The addresses of the latter include FARKAS BOLYAI, WACHTER, OLBERS, SCHUMACHER, GERLING, TAURINUS and BESSEL. These letters were published in the 8th volume of GAUSS's collected works. Other information can be gleaned from STÄCKEL's article *Gauss als Geometer* and from several recently published papers.

The data these sources afford us reveal that GAUSS had a profound knowledge of the investigations of his predecessors and contemporaries concerning Axiom XI. Upon the influence of the frustrated attempts of others at providing a proof he gradually awoke to the possibility of a non-Euclidean system. A letter by WACHTER gives us the impression that around 1816 he already knew the paracycle and the parasphere. A somewhat later manuscript of his discloses that he defined parallelism in essentially the same way as LOBACHEVSKY and JÁNOS BOLYAI. Around the same time he also defined the notation of corresponding points.

He wrote to TAURINUS in a letter of 8 November 1824:

"The assumption that the sum of the three angles [in a triangle] is less than 180° leads to a curious geometry, quite different from ours [the Euclidean] but thoroughly consistent, which I have developed to my entire satisfaction, so that I can solve every problem in it excepting the determination of a constant, which cannot be fixed a priori."

Elsewhere in the same letter he wrote:

". . . the three angles of a triangle become as small as one wishes, if only the sides are taken large enough, yet the area of the triangle can never exceed, or even attain a certain limit, regardless of how great the sides are."

That clearly shows that at the time of writing the letter he already knew that in hyperbolic geometry the area of the triangle had an upper bound. In a letter to GERLING dated 16 March 1819 he also put down the expression giving the area.[7]

Some later notes by GAUSS, probably dating from 1840—1846, contain the proofs — obtained by more modern tools than those used by LOBACHEVSKY or BOLYAI — of certain formulae of hyperbolic trigonometry.

GAUSS, however, failed to synthesize his relevant results in an integrated system and forbade his corresponding partners to disseminate them.

What must have been underlying GAUSS's caution was obviously the oppressive authority of the Kantian philosophy at the time. He probably saw more clearly than anyone else that the creation of hyperbolic geometry would rock the foundations of the view that geometry as the science of space could be built exclusively on *a priori* principles. So he was afraid of being attacked, not by the mathematicians but by the philosophers. This may explain why he refused to use his enormous international prestige to promote the recognition of either LOBACHEVSKY's or JÁNOS BOLYAI's discovery. His attitude towards the latter was the colder: he only put down his name in two private letters (to FARKAS BOLYAI and GERLING) while he had LOBACHEVSKY at least coopted into the Scholarly Society of Göttingen as a corresponding member in 1843. The bigger eyesore for the philosophers of the age would have been JÁNOS BOLYAI who included even in the title of his work the anti-Kantian words "a priori haud unquam decidenda" (=a priori undecidable for ever).

JÁNOS BOLYAI himself gradually recognized the philosophical consequences of his theory of space. Let us quote, in evidence, two passages from his posthumous papers:

". . . the otherwise highly meritorious and sharpwitted Kant happened to declare the highly unwarranted and erroneous tenet that the void (=space) and the time are not independently existing things but mere appearances or *figures* of our imagination(!)." *(TK)*

[7] *Briefwechsel zwischen C. F. Gauss u. C. L Gerling.* Berlin, 1927, p. 195.

"... the incorrect opinion or view of the renowned and sagacious good Kant about time ... was born of total ill-conception ... Thus (viz. in system *S*) the notions of time and space are clear and evident, or clarified: and the presentiment of the so-called idealist phil[osophers] that perhaps only their souls exist and nothing beyond has vanished in thin air." *(TK)*

The passages of rather archaic wording in Hungarian reveal that JÁNOS BOLYAI discarded KANT's idea that time and space were only reflections on our minds.

JÁNOS BOLYAI described the relationship between space and gravitation as follows:

"... the law of gravitation appears to be in close connection with, or continuation of the size and actually the quality (=constitution) of space." *(TK)*

Such and similar remarks of BOLYAI have led many researchers to regard him as one of the pioneers of the geometrization of physics.

IMRE TÓTH has analyzed JÁNOS BOLYAI's theory of space from the aspect of philosophy in several of his studies. Especially noteworthy is the one [37] in which he gives a philosophical examination of several theorems of the *Appendix*.

It is true in general that JÁNOS BOLYAI was clear about the immense significance of his geometrical system. Let us quote a few sentences in original German to bear this out. By way of explanation we note that on 3 May 1832 he submitted a request to JOHANN VON HABSBURG for a three-month furlough for rest and research work. He enclosed the first 33 sections of the *Appendix*, though in a somewhat modified form. STÄCKEL published this altered text. Some sentences read as follows:

"... es *lebt* in dem Verfasser die (vollkommen geläuterte) Überzeugung (desgleichen er auch von jedem einsichtsvollen Leser erwartet), dass durch Aufklärung des Gegenstandes Einer der *allerwichtigsten* und *allerglänzendsten* Beiträge zur wahren Bereicherung der Wissenschaft, zur Bildung des Verstandes und somit zur Hebung des menschlichen Schicksals gemacht wurde" ([31], II., p. 202).

*

To conclude, let us adduce some remarks of KAGAN concerning JÁNOS BOLYAI's views on LOBACHEVSKY's theory [23]. To shed light on the historical background it has to be noted that in 1848 JÁNOS BOLYAI received LOBACHEVSKY's fundamental treatise *Geometrische Untersuchungen zur Theorie der Parallellinien* from his father and soon attached comments to it in Hungarian (almost the entire text can be found in [31], I. 140—160).

"JÁNOS — KAGAN writes — studied LOBACHEVSKY's work carefully and analyzed it line by line, not to say word by word, with just as much care as he administered in working out the *Appendix*. The work stirred a real storm in his soul and he gave outlet to his tribulations in the comments added to the 'Geometrical examinations'.

The 'Comments' to the 'Geometrical examinations' are more than a critical analysis of the work. They express the thoughts and anxieties of JÁNOS provoked by the perusal of the book. They include his complaint that he was wronged, his suspicion that LOBACHEVSKY did not exist at all, and that everything was the spiteful machination of GAUSS: it is the tragic lament of an ingenious geometrician who was aware of the significance of his discovery but failed to get support from the only person who could have appreciated his merits."

"In spite of the mental agitation amidst which JÁNOS put his observations to paper, he preserved enough objectivity to highly appreciate the work of his rival. In his comment to Theorem 35 he remarks that the proofs of LOBACHEVSKY concerning spherical trigonometry bear the impress of a genius and his work should be esteemed as a masterly achievement."

"Highly interesting are BOLYAI's theoretical considerations of the question as to what extent it can be declared that non-Euclidean geometry does not certain intrinsic contradictions. This tough question engaged LOBACHEVSKY's mind all through his life; the main of his long computations of integrals was to find regular coincidence in the results and thereby exhibit the firmness of the new geometry. The brief considerations related to this subject in the 'Geometrical examinations' did not satisfy JÁNOS, certainly with good reason. Here JÁNOS claims to have obtained conclusive proof of the 'consistency' of hyperbolic geometry, as it is termed today. But he left no sign of this proof behind to posterity,[8] so it has remained under veil how he arrived at this proof."

It clearly shows the human greatness of JÁNOS BOLYAI that he wrote in what we may consider his scientific testament, referring to LOBACHEVSKY: ". . . I am sharing the finder's merit with pleasure." *(TK)*

The volume of the tetrahedron in absolute geometry

Having built up the absolute trigonometry of the plane, one would logically be urged to address the question of the volume of the tetrahedron. JÁNOS BOLYAI devoted much attention to this problem; his relevant manuscripts were published by STÄCKEL ([31], I., pp. 109—118). There one can find the complicated formulae (not reproduced here) resulting from BOLYAI's calculations.

GAUSS called JÁNOS's attention to the problem in a letter to FARKAS BOLYAI dated 6 March 1832 ([12], p. 112), but some sources attest that the author of the *Appendix* had already performed such computations by that time. The notes themselves, very sketchy and sparse, are of a later date, however, GAUSS and LOBACHEVSKY have also dealt with the problem and the ways they followed intersected that of JÁNOS BOLYAI at several points. Undoubtedly, all three worked independently of one another.

Both LOBACHEVSKY and BOLYAI have chosen a special tetrahedron. Its base is a right triangle, the fourth vertex being on the line perpendicular to the plane and expected at one of the vertices with base from any vertex at an acute angle. Let us call it *normal tetrahedron*. Note that all its faces are right triangles. Let us also introduce the *asymptotic normal tetrahedron:* a normal tetrahedron whose lateral edges are parallel in the non-Euclidean sense.

---

[8] This statement by KAGAN does not entirely comfort to the facts. See the discussion of *Raum-Lehre*.

The normal tetrahedron plays the same role in computing the volumes of polyhedra as does, e.g., the right triangle in calculating the areas of polygons. Its introduction is useful since it is determined by only three data. Depending on the choice of the three data, the volume of the normal tetrahedron can be computed in different ways.

To compute the volume, we have to decompose the tetrahedron into elementary bodies by some procedure, after which the required formula is provided by a definite integral.

János Bolyai proposed three procedures to dissect the normal tetrahedron; in addition to the corresponding methods of cubature there is a fourth one that rests on a different principle using the asymptotic normal tetrahedron.

1. Drop a perpendicular to the edge $\overline{AC}$ from any point G of the edge $\overline{AD}$ of the normal tetrahedron *(Figure 49)*, and from the foot F drop a perpendicular to the edge $\overline{AB}$. Repeat this process starting from a point G' which is infinitesimally close to G. The resulting plane triangles EFG and $E'F'G'$ enclose the volume element dK. Its value, however, can only be expressed by a very complicated formula and even for an asymptotic normal tetrahedron the problem leads to an elliptic integral.

A later note of János Bolyai includes his trials to express this integral with elementary functions in a closed form. Naturally, these attempts were in vain.

This method of János Bolyai for the dissection of the tetrahedron tallies with one proposed by Gauss in 1832 *(Werke,* Vol. 8, p. 228).

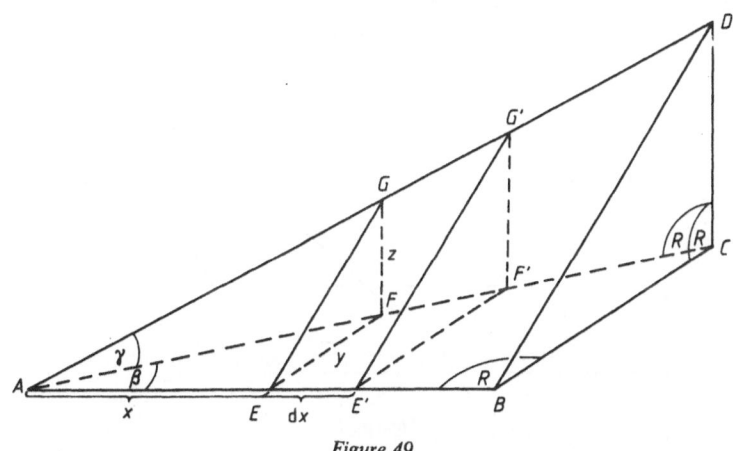

Figure 49

2. Let us turn the normal tetrahedron by 360° around an edge, say $\overline{AB}$ *(Figure 49)* as axis. In this way the cubage of the tetrahedron can be reduced to that of straight cones. The volume element in this case can be obtained with the help of planes at right angle to the axis of rotation.

János Bolyai made mention of this procedure at several places as he thought that computating the volume elements of the solid of revolution was easier than the same for the normal tetrahedron. But after passing from the volume element of the solid of revolution to that of the normal tetrahedron, the expression of dK was more complicated than it had been in the first approach, so János did not even try to integrate it.

Interestingly enough, LOBACHEVSKY also used the idea of reducing the cubature of the tetrahedron to the cubature of the cone in his treatise of 1829, *On the Foundations of Geometry*. The formula LOBACHEVSKY devised for the volume element is correct, whereas BOLYAI's has an error, as STÄCKEL pointed out.

3. In a note of JÁNOS BOLYAI probably from 1856, the dissection of the normal tetrahedron with the hyperspheres of the base plane ABC is expounded *(Figure 50)*.

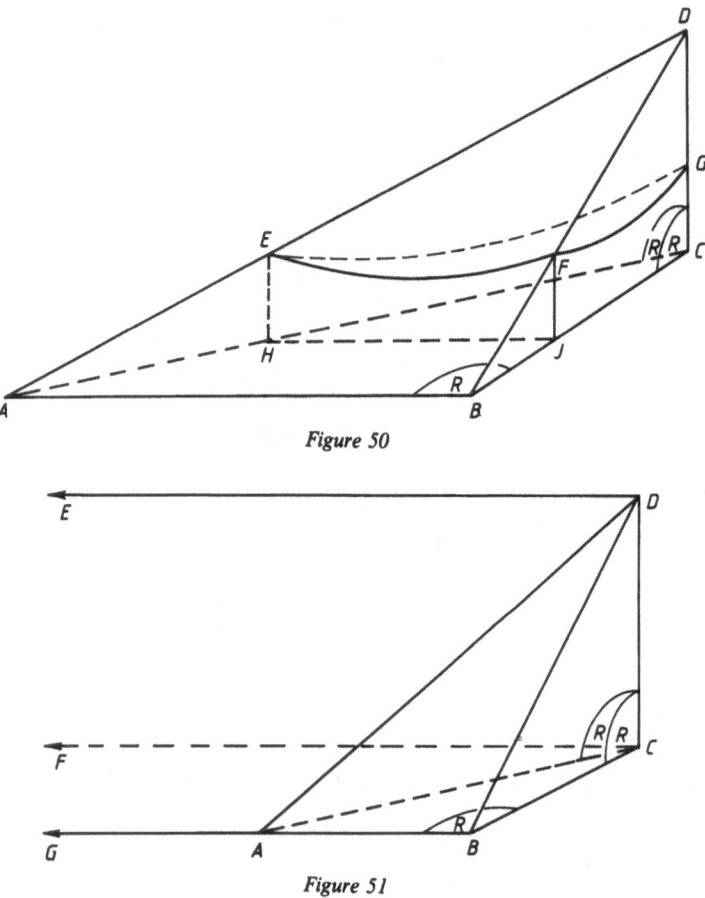

Figure 50

Figure 51

In this case the volume element is bounded by two hyperspheric triangles. BOLYAI, however, could not go beyond the computation of the data of one hyperspheric triangle: he shrank from further investigations seeing the overcomplicated formula obtained.

It is intriguing to recount the subsequent fate of this way of decomposition. In his *Elemente der absoluten Geometrie* published in 1876, FRISCHAUF mentioned the method independently of BOLYAI, but he, too, refrained from performing the computations. When, however, STÄCKEL published BOLYAI's notes in 1901, FRISCHAUF returned to the problem and managed to prove that this way of dissection produced the same integral as the first method of decomposition.

4. Finally, an early note of a few lines by JÁNOS BOLYAI features an idea of cubature based on the asymptotic tetrahedron. Namely, if we determine the volumes of the asymptotic normal tetrahedra of base *BCD*, and *ACD*, respectively *(Figure 51)*, then their difference yields the volume of the normal tetrahedron *ABCD*.

JÁNOS BOLYAI's note breaks off here, without giving any computation.

The Responsio

As his treatise submitted to the *Jablonowski Society* in Leipzig under the title *Responsio* suggests, JÁNOS BOLYAI also entertained modern ideas in the field of complex numbers ([31], II., pp. 223—233). As was mentioned elsewhere (see note 11 on page 190), the problem to be considered in the tracts submitted was whether imaginary quantities involved in geometry could be constructed or not. The introductory lines claim that the *Responsio* seeks to answer this question, but in actual fact the treatise is an indictment against the incorrect formulation of the question. In JÁNOS BOLYAI's view the possibility of geometrical construction was a secondary concern have, while the exact definition of complex numbers and the determination of their role in geometry were of paramount importance.

The *Responsio* has contributed new results in regard to the latter two problems, improving on certain ideas of FARKAS BOLYAI in the first and relying on absolute geometry in the second.

In the theory of complex numbers JÁNOS BOLYAI postulates four units to be given as a start,[9] two of them positive and two negative. Their symbols are:

Despite the tortuous writing, we learn from a later table that actually these units are:

$$+1 \quad -1 \quad +i \quad -i$$

The father's influence is unmistakable here: JÁNOS BOLYAI, just like his father, wished to avoid the speculations necessary for the introduction of negative numbers, so they regarded them as given *a priori*. What is new in his conception, as § 11 of the *Responsio* bears out, is that he did not find any obstacles of principle to postulate *an arbitrary number* of units. He writes:

"... *any number* of quantities (you like) can be introduced; but this is *not necessary*" ([31], II., p. 233).[10]

---

[9] LAJOS SCHLESINGER writes that JÁNOS BOLYAI "like Hamilton in 1837 ... regards the complex quantity as a quadruple of numbers and for this quadruple he lays down the formal rules ..." *(Acta ... Kolozsváriensis*, 1902—3, p. 17). This sentence may lead to the misconception as if JÁNOS BOLYAI were a forerunner of the theory of quaternions. This is misleading because the four units he started from did not satisfy the operational rules imposed on the units defining the quaternions.

[10] This idea of BOLYAI was again raised by H. GRASSMANN (1855) who defined hypercomplex numbers by the formula

$$x_1 e_1 + x_2 e_2 + \ldots + x_n e_n,$$

where the e_i-s denote arbitrary units. See, e.g., KLEIN, F.: *Vorlesungen über die Entwicklung der Mathematik* ... Berlin, 1926, Part I, 179.

It aggravates the understanding of the *Responsio* that in addition to the symbols used for the units, JÁNOS BOLYAI also applies four similar signs:

$$+ \quad - \quad \phi \quad \ominus$$

Their role is at first not quite clear in the paper, partly because of the several slips of the pen. Perhaps this is caused by the similarity of the symbols.

BOLYAI felt obliged to give an explanation at another place ([31], I., p. 105). There he states that the latter signs

"... do not mean anything else but the mode in which — setting up some conventions, a step that should be done — ... we have to treat the quantities in algebraic computations."

Thus the latter are symbols of operation and denote that one adds a term multiplied by $+1$, -1, $+i$ or $-i$. This again suggests the influence of FARKAS BOLYAI who used different signs, as we saw earlier, to designate the *qualities* of being positive or negative and the *operations* of addition and subtraction in the course of laying the foundations.

§ 6 of the *Responsio* also deserves special attention. But before evaluating it, we adduce some historical data.

The theory of complex numbers began to be worked out at the end of the 18th century. There were two main trends in defining and using them. Chronologically, first was the *geometrical* trend (WESSEL 1797, ARGAND 1806 and, in certain respect, GAUSS) which took different shapes from mathematician to mathematician but was uniform in treating complex numbers as segments of given length and direction. The attempt to define the complex numbers independently of geometrical interpretation and build up their algebra by observing the principles of permanence and consistency was of a somewhat later date. The first representative of the *arithmetical* trend was HAMILTON (1837). In his view the complex number was an ordered pair of numbers satisfying some operational suitably chosen.

According to § 6 of the *Responsio*, JÁNOS BOLYAI was also one of the early advocates of the *arithmetical* trend. Though, as STÄCKEL contended,[11] BOLYAI's definition of the complex number as an ordered pair of numbers was not so complete as HAMILTON's, but a formula of the *Responsio* shows that he did regard the complex numbers as ordered pairs of numbers and postulated the formal rules of addition and multiplication among them.

However, ponderous the wording of § 8 of the *Responsio* may be, the point of it cannot be mistaken. Already FARKAS BOLYAI defined the logarithm of a real number in a novel way, with the help of an infinite series. JÁNOS adopted this definition but generalized it to complex numbers:

Let

$$f(z) = e^z = 1 + z + \frac{z^2}{2!} + \frac{z^3}{3!} + \ldots;$$

---

[11] JOHANN BOLYAI's Theorie der imaginären Grössen. *Math. u. Naturw. Berichte aus Ungarn*, Vol. 6 (1898), pp. 263—297.

190

then we call z the logarithm of $f(z)$. JÁNOS BOLYAI must have been unaware of CAUCHY's investigations into function theory, so the above definition of the logarithm is to be considered original.

Undoubtedly the most valuable part of the *Responsio* is § 9, which remained perfectly incomprehensible for the judges of the competition. By simply referring to the *Appendix*, JÁNOS BOLYAI presented here two formulae of absolute trigonometry, mentioned the concept of the hypersphere and remarked that, formally, absolute trigonometry coincides with the trigonometry of the sphere of radius $\dfrac{k}{i}$. These sketchy observations were meant to show that the imaginary unit i had an important and till then unknown geometrical application in his system. Probably he entertained the quite naive hope that the judges would recognize the profundity of his arguments and bestow upon him, besides the meagre award, true recognition.

Vain hope it was — how could the judges, having a rather modest place in the history of mathematics, realize the value of the laconically worded notes if they had possibly never seen the *Appendix*.

In JÁNOS BOLYAI's view the geometrical representation of imaginary quantities is merely a question of agreement. In § 10 of the *Responsio*, we read the following:

"Let me only remark here that D'ALEMBERT's reasoning which he put forth in an attempt to prove that in the ordinary construction of equations the positive and negative values (of the ordinates and abscissas) must be measured on opposite sides of the coordinate axes ill-founded and wrong, as it comes from the nature of the thing that in this question there is no place for any compulsory rule." ([31], II., p. 231).

JÁNOS BOLYAI took up arms even against GAUSS in this question:

"... the notions of *right* and *left*, *up* and *down*, etc. are not determined, and must — and can — be avoided here as *relative* notions alien to geometry." (*Ibid.*, p. 233)

Raum-Lehre oder Geometrie[12]

Besides creating absolute geometry, JÁNOS BOLYAI was also captured by the idea of "constructing the entire mathematics perfectly and lucidly from the beginning to the farthest possible limit". More information about this can be gleaned from his project *Reformation der Elemente der Mathematik*, which he began to write around 1832 but abandoned after completing the preface. He envisioned an encyclopaedic work constructed quite of his own ideas including the branches of mathematics that he was fond of (arithmetic, number theory, geometry, formal logic, mechanics).

---

[12] As far as we know, JÁNOS BOLYAI wanted to write a book on geometry with that title. The following pages contain a few of the novel ideas put down in notes for that book.

After a long and relatively fruitless period, he resumed work on this project around 1850. Yet the treatise written in German and intended for publication remained unfinished: he promised to solve the new problems that arose incessantly while writing the *Raum-Lehre* but he did not have enough strength to keep his promise and complete the work. Nevertheless, the *Raum-Lehre* and the numerous notes added to it in subsequent years contain several ideas that are not unworthy of the author of the *Appendix*.

In the following we summarize these results, without separating the material of the *Raum-Lehre* from that of the notes attached.

The material can be divided into four parts. The first part ("Foundations") discusses problems, in immature form occasionally incomplete proofs, which today belong to the foundations of geometry. The following remark of JÁNOS BOLYAI refers, first of all, to this area:

"Many things which are usually proved tortuously are almost obvious here, while several things usually ignored are developed most carefully." ([31], II., p. 249).

It is quite original in JÁNOS BOLYAI's set of postulates that, unlike the Euclidean system, he tries to built up the whole of geometry on the notions of *plane* and "*ring*". When defining these concepts, he also uses some of FARKAS BOLYAI's definitions and, just as his father, admits the concept of motion to geometry. Thus, for example, for JÁNOS the *point* is a "partless part", the *ring* is a "simple, uniform, closed curve".

To correctly understand the attribute *simple*, one has to know that curves are either "simple" or "knotty" (i.e., having double or multiple points).

"Only the collection of those points constitutes a *simple curve* from any point of which there is either a single path — that can be run through by a material point — to any other or there always exist two such pathes at most". ([31], I., p. 181).

The latter is obviously the case of a *closed* simple curve.

The involvement of physical motion in the definition implies the *continuity* of the curve. JÁNOS BOLYAI must have adopted FARKAS's opinion that in geometry the temporal aspect of physical motion need not be taken into account. In other words, in JÁNOS BOLYAI's definition the curve is *continuous but not necessarily differentiable*. Thus the "simple" curves — however intuitive their definition — belong to the class of *simple Jordan curves*.

Some of the sections of *Raum-Lehre* give a step-by-step verification that the "ring" is identical with the circles and some remarks are made to the effect that the *plane* and the *ring* are suitable notions to build a rigorous geometry on them. The question was thoroughly probed into by ISTVÁN SPRINGER [30] who concluded that although this idea of JÁNOS BOLYAI was fertile, the axioms presented in the *Raum-Lehre* were insufficient for a perfect construction of geometry and had to be complemented by some axioms of congruence and continuity.

Disregarding the methods of analysis and relying solely on intuition, JÁNOS BOLYAI gave a plausible definition and classification of surfaces. He divided the simple surfaces into two groups: "complete" (simply connected) ones and "punctured" ones. The following sentence has salience in JÁNOS BOLYAI's rather laboured text:

"From any simple surface we may remove an arbitrary number of holes, replace them with tubes, and connect the tubes pairwise." ([31], I., p. 181).
And, without giving a proof, he goes on:
"In the most general case the simple surface is like that." *(Ibid.)*

These sentences reflect a geometrical thinking far ahead of the age: they involve the ideas of homeomorphic mapping and topological invariance, and give a construction which still appears as an example in "descriptive topology".

The second part of *Raum-Lehre* ("The theory of constructions") describes some problems of construction for creation of geometric loci "of which people have only had a vague idea so far, merely assuming rather than proving their existence, not to say find there a priori" ([31], II., p. 251).

In JÁNOS BOLYAI's opinion, such problems are: given two points, to construct a sphere centred in one and passing through the other; or, given three (four) points, to decide if they are collinear (coplanar) or not. BOLYAI seeks to solve to these problems with the help of the compasses alone. This suggests that he tried to set up new problems inspired by the MOHR (1672)—MASCHERONI (1797) theorem. According to this theorem, any construction that can be performed with the compasses and the straightedge can be carried out with the compasses alone. He must have been influenced by MASCHERONI's work *Geometria del compasso* (Pavia, 1797) which he praised in one of his notes.

Part III of *Raum-Lehre*, a rather defective and unfinished part, is concerned with angles and polygons. In it, JÁNOS BOLYAI pays particular attention to certain problems; e.g., he proves meticulously that a polygon can be divided into triangles, defines the notation of interior angle of a polygon, etc.

Probably Part IV of *Raum-Lehre* was intended to become a revised and enlarged version of the *Appendix*, but of this section unfortunately only few notes have survived; they were published by STÄCKEL ([31], I., pp. 185—187). In them, JÁNOS BOLYAI returns to the question whether absolute geometry contains intrinsic contradictions or not. He saw consistency proved for the plane by the fact that, formally, the relations of the absolute trigonometry of the plane coincide with the corresponding formulae for the spherical triangle of imaginally radius. Yet he left from the beginning that this did not decide the question of consistency *in space*.

The method he figured out to answer this question reduces the problem of consistency of absolute geometry to that of Euclidean geometry, but despite all its ingenuity it is not adequate to give an answer of universal validity.

JÁNOS BOLYAI's conception was the following. If the six edges of the tetrahedron are given, we can compute the dihedral angles in two ways (see below). Using the formulae of Euclidean trigonometry, the two ways of computation yield identical results, as expected. If we also obtain identical results when computing in two ways with the formulae of hyperbolic geometry, then — according to JÁNOS BOLYAI it is likely that the absolute geometry of space is consistent.

Let us go into some detail here. Let the six edges a, b, c, d, e, f of the tetrahedron (to use BOLYAI's term: "system of four points") be given. Let (ae) denote the angle subtended by the edges a and e, let (af) be the angle formed by a and f, etc. *(Figure 52)*. Let (aec) stand for the dihedral angle at edge e.

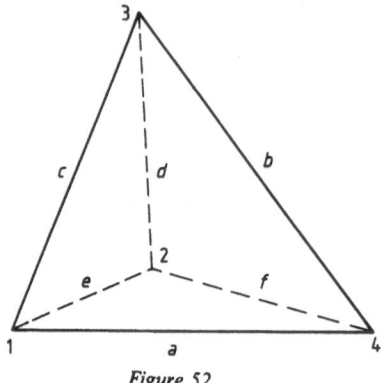

Figure 52

Now, the angles can be computed with the cosine law of hyperbolic geometry. The three angles at vertex 1 are (ac), (ae), (ce); their cosines are given by the relations

$$\cos(ac) = \frac{ac-b}{AC}; \quad \cos(ae) = \frac{ae-f}{AE}; \quad \cos(ce) = \frac{ce-d}{CE}, \tag{I}$$

on the right-hand side of which we have used the notation

$$a = \operatorname{ch}\frac{a}{k}; \quad b = \operatorname{ch}\frac{b}{k}; \quad \ldots; \quad A = i\operatorname{sh}\frac{a}{k}; \quad B = i\operatorname{sh}\frac{b}{k}; \ldots$$

Similarly, the angles at vertex 2 are (df), (de), (fe), their cosines being

$$\cos(df) = \frac{df-b}{DF}; \quad \cos(de) = \frac{de-c}{DE}; \quad \cos(fe) = \frac{fe-a}{FE}. \tag{II}$$

The twofold way of computation means that the dihedral angle (aec) can be calculated either with the help of the angles at vertex 1 or with those at vertex 2.

For vertex 1:

$$\cos(aec) = \frac{\cos(ac) - \cos(ae)\cos(ce)}{\sin(ae)\sin(ce)}. \tag{III}$$

Using the angles at vertex 2:

$$\cos(aec) = \frac{\cos(df) - \cos(de)\cos(fe)}{\sin(de)\sin(fe)}. \tag{IV}$$

Substituting (I) and (II) into the right-hand side of (III) and (IV) we obtain two rather involved formulae, of which JÁNOS BOLYAI was able to probe through elementary transformations that they were identical.

194

Thus, for four points JÁNOS BOLYAI found no inconsistency. So he went on and extended the "system of four points" by a fifth point not coplanar with any three of the previous ones. Knowing the nine edges of this body, we can compute a dihedral angle *in three ways*. Here, JÁNOS BOLYAI at first failed to arrive at the required coincidence; this led him to the conclusion that his system was still inconsistent, which he believed to be an indirect proof of EUCLID's Postulate 5. But checking his computations he found a mistake; so the "system of five points" did not disclose any contradiction in absolute geometry either.

JÁNOS BOLYAI would have liked to continue, but he was frightened off by the complexity of the calculations for *six* points.

So, when JÁNOS BOLYAI's notes broke off for ever, he did not know for sure whether his geometric system was devoid of contradictions or not. Only a few years had to pass before the question was answered conclusively. The geometry of BELTRAMI's pseudosphere, and shortly afterwards the CAYLEY–KLEIN model (1871) proved the *consistency* of absolute geometry relative to the arithmetic of real numbers.

Bibliography

[1] ALEXITS, GYÖRGY: *The World of János Bolyai.* Budapest, 1977. (In Hung.)

[2] BEDŐHÁZI, JÁNOS: *The Two Bolyais. An account of their lives and characters.* Marosvásárhely, 1897. (In Hung.)

[3] BEKE, MANÓ: Bolyai's trigonometry. *Math. és Phys. Lapok,* Vol. 12 (1903), pp. 30—49. (In Hung.)

[4] BENKŐ, SAMU: *The Confessions of János Bolyai.* Bucharest, 1968. (In Hung.)

[5] BOLYAI, FARKAS: *The beginnings of arithmetic.* Marosvásárhely, 1830. (In Hung.)

[6] — Tentamen juventutem studiosam in elementa matheseos purae, elementaris ac sublimioris, methodo intuitiva, evidentiaque huic propria, introducendi. Cum Appendice triplici. I—II. Marosvásárhely, 1832, 1833. The second edition was prepared by GYULA KŐNIG, JÓZSEF KÜRSCHÁK, MÓR RÉTHY and BÉLA TŐTÖSSY, also in two volumes (three parts). Budapest, 1897, 1904.

[7] — *The beginnings of arithmetic, geometry and physics* ... Marosvásárhely, 1834. (In Hung.)

[8] — A partly abridged and partly enlarged, generally improved and corrected edition of *The beginnings of arithmetic,* first printed at Marosvásárhely in 1829. (In Hung.) 1843.

[9] — *The beginnings of arithmetic, for beginners.* Marosvásárhely, 1850. (In Hung.)

[10] — *The elements of the theory of space.* Marosvásárhely, 1851. (In Hung.)

[11] — *Kurzer Grundriss eines Versuchs* ... Marosvásárhely, 1851. Also published in [31], Vol. II., pp. 119—179.

[12] *Briefwechsel zwischen C. F. Gauss und Wolfgang Bolyai.* Leipzig, 1899.

[13] BOLYAI, JÁNOS: *Appendix. The Theory of Space.* With introduction, comments, and addenda edited by FERENC KÁRTESZI. Suppl. by BARNA SZÉNÁSSY. Akadémiai Kiadó, Budapest, 1987.

[14] BONOLA, R.: Index operum ad geometriam absolutam spectantium. In: *"Joannis Bolyai in memoriam."* Kolozsvár, 1902, pp. 83—149.

[15] BONOLA, R.—LIEBMANN, HEINRICH: *Die nichteuklidische Geometrie.* Leipzig—Berlin, 1908.

[16] BRASSAI, SÁMUEL: Commemorative oration over Farkas Bolyai. *Publications of the department of philosophy, linguistics and history of the Transylvanian Museum Society,* Vol. 2, Kolozsvár, 1886, pp. 209—247. (In Hung.)

[17] CSADA, IMRE: Farkas Bolyai's equivalents of the 5th postulate. *Közlemények a Debreceni Tudományegyetem Matematikai Szemináriumából,* III, Sárospatak, 1929. (In Hung.)

[18] DÁVID, LAJOS: *The life and work of the two Bolyais.* Budapest, 1923, 1979. (In Hung.)

[19] — *The Bolyai geometry on the basis of the Appendix.* Kolozsvár, 1944. (In Hung.)

[20] — In memoriam Wolfgangi Bolyai. *Az MTA III. Osztályának Közleményei,* Vol. 9 (1959), pp. 215—236.

[21] FRISCHAUF, J.: *Absolute Geometrie nach Johann Bolyai.* Leipzig, 1872.

[22] GERGELY, JENŐ: *An introduction to non-Euclidean geometry on the basis of János Bolyai's Appendix.* In: [42], pp. 179—226. (In Hung.)

[23] KAGAN, V. F.: The construction of non-Euclidean geometry by Lobachevsky, Gauss and Bolyai. *Proc. Inst. History of Science*, 1948, II., pp. 323—399. (In Russian.)

[24] KALMÁR, LÁSZLÓ: The impact of the Bolyai—Lobachevsky geometry on the development of the axiomatic method. *Az MTA III. Osztályának Közleményei*, Vol. 3 (1953), pp. 235—242. (In Hung.)

[25] KÁRTESZI, FERENC: The antecedents and scientific results of the Appendix. *Matematikai Lapok*, Vol. 31 (1978—1983), pp. 15—22. (In Hung.)

[26] RAPCSÁK, ANDRÁS: The effect of the Appendix on modern mathematics. *Matematikai Lapok*, Vol. 31 (1978—1983), pp. 23—28. (In Hung.)

[27] RÉNYI, ALFRÉD: The importance of the Bolyai—Lobachevsky geometry in the transformation of world view. *Az MTA III. Osztályának Közleményei*, Vol. 3 (1953), 253—273. (In Hung.)

[28] RÉTHY, MÓR: An introduction to János Bolyai's "different, new world". *Math. és Phys. Lapok*, Vol. 12 (1903), pp. 1—29 and 303—320. (In Hung.)

[29] SCHLESINGER, LAJOS: János Bolyai. *Math. és Phys. Lapok*, Vol. 12 (1903), pp. 57—88. (In Hung.)

[30] SPRINGER, ISTVÁN: A supplement to the axiomatic of János Bolyai's geometry. *Közlemények a Debreceni Tudományegyetem Matematikai Szemináriumából*, 1, Budapest, 1927. (In Hung.)

[31] STÄCKEL, PAUL: *Wolfgang und Johann Bolyai. Geometrische Untersuchungen I—II.* Leipzig—Berlin, 1913. In Hungarian: Budapest, 1914. (Transl. by IGNÁCZ RADOS.)

[32] SZÁSZ, PÁL: *An Introduction to the Bolyai—Lobachevsky Geometry.* Budapest, 1973. (In Hung.)

[33] SZÉNÁSSY, BARNA: Farkas Bolyai's ideas on infinitesimal quantities. *Közlemények a Debreceni Tudományegyetem Matematikai Szemináriumából*, XIII, Debrecen, 1937. (In Hung.)

[34] — *Farkas Bolyai.* Budapest, 1975. (In Hung.)

[35] — *János Bolyai.* Budapest, 1978. (In Hung.)

[36] — The lives of the two Bolyais and the scientific significance of the Tentamen. *Matematikai Lapok*, Vol. 31 (1978—1983), pp. 3—14. (In Hung.)

[37] TÓTH, IMRE: The philosophical implications of the Bolyai geometry. In: [42], pp. 257—340. (In Hung.)

[38] — Wann und von wem wurde die nichteuklidische Geometrie begründet? *Archives Internat. Hist. Sciences*, Vol. 30 (1980), pp. 192—205.

[39] WESZELY, TIBOR: *Farkas Bolyai.* Bucharest, 1974. (In Hung. and Romanian)

[40] — *The mathematical work of János Bolyai.* Bucharest, 1981. (In Hung.)

[41] — On the posthumous mathematical manuscripts of János Bolyai. *Matematikai Lapok*, Vol. 31 (1978—1983), pp. 29—37. (In Hung.)

[42] *The life and work of János Bolyai.* (Collected studies.) Bucharest, 1953. (In Hung.)

16. The mathematical activity
of the Hungarian Scholarly Society
in its first decades

As the only concrete achievement of the 1825 Diet, the Estates succeeded in getting permission to establish the Hungarian Scholarly Society (later called the Hungarian Academy of Sciences), and five years later our first truly scientific organization began work. While the earlier writings arguing for the need of an academy (DÁNIEL TERS-TYÁNSZKY, MIKLÓS RÉVAI, SÁMUEL DÉCSI and others) reasoned that agriculture, stock breeding, trade and industry must be developed, ISTVÁN SZÉCHENYI's writings and speeches reveal that what he envisioned when setting up the Academy was a sort of society for the cultivation of the language. Indeed, the highlight of the Academy's activities in the first three or four decades was the promotion of the language, and the endeavours in the exact sciences — few in number — were nowhere near the European standards.

The first charter of the Academy reflects both the conservative position of the ruling house and the aristocracy, and the progressive, reformist zeal of the lesser nobility. The management of the Academy was in the hands of the "council of directors" and the "honorary members" mainly coming from the aristocracy and high clergy, but the ordinary and corresponding members, mostly lesser nobles, also had some rights. The proposals put forth by the members could, however, easily get stranded by the resistance of the honorary members, the directorial council or ultimately the Vienna Court, if the initiatives were seen as jeopardizing the status quo. The charter spelt out that the upper limit of the number of ordinary members was 42, including 18 from the capital and 24 from the provinces. This provision may have been aimed at preventing the capital from overpowering the country by its scientific weight. But it must have been a far more important consideration that the "rebellious" ordinary members concentrated in Pest should not represent a greater force than the rural members scattered all over the country and having only loose contact with the capital because of the conditions of transportation.

Neither was the division of the Scholarly Society into departments more favourable. Future economic prosperity would first of all have required the promotion of natural sciences, yet they were very poorly represented in the six departments (linguistics, philosophy, history, mathematics, law and natural science). In the first times, members dealing with questions of the military sciences, and later (from 1858) also those working in mechanics, theoretical physics, astronomy, geodesy and various kinds of engineering also belonged to the department of "mathesis". At the time of the foundation, this department had only three seats for ordinary members from Pest and three for those from the provinces. It is most revealing of the relative backwardness of mathematical life in the country and of the haphazard way of choosing the members that in the beginning not even these six places were filled, leaving three or four places vacant until the Austro—Hungarian Compromise of 1867.

Fairly ample information of the activity of the Hungarian Scholarly Society can be gleaned from its publications. The first series launched was the *Yearbooks of the Hungarian Scholarly Society* that wished to inform the reader of the proceedings of the Academy and carried a few studies of various character. The *Yearbooks* came out biannually, until the crushing of the war of liberation, and at longer and longer intervals in the subsequent years of Austrian autocracy that also gravely afflicted the life of the Academy.

The question of *cultivating and popularizing* the sciences in the Hungarian language was a central concern of the members of the Academy from the very beginning, so they launched a periodical in anticipation of a wide readership in 1834. The thick volumes of *Tudománytár* were filled with easily understandable original and translated articles, all in Hungarian.

The Academy started a scientific journal in 1840 with the title *Magyar Académiai Értesítő* published in ten issues a year. At the beginning it carried the communications and treatises read at the department meetings as well as other major results of the members. In 1859 the *Értesítő* was divided into two parts and a year later into three, with separate issues devoted to communications read in the mathematics and natural science departments.

There was another modification in 1860 in order to promote the natural sciences. Upon the proposal of ANTAL CSENGERI (1822—1880) a separate *committee of mathematics and natural science* was set up. It was the task of this committee to find problems related to Hungary in these sciences and coordinate their elaboration, lest "... foreign scholars continue to our country on expeditions just like uncultured, barbarous lands" ([19], 1. VIII). Although the committee included scholars interested in mathematical problems as well (OTTÓ PETZVAL, ISTVÁN KRUSPÉR, JÓZSEF SZTOCZEK), the journal they launched in 1861 under the title *Mathematikai és Természettudományi Közlemények* did not contain any articles on mathematical subjects. There were two reasons for that. One was that despite the epithet "mathematical", the scope of the committee's activity encompassed almost exclusively problems of natural science. The other was that the *Értesítő* could publish all mathematical results of that period.

Still, these publications caused much headache to the Academy that was sustained from public donations without any official support. *Tudománytár* had to be stopped in 1844 after several unanswered calls to the nation for more efficient support. The limited readership of the only scientific periodical of the time could not ensure its future, and the editors failed to attain their goal because of restricted circulation. The same fate befell the *Mathematicai Pályamunkák* after three volumes. In the series, the Academy had wanted to publish prize essays in mathematics regularly.

After 1849 also other reasons were added to aggravate the position of mathematics within the Academy. KOSSUTH's programme for industrialization and the establishment of the *Protective Society* (1844—1846, for the protection of Hungarian products against foreign goods) kindled great enthusiasm in the majority of academicians committed

198

to the promotion of natural science in the country, and they rallied under KOSSUTH's banner. But after the suppression of the War of Independence (1849), Hungary had to pay a great price for the glorious achievements of the few preceding years: the members of the Academy engaged in mathematics were most severely persecuted. SÁMUEL BRASSAI, a valiant supporter of the war of independence, had to go into hiding and later had to work for years as a badly paid tutor at private schools. In 1850 the Vienna government stripped ANTAL VÁLLAS, one of our hardest-working and most versatile mathematicians, of his professorship for his pamphlet including reformatory ideas and calling for the establishment of a Hungarian technical university [33]. Deeply hurt in his dignity, he chose voluntary exile and died after a life of vicissitudes in New Orleans. After the fall of the war, KÁROLY NAGY was arrested for having been the manager of the progressive thinker KÁZMÉR BATTHYÁNY's estate. After his release he had to sell his modest landed property to get a passport and go to Paris, never to see his native country again. The heroic defender of Pétervárad, ERNŐ HOLLÁN was forced to surrender to the numerically superior enemy and was imprisoned before he could emigrate. DÁNIEL CSÁNYI, a professor of mathematics at the Debrecen college had organized the defence of Komárom. His valiance earned him six years in a dungeon.

And the list could be continued. It can fill us with due pride that the great majority of Hungarian mathematicians of the age were firmly pledged to progress and to the ideal of freedom. On the other hand, it is a great shame that despotism managed to paralyze the fledgling mathematical life of Hungary just when it started to take wing.

Absolutism hindered the work of the Academy in several other ways as well. For some time after the war of liberation, no meeting of the Academy was permitted. Though in the mid-1850s members began to convene again under heavy security control, no new member could be coopted before 1858. During these years the Academy seemed to be doomed to slow death; by its disheartening dilettantism typical in those days, it was more an obstacle than a helper of our scientific life.

*

This was the background against which all the noteworthy events of the "mathesis" department of the Academy, and indeed, the whole mathematical life of the country took place. After the foundation, the only ordinary member of the mathematical department from Pest was PÁL TITTEL, professor of astronomy at the University of Pest, and the only member from the provinces was LAJOS BITNICZ, a priest teacher at Szombathely. The former died shortly after his admission, without having time to contribute significantly to the life of the Academy. In 1831 ISTVÁN NYÍRY, a teacher in Sárospatak, and in 1832 SÁNDOR GYŐRY a "chartered surveyor" was elected ordinary member, while FARKAS BOLYAI, KÁROLY NAGY and PÁL SÁRVÁRI became corresponding members. Of them, only GYŐRY's name can be encountered in the mathematical literature of the subsequent, relatively long period: at the time of Austrian absolutism (1849—1867) he represented the mathematical department of the Academy almost alone, FARKAS BOLYAI's membership meant no more than a few book reviews for the *Science Digest* and an article of one and a half pages on ethnography [4]. JÁNOS BOLYAI's name never appeared in the publications of the Academy in his lifetime.

MATHEMATIKAI

MŰSZÓTÁR.

KÖZBE BOCSÁTJA

A' MAGYAR TUDÓS TÁRSASÁG.

BUDÁN.

A' MAGYAR KIR. EGYETEM' BETŰIVEL.

1834.

35. The title page of the *Dictionary of Mathematics*

Celeritas media, közép se-
besség. *Győry.*

Celeritas virtualis, sebes-
ség törekedés' nagysaga.
Bitnicz.

Cella vinaria, pincze. *Jó.*

Centrum aequilibrii, sulye-
gyen' középpontja. *Bitnicz.*

Centrum circuli, kerekszek.
Dugonics. Kellőközép. *Pethe.*
Középpont. *Bolyai. Bresz-
tyenszky. Beregszaszi.* Szek-
pont. *Lexicon.* Kullóközép.
Udvardy. Középpont. *Bedeus.*
Ker. *Részédes.*

Centrum gravitatis, nehez-
ség középpontja. *Varga M.*
Sulypont. *Közhasznu.*

Centrum magnitudinis, nagy-
ság középpontja. *Bitnicz.*

Centrum motus, mozgás'
középpontja. *Bitnicz.*

Centrum oscillationis, lóg-
gas' középpontja. *Varga M.*

Centrum percussionis, ü-
tes középpontja. *Nyiry.*

Centrum quietis, nyugvás'
középpontja. *Bitnicz.*

Centrum sectionis conicae,
szek. *Dugonics.*

Centrum virium, erők kö-
zeppontja. *Varga Márt.*

Character numericus, szám-
tag. *Pethe.* Számjegy. *Bresztyenszky.*

Characteristica loga-
richmi, tagjegy. *Pethe.* Bé-
lyeg. *Dugonics.* Esmertető-
jegy. *Bresztyenszky.*

Chaussee, töltött ut. *Varga
Ján.* Csinalt ut. *Közhasznu.*

Chelonium, fogantyu.
Ameth.

Cheminements, utayita-
sok. *Jelenkor.*

Chevalets, raketabak. *Kis.*

Chorda circuli, kerekhur.
Dugonics. Hur. *Pethe. Bresztyenszky. Arithm. Bolyai.*

Chorda munimenti, közep-
gat, bástyaküz. *Bitnicz.* De-
reklal. *Kovics.*

Chorographia, tajleiras.
Bitnicz.

Choroidea oculi, szem
fekete hartyaja. *Varga Márt.*
Eres hartya.

Chors, ol. *Közszó.*

Chronologia, idötudomány.
Lexicon. Idöszamlalás. *Lux-
ghy.* Idöszamolás *Közhasznu.*

Chronometer, idömérő.
Közhasznu. Hosszasagmérő
óra. *Nyiry.*

Chrymotheca, jegverem.
Közszó.

Cincta, pántlag. *Kresz-
nics.*

Circinnus, czirkalom. *Du-
gonics. Pethe. Bresztyenszky.
Sarvary. Arithmet.* Körítő
Güezei szó.

Circinnus, proportionalis,
szeres czirkalom. *Pethe.*

Circulus, kerület. *Czere.*
Kerekded, czirkalom. *Cisio.*
Kerek, kür. *Dugonics. Bresz-
tyenszky.* Karika, kerek. *Var*

36. A page from the *Dictionary of Mathematics*

Following the objectives of the Academy, members first strove to create the Hungarian terminology of mathematics as their main task. They set out with great enthusiasm and little competence to collect the mathematical vocabulary of the Hungarian language. Ahead of all other departments, they published the first Hungarian *Mathematikai Műszótár*[1] as early as 1834. A book of 110 pages, it comprises more than the title suggests as it contains the most frequent Hungarian terms of navigation, architecture, painting, mining, forestry and warfare together with their Latin, German and French equivalents.

The publication of the dictionary was most urgent indeed as the Hungarian mathematical language was perfectly chaotic and incomprehensible. The preface to the book lists a meagre bibliography of 35 books and treatises from which the authors (TITTEL, GYŐRY, BITNICZ and NYÍRY) compiled the material of the dictionary. Strictly speaking, only 4 or 5 of the sources are mathematical works; most of the Hungarian mathematical terms in the dictionary originate from JÁNOS APÁCZAI CSERE, ANDRÁS DUGONICS, FERENC PETHE, FARKAS BOLYAI, BÉLA BRESZTYENSZKY and SÁNDOR GYŐRY. The list by itself proves that the editors did a very incomplete job as they ignored the mathematical vocabulary created in the previous centuries, the credit for which is due to the authors of the *Arithmetics of Debrecen* and *of Kolozsvár* as well as FERENC MENYŐI TOLVAJ, JÁNOS ONADI and, above all GYÖRGY MARÓTHI.

To a certain extent, the dictionary helps us reconstruct the stage of development of some scientific disciplines in those days. Apparently, the most felicitous expressions are those of the art of war, the bulk of which remained unchanged to this day both in form and meaning. By contrast, the mathematical lingo abounds in misconceived terms. It would not be hard to cull those 30 or 40 words in the dictionary that are still used in mathematical literature today. The more successful words chiefly belong to elementary arithmetic, while the terms of analysis and geometry fell victim to forced Magyarization.

The impact of the dictionary upon the Hungarian mathematical language is negligible, having fallen short of its express aim of creating a unified Hungarian mathematical vocabulary. This was caused not so much by the defectiveness of the terminology as by the editors' reluctance voiced in the preface to take sides and to specify the principles to be followed in coining Hungarian terms. In this way the dictionary enhanced the fervour of haphazard Magyarization as after its publication every mathematician felt it his patriotic duty to coin one or more Hungarian equivalents to international mathematical terms. This tendency often made otherwise valuable works incomprehensible. Frequently, the meaning of the tortuous words and phrases can only be made out from the context, or from the Latin original in brackets. Not even those mathematicians who chastised forced Magyarization (GYŐRY, NYÍRY, VÁLLAS, OTTÓ PETZVAL) were exempt from exaggerations either. E.g. ANTAL VÁLLAS admitted in an address to the academic committee that "Magyarization is a fine and noble deed" and found such coinages as *kebel* 'lit. bosom' for sine and *pótkebel* 'lit. substitute bosom' for cosine quite acceptable, and only raised his voice against abbre-

---

[1] For a detailed linguistic analysis see: GÁLDI, LÁSZLÓ: *Hungarian dictionaries in the Age of Reason and the Age of Reforms.* Budapest, 1957, pp. 450—454. (In Hung.)

viating these circular functions in several textbooks as *keb x* and *pkeb x* (*Gazette*, Vol. 7, p. 13).

Though not as a result of the dictionary, our mathematical language made rapid progress during the first decades of the Academy. This progress was closely related to the development and refinement of the entire Hungarian language and literature. In this field the role of the Academy was indubitably significant. Some of the credit must go to those members (KÁROLY NAGY, OTTÓ PETZVAL, ÁRMIN JÁNOS VÉSZ) who tried to gradually purge the language of the forcefully Magyarized terms. At the end of his two-volume textbook [36], of great popularity for a long time, ÁRMIN JÁNOS VÉSZ listed the mathematical terms he used — some one hundred and fifty in all —, most of which are used unchanged today.

In the above sense, the first decades of the Academy eventually proved to be decisive in creating the Hungarian mathematical language. The emphasis on the Hungarian language was a motivating force for the mathematicians, and even if they often carried things to extremes, several of the terms they coined have survived in our mathematical literature. The botched terms coined contrary to the laws of the Hungarian language warned our mathematicians later that it was more appropriate also from the didactical point of view to use the internationally accepted terms than bad Hungarian ones. That is how the mathematical vocabulary of the Hungarian language had evolved by the end of the last century, blending international and Hungarian words.

A very important, perhaps the most important, contribution of the first mathematicians of the Academy was their untiring struggle to have the significance of the mathematical sciences acknowledged and the standards of teaching mathematics raised. The engineer members of the Academy, SÁNDOR GYŐRY, JÓZSEF BESZÉDES, VILMOS FEST, PÁL VÁSÁRHELYI, OTTÓ PETZVAL, played an active role in SZÉCHENYI's projects to regulate the rivers and build bridges. In the course of their work they must have experienced day by day that mathematics was central to boosting the economy of the nation, but they must also have realized how backward the country was. That might be the reason why these years saw the publication of several treatises in various journals of the Academy that underlined the role of mathematics and the need to improve its teaching. Envisaging a better future, they tried to stir the conscience of the nation.

At some place SÁNDOR GYŐRY posed the following question:

"Can we ever overcome those way ahead us, and when? The distance and gulf separating us from the higher educated nations are gradually increasing."[2]

He complained that in Hungary even the natural sciences were controlled by lawyers, that the money tended to run short when it came to improving the level of cultural life, and that two-thirds of young university graduates were living from hand to mouth without employment. We had fallen behind, he said, first of all in the natural sciences and mathematics because

---

[2] On the regulation of the Danube. *Évkönyv*, Vol. 2, Part 2, p. 120. (In Hung.)

"... those who have prospects of getting more brilliant and glamorous offices will surely not choose the despised mathematical sciences".[3]

LAJOS BITNICZ also complained that the Hungarian youths were "somewhat reluctant to take up mathematics."[4] And ANTAL VÁLLAS noted bitterly in an essay that while nearly all the universities abroad had several chairs for advanced mathematics, in Hungary the government only allowed a single extraordinary chair to be set up for analysis for the whole of 14 million inhabitants ([32], p. 166). It was again he who wrote in another paper that the Hungarian mathematicians had no scientific instruments at their disposal, they lacked professional books, and they had to waste their powers in the most diverse odd jobs to make ends meet. "We can hardly wait to see the authorities ... remove the obstacles." A great many places could be quoted, however, to prove that his words were not given a hearing and after the fall of the 1848 war of independence the situation deteriorated further. In the preface of his book published as late as 1867, OTTÓ PETZVAL [25] still felt obliged to state that the number of those engaged in mathematics was very low in Hungary although the nation was famous for its mathematical disposition. "The scientific culture of Hungary is like a dwarf as compared to foreign countries."

There was much discussion in those years about the need to set up a Hungarian technical university. In his addresses to the Academy, ISTVÁN SZÉCHENYI also raised the problem several times asking whether it would not have been more expedient to set up an institution of higher education in technology instead of the Academy. But the ultimate conclusion he arrived at every time was that the problem of paramount importance was the cultivation of the Hungarian language, and only when this aim had been achieved could the improvement of the technical sciences come to the fore. SÁNDOR GYŐRY took a firm stand on the establishment of the technical university in several writings, while ANTAL VÁLLAS lashed out against the current situation in very biting terms in his boldest pamphlet [33]. The basic tone of the treatise is epitomized by the following disheartening sentence: "The guardian spirit of the future sits brooding over the ruins of a country once envisioned as a fair land" ([33], p. 58). This bitterness comes up elsewhere as well: "The easiest way to gain a Hungarian's confidence is to ridicule the sciences" (*Idem.*, p. 13), the author complains and goes on to pass sharp criticism on the monopoly of the liberal arts; he blames the misery of the country exactly on the backwardness and neglect of mathematics and the technical sciences, and urges for the setting up of a Hungarian politechnical university to remedy these maladies. Although his writings did not bring immediate results, the proposals of VÁLLAS came in handy when the *Joseph Industrial School* was promoted to the rank of polytechnical school (1856) and later to that of technical university (1871).

In order to attenuate the cultural backwardness of the country, the Academy strove from the beginning to put across some organizational measures. These measures aimed at improving the scientific life as a whole, thus they were beneficial for mathematics as well. Such moves included the gradual establishment of contacts with

---

[3] *Idem*, p. 137.

[4] Commemorative address for ordinary member PÁL TITTEL. *Évkönyv*, Vol. 2, Part 2, p. 9. (In Hung.)

foreign scholarly institutions, first of all with academies of sciences abroad. This activity began in the very first years of the Academy and soon encompassed the noted scholarly institutes of Germany, France, Italy, Russia, England and the United States. At the time of the Austro—Hungarian Compromise of 1867, personal contacts were being nursed with 25—30 foreign academies and an exchange of publications was also going on. The growth of the library of the Academy was also comparatively fast, especially from donations by private persons — a survey set the stock of books at 50—60,000 already in 1844 (of which some 500 were on mathematics) — but at the same time the total annual attendance of the library was as low as 7,000!

The Academy profited little from its attempts to extend the contacts by coopting foreigners into its membership. The selection of foreign scholars was being done rather randomly. In many cases it was personal affection rather than scholarly merit that underlay the decisions of the Academy.

The first foreign mathematician to be coopted was the Cambridge professor CHARLES BABBAGE (1792—1871) who became a member of the Scholarly Society in 1833. BABBAGE was a forerunner of modern computer science and a pioneer of the theory of functions. His logarithm table, published with a Hungarian foreword, largely contributed to popularizing the idea of computing with logarithm. GAUSS and PONCELET became foreign members of the Academy in 1847. GAUSS exerted a great influence on our scientific development, but the Hungarian scholars of the period, except for the two BOLYAIS, held him in high esteem not so much for his mathematical exploits as for his results in astronomy and geodesy.

In 1858 apart from the prominent astronomer JOHN HERSCHEL, the Viennese mathematician ANDREAS ETTINGSHAUSEN (1796—1887) and a Belgian astronomer ADOLPH QUETELET (1796—1874) were elected into the Academy. ETTINGHAUSEN's name has survived mostly for his investigations in combinatorics, while QUETELET earned fame for his statistical research and geometrical results. The latter was among the most eminent practitioners of statistics, advocating the view that the social phenomena were independent of the forms of social life, consequently, the social system could not be blamed, e.g., for crime. What accounts for his cooption was most probably the growing interest in data collection; at the same time, QUETELET was the editor of the Belgian journal *Correspondence Mathématique et Physique* for a long time, so another reason for his election must have been the desire to get into contact with a well-known foreign mathematical periodical. This motivated the cooption of the mathematician AUGUST GRUNERT of Greifswald in 1860. The *Archiv der Mathematik und Physik* he founded in 1841 (ceased in 1920) had a good reputation and published several new mathematical results. Indeed, after the mid-19th century one can come across Hungarian names on the pages of GRUNERT's *Archiv* more and more frequently.

*

Despite its limited means, the Scholarly Society ambitioned to encourage the publication of good textbooks of mathematics from the very beginning. There was pressing need for the interference of the Academy in this regard, for the bulk of the profuse output of mathematical textbooks were substandard, fraught with errors. Let us cite but a few of the mistakes and naiveties of the books. ANDRÁS TATAI, for instance, illustrated the multiplication and division of common fractions with the following:[5]

$$\frac{2}{3} \times 4 = \frac{8}{12}; \quad \frac{8}{10} : 2 = \frac{4}{5}.$$

GÁBRIEL CSEPCSÁNYI made no scruples about concluding from $\frac{a}{a} = 1$ that zero divided by zero was also equal to the unit.[6] One can imagine how gross the mistakes committed by the authors in the course of presenting algebra were, not to speak of the daring new symbols, tortuous technical terms and the eulogies of primitive relations as new results. Even respected mathematicians could be reproached with such mistakes. For example, FERENC KEREKES, after presenting the identity $-\sqrt[\frac{1}{3}]{a} = \frac{1}{a^3}$, proudly announced that as far as he knew, no one before him had noticed it.

Most regrettably, however, the intervention of the Academy was slow to bear fruit. KÁROLY NAGY's *Arithmetic* was published in 1835, which together with his subsequent *Elementary algebra* [22] provided the primary and secondary schools with textbooks undeniably better than the ones formerly in use. The Academy rewarded the author of the textbooks and ANTAL VÁLLAS wrote an appreciative review of them [32]. However, it would be mistaken to believe that the two books would pass a more severe test of proficiency. In certain fields (e.g. the definition of logarithm) it is teeming with grave mistakes. In a letter to GAUSS, FARKAS BOLYAI said he regarded as the only merit of the book that it was correctly printed in Vienna, but he subjected its language and method to severe criticism ([5], p. 123). Much later SÁNDOR GYŐRY also criticized the book's misconceived section on interpolation. And indeed, how can one compare this book to the profound and original thoughts of the *Tentamen!* The fact that even our outstanding mathematicians commented on the works of KÁROLY NAGY proves that these works rose above the average of those days.

SÁNDOR GYŐRY's calculus [8] also earning a prize from the Academy was designed to raise the standards of university education. The collection of the material for the book reflects the author's individual teste and hence it cannot be regarded as a systematic text in calculus. In spite of the professional flaws and the laboured technical terms the book deserves credit for being the first to emphasize the importance of the notion of function. By the way, the Hungarian term *függvény* for function also comes from GYŐRY.

Somewhat more meritorious is OTTÓ PETZVAL's *Elementary arithmetic* of a later date [24]. As the preface to the book reveals, the aathor wrote his book in view of the needs of students and of the specific conditions of Hungarian education. Apart from today's secondary school material, the voluminous book also touches on cubic and quartic equations. The layout of the book is systematic, its language and style pliant and unlaboured.

---

[5] TATAI, ANDRÁS: *The beginnings of pure mathesis.* Pest, 1836. (In Hung.)
[6] CSEPCSÁNYI, GÁBRIEL: *Elementa matheseos purae.* Pozsony, 1824.

In the sixties, the Academy also promoted the publication of systematic works for higher education. The two-volume work of ÁRMIN JÁNOS VÉSZ (WEISZ) [36] and the *Higher mathematics* by OTTÓ PETZVAL [25] elaborated the areas of calculus as accepted today; the former also addressed — for the first time in Hungarian mathematical literature — differential equations and the calculus of variations, while the latter also dealt with functions of several variables.[7]

*

Less inaccurate than the textbooks were the diverse trigonometric and logarithm tables of the age. Their principal aim was to lend assistance to astronomical and geodesic computations, hence their editors included scholars with affiliations to astronomy and geodesy (JÁNOS PASQUICH, KÁROLY NAGY, ANTAL VÁLLAS and others). NAGY's commitment to the cause was so great that he edited the well-known British mathematician BABBAGE's book in colour print in London so as to popularize computations with logarithm. Besides astronomy and geodesy, interest in commercial arithmetic also began to grow at the beginning of the last century. FERENC KEREKES compiled his logarithm table titled *Multiplications* since "... our country is slowly waking to the financial matters" ([16], p. 5). One peculiarity of the book is that its author took up cudgels with weighty arguments against *malthusianism*, which warned of the danger of the overpopulation of the planet ([16], p. 103). He declared that the threat of overpopulation was unfounded as the sciences helped industry to develop and the size of arable land to increase at an ever growing pace: besides, in the future the birthrate was likely to become lower. The time would come, he maintained, when a lasting equilibrium between production and population would ensure the required standard of living.

As for reliability, the best logarithm table was compiled by NÁNDOR LUTTER; this table has appeared in many editions from the 1860s to the first half of the 20th century.

*

The value of the first academic publications aspiring at *scholarly merits* was next to nothing. Some progress was perhaps made in the area of the numerical and graphical solution of equation. Our academicians seem to have been most fascinated by the HORNER scheme (1819) and the GRAEFFE procedure (1837). SÁNDOR GYŐRY devoted a whole series of articles to analyzing HORNER's scheme and applied the method, often demanding lengthy computations, to many concrete equations [9—14]. KÁROLY TAUBNER devoted a separate paper to present the BUDAN–FOURIER theorem [31]. As far as we know, ANTAL VÁLLAS was the first in the world to give a detailed exposition of the algebraic results of FOURIER, HORNER and GRAEFFE in his nearly 600-page book which won a prize of the Academy [35]. Unfortunately, the book is

---

[7] For the more outstanding achievements of the members of the Academy (up to 1975) see also the chapters by LÁSZLÓ VEKERDI in *150 years of the Hungarian Academy of Sciences*. Budapest, 1975. (In Hung.)

laden with so many dull and erroneous sections that all we can actually appreciate in it is the good intention of the author.

The popularization of the theory of probability begun by ISTVÁN HATVANI was resumed by LAJOS BITNICZ in his work entitled *The principle of the least squares* [2]. The author had a reliable knowledge of the relevant results achieved previously; after defining several concepts of the theory of probability, he explained GAUSS's normal distribution in a comprehensible way, deducing from it the principle of the least squares. ISTVÁN NYIRY used three tables of mortality published abroad to present the concept of life expectancy and, on the basis of the results thus gained, proceeded to determine the expected number of the population and the possible number of conscripts [23]. Another of BITNICZ's essays [3] is similar in nature: in it, he explains the *law of large numbers* on the basis of data concerning criminals.

These mathematically humble attempts were not perfectly independent of the fact that various insurance activities were spreading abroad and since the general census ordered by JOSEPH II (1784) it had become necessary in Hungary, too, to collect various statistical data. Based on plans of the renowned writer ANDRÁS FÁY, the *National Savings Bank of Pest* began operation in 1840. FÁY soon compiled the first Hungarian table of mortality to promote the bank's life insurance activities.

The first systematic Hungarian treatment for educational purposes of the elements of probability theory was completed by OTTÓ PETZVAL ([25], pp. 373—396). His work deserves special attention because several concepts (probability, certainty, improbable, doubtful event, etc.) struck root in the Hungarian mathematical literature with the terms he coined.

Apart from the work of the two BOLYAIS, the literature on geometry was very poor in the studied period. The first and possibly most significant initiative of the Academy in this respect was the decision in 1832 to publish EUCLID's *Elements* in Hungarian. SÁMUEL BRASSAI was assigned the task of translation who went to great lengths to compare various foreign EUCLID editions and annotated the Hungarian translation with a detailed foreword and explanatory comments [6]. That later it was felt necessary to translate it again was not so much for professional as for linguistic reasons (*The first six books of the Elements.* Budapest, 1905. Transl. ALAJOS BAUMGARTEN; Contains only the planimetry part of EUCLID's work). An up-to-date translation of the Elements appeared recently prefaced by ÁRPÁD SZABÓ and translated by GYULA MAYER (Budapest, 1983).

The subject of the competition advertised by the Academy in 1838 was also geometry. Invited were studies on first and second-order planar curves as they appear in practice. The first prize was won by KÁROLY TAUBNER, the second by VILMOS FEST (*Mathematicai Pályamunkák*, Vol. 1, Buda, 1844). Neither submission contained more than what is usually included today in the secondary schools curricula of analytic geometry, and unfortunately they abound in errors, misprints and laboured technical terms to boggle the reader's mind. TAUBNER devoted another paper to conic sections [28], and in a study he analyzed the curves frequently occurring in geometrical optics [30].

Bibliography

[1] ARENSTEIN, JÓZSEF: The properties of imaginary quantities. *Mathematicai Pályamunkák*, Vol. 2. Pest, 1847. (In Hung.)

[2] BITNICZ, LAJOS: The principle of the least squares. *Évkönyv*, Vol. 3, 1834—1836, pp. 42—66. (In Hung.)

[3] — On the law of large numbers. *Magyar Academiai Értesítő*, 1851, pp. 241—249. (In Hung.)

[4] BOLYAI, FARKAS: Wedding rituals in Marosszék county. *Tudománytár*, Vol. 2, 1834, pp. 221—222. (In Hung.)

[5] *Briefwechsel zwischen C. F. Gauss und Wolfgang Bolyai.* Leipzig, 1899.

[6] BRASSAI, SÁMUEL: *Euclid's Elements.* Pest, 1865. (In Hung.)

[7] FEST, VILMOS: Curves of the first and second order ... *Mathematicai Pályamunkák*, Vol. 1, Buda, 1844. (In Hung.)

[8] GYŐRY, SÁNDOR: *The elements of advanced calculus.* I—II. Buda, 1836, 1840. (In Hung.)

[9] — Higher-degree numerical equations. *Évkönyv*, Vol. 6, 1840—1842, pp. 395—403. (In Hung.)

[10] — On the reduction of equations. *Magyar Academiai Értesítő*, 1847, 12. (In Hung.)

[11] — On the solution of quartic equations. *Magyar Academiai Értesítő*, 1848, 61. (In Hung.)

[12] — On the general solution of equations. *Magyar Akadémiai Értesítő*, Vol. 2, 1860—1861, pp. 96—122. (In Hung.)

[13] — On the roots of equations of higher-degree. *Az MTA Értesítője*, Vol. 1, 1867, pp. 169—170. (In Hung.)

[14] — On the solution of equations. *Az MTA Értesítője*, Vol. 2, 1868, pp. 291—293. (In Hung.)

[15] — The mathematical sciences' ... *Évkönyv*, Vol. 2, 1832—1834, part 2, pp. 23—43. (In Hung.)

[16] KEREKES, FERENC: *Multiplications.* Debrecen, 1845. (In Hung.)

[17] KERESZTESI, MÁRIA: *A history of the Hungarian mathematical language.* Debrecen, 1935. (In Hung.)

[18] KÜRSCHÁK, JÓZSEF: The last hundred years in the history of Hungarian mathematics. In: *The first century of the Hungarian Academy of Sciences.* Budapest, 1926, pp. 451—459. (In Hung.)

[19] *Mathematikai és Természettudományi Közlemények* (Bulletin of Mathematics and Natural Sciences) (In Hung.)

[20] *Mathematikai Műszótár* (Dictionary of Mathematics). Buda, 1834. (In Hung.)

[21] NAGY, KÁROLY: *Arithmetic.* Vienna, 1835. (In Hung.)

[22] — *Elementary algebra.* Vienna, 1837. (In Hung.)

[23] NYIRY, ISTVÁN: A history of calculating human mortality. *Tudományos Gyűjtemény*, Vol. 5, 1821, pp. 49—69. (In Hung.)

[24] PETZVAL, OTTÓ: *Elementary mathematics.* Budapest, 1856. (In Hung.)

[25] — *Higher mathematics.* Budapest, 1867. (In Hung.)

[26] SZÉNÁSSY, BARNA: The mathematical activity of the Hungarian Academy of Sciences up to 1867. *Acta Universitatis Debreceniensis*, Vol. 1, 1954, pp. 5—28. (In Hung.)

[27] SZTOCZEK, JÓZSEF: Mathematics and natural science. In: *Sketches from the 50 years of the Hungarian Academy.* Budapest, 1881, pp. 86—94. (In Hung.)

[28] TAUBNER, KÁROLY: A contribution to the theory of conic sections. *Tudománytár*, Vol. 11, 1842, pp. 245—253 and 275—307. (In Hung.)

[29] — First- and second-degree curves, etc. *Mathematicai Pályamunkák*, Vol. 1, Buda, 1844. (In Hung.)

[30] — Notable optic curves ... *Magyar Académiai Értesítő*, 1847, pp. 375—393. (In Hung.)

[31] — Fourier's method for solving equations of higher degree. *Tudománytár*, 1841, pp. 7—27 and 67—81. (In Hung.)

[32] VÁLLAS, ANTAL: Recent Hungarian literature in mathematics, etc. *Tudománytár*, Vol. 4, 1836, pp. 143—172. (In Hung.)

[33] — *On the prospects of a Hungarian central technical university.* Pest, 1841. (In Hung.)

[34] — *Beitrag zur Auflösung der höheren Gleichungen.* Wien, 1843.

[35] — *Higher-degree equations in one unknown quantity.* Buda, 1848. (In Hung.)

[36] VÉSZ, JÁNOS ÁRMIN: *The outlines of higher mathematics.* Vols 1—2, Pest, 1861—1862.

V. The expansion
of mathematical research

17. Rising demands towards mathematics

As is well known, although the Compromise between Hungary and Austria (1867) failed to bring the long-desired liberation for the Hungarian masses as it was a compromise between the Hungarian owners of large and middle-size estate on the one side and the ruling house and the ruling classes of Austria on the other, yet it mean a great stride forward in overall middle-class development. The relative consolidation of the country's economic and political situation encouraged the investment of capital, which boosted the entrepreneurial spirit in commerce, resultated in setting up a lot of factories (in heavy industry, distilling, milling), increased the level of the mechanization of agriculture, etc. In short, the spreading of capitalism accelerated despite the vestiges of the feudal and colonian system. And although economic crises typical of capitalistic development often disrupted the process of advancement, toward the end of the 19th century the country reached the state of monopoly capitalism.

Let us quote but a few data which probably are little known to mathematicians in order to illustrate the pace of progress: the number of power machines in the factories grew 30-fold between 1863 and 1898, and the number of credit banks rose from 107 in 1867 to 4954 at the outbreak of World War I.

Under such circumstances it was inevitable to streamline the Hungarian educational system and academic life so as to enable them to meet the requirements of a more advanced social order. The main events of this process with relevance to mathematics are summarized below.

*

Let us illustrate on a typical — and scientifically significant — example how a technical problem of the period considered led to abstract and profound mathematical results within a short time. Such examples are large in number, testifying to the fact that highly abstract mathematical investigations were very often prompted by practical needs. Later, however, what KOLMOGOROW said also proved valid on the Hungarian scene: "... great and new theories are not only generated directly by problems arising in the natural sciences and engineering but also by the intrinsic needs of mathematics."

The Hungarian rivers, first of all the Danube, having been rendered navigable, from the middle of the last century interest focussed on the technical problems of navigation including the problem of the *screw propeller* of optimal efficiency. A kindred problem was raised by the development of flour-milling and by irrigation for intensive farming: that of the *air vane*.

Inquiries into the secrete of the propeller began much earlier: MACLAURIN and EULER described the optimal bend of the propeller blades by the following formula:

$$\operatorname{tg} \alpha = \frac{3c}{4\omega r} \pm \sqrt{\left(\frac{3c}{2\omega r}\right)^2 + 2}.$$

In the formula, c is the speed of advance, ω is the angular velocity of the propeller, r is the distance from the centre of the axis of rotation to the surface element considered, and α is the angle that is to be subtended by the normal of the surface element with the direction of advance at distance r so as to obtain optimal efficiency. Most researchers tried to find the equation of the propeller surface with the help of this formula.

After several dilettante attempts, the first scientific treatment of the problem of the most efficient marine screw and the structurally similar air propeller was accomplished by LAJOS MARTIN, a professor of mathematics at the university of Kolozsvár. A special incentive for him was the competition announced by the Hungarian Ministry of Commerce inviting constructions of steam, water, wind or horse-powered devices that would have solved the problem of irrigation until the extension of the system of canalization. In his theoretical research, MARTIN accepted the MACLAURIN–EULER formula but also assumed that the best ship screw and air vane must be some sort of *conoid*.[1] Making physically and technically untenable simplifications, he put down the equation he had figured out, and although the contraptions constructed on its basis all failed at the testing, MARTIN deemed it justified to publish his theoretical findings.

That launched one of the fiercest and at the same time most fruitful literary debates in the history of Hungarian mathematics involving many of our outstanding scholars. The first to point out the errors in MARTIN's lengthy treatises were his official critics KÁLMÁN SZILY and ISTVÁN KRUSPÉR, who declared that the essays [9, 10] should only be published with the far-from-flattering critical comments ([9], I—IX). SZILY, for example, revealed with scathing sarcasm that MARTIN's surface did not satisfy either of the hypotheses: it was not a conoid and it did not fulfil the MACLAURIN–EULER formula. Later SZILY took an even more active share in the investigation of this question: in two studies [14, 15] he first of all refuted MARTIN's assumption that the sought-after propeller should necessarily be a conoid because such surfaces did not meet the MACLAURIN—EULER requirement. Then he proceeded to demonstrate that there was only one conoid that could be placed symmetrically to the axis of rotation and used as a propeller: the Archimedean helix.

After this discovery, SZILY dropped the MACLAURIN—EULER restriction and reformulated the problem in a less pretentious way: does an Archimedean helix of a given

---

[1] Those surfaces are called conoids which satisfy the first-order homogeneous linear partial differential equation $x\dfrac{\partial z}{\partial x} + y\dfrac{\partial z}{\partial y} = 0$, that is, which can be defined by the equation $z = f\left(\dfrac{y}{x}\right)$.

size that can be placed in a straight cylinder (fixed below a ship in the usual way) produce a larger propelling force than any other surface of the same area? Expressing the component of force acting in the direction of advance with a rather complicated formula (a double integral) he proved that of all conoids, under certain restrictions, the Archimedean helix was the most appropriate for the required purpose. We mention that SZILY already, applied the calculus of variations in his computations but gave up the tiresome examination of the second variation.

Having noticed this, MARTIN did not rest with arms folded. In his reply of a highly ironical tone [11] he called SZILY's variational method an unnecessary and erroneous extravagance. MÓR RÉTHY managed to replace SZILY's pressure formula by a simpler one, and gave an exact solution of the variational problem so obtained [12, 13].

Leaving the more detailed exposition of the relevant results to a later chapter, let us only note here that this problem initiated the famous doctoral dissertation of GYULA VÁLYI (see Chapter 22) in which he gave the criterion for the integrability of the partial differential equation corresponding to the variational problem. The same problem stimulated GYULA KŐNIG's fundamental treatises related to the general theory of partial differential equations (see Chapter 20).

JÓZSEF KÜRSCHÁK continued the VÁLYI—KŐNIG investigations in a whole series of papers [1—8]. First of all, he studied the problem in greater generality than VÁLYI had done; namely, assumed that the variational integral depended not only on the direction of the surface element but also on its place. Then, by the application of the *contact transformation*[2] he presented a simpler proof of VÁLYI's results and announced the important theorem that the second or higher-order partial differential equations arising from variational problems remain differential equations of variational problems even after the application of the contact transformation.

So, this field of problems with meagre initial results became the motive force and point of departure for Hungarian research into the calculus of variations and the theory of differential equations which can pride itself on respectable traditions now.

Bibliography

[1] KÜRSCHÁK, JÓZSEF: On the partial differential equations of the second order occurring in the variation of double integrals. *Math. és Term. tud. Értesítő*, Vol. 7, 1889, pp. 296—307. (In Hung.)

[2] — Über die partiellen Differentialgleichungen, etc. *Math. u. naturw. Berichte aus Ungarn*, 7, 1889, pp. 263—275.

[3] — Über eine besondere Classe der partiellen Differentialgleichungen des Variationscalculus. *Math. u. naturw. Berichte aus Ungarn*, 8, 1890, pp. 35—50.

[4] — On a special class of partial differential equations in the calculus of variations. *Math. és Term. tud. Értesítő*, 8, 1890, pp. 60—75. (In Hung.)

[5] — Über partielle Differentialgleichungen zweiter Ordnung mit gleichen Charakteristiken. *Math. Annalen*, 57, 1890, pp. 317—320.

---

[2] A contact transformation takes two mutually tangent surfaces into another two mutually tangent surfaces. The germ of this idea can be found in EULER and JACOBI, but its precise analytic definition was given by SOPHUS LIE (Zur analytischen Theorie der Berührungstransformationen. *Forhandlinger i Videnskabe Selskabet i Christiania*, 1873, pp. 237—262).

[6] — Über eine Classe der partiellen Differentialgleichungen zweiter Ordnung. *Math. u. naturw. Berichte aus Ungarn*, 14, 1896, pp. 285—318.

[7] — On a class of partial differential equations of the second order. *Math. és Term. tud. Értesítő*, 15, 1897, pp. 225—256. (In Hung.)

[8] — On the transformation of the partial differential equations appearing in the calculus of variation. *Math. és Term. tud. Értesítő*, 17, 1899, pp. 457—466. (In Hung.)

[9] MARTIN, LAJOS: Helical surfaces in mechanics. *Értekezések a Math. Tud. Köréből*, Vol. 3, 1874—1875. (In Hung.)

[10] — The theory of the horizontal air vane. *Ibid.*

[11] — Application of the differential coefficient for solving the equation of the propeller surface. *Értekezések a Math. Tud. Köréből*, Vol. 5, 1877. (In Hung.)

[12] RÉTHY, MÓR: A contribution to the theory of propeller and peripeller surfaces. *Értekezések a Math. Tud. Köréből*, Vol. 4, 1876. (In Hung.)

[13] — To the theory of the screw propeller. *Műegyetemi Lapok*, Vol. 2, 1877, pp. 257—271. (In Hung.)

[14] SZILY, KÁLMÁN: On the theory of the propeller. *Az MTA Értesítője*, Vol. 9, 1875, pp. 237—238. (In Hung.)

[15] — Zur Theorie der Propellerschraube. *Der Civilingenieur*, Vol. 23, 1877, pp. 177—186.

18. External factors of the development of mathematics in Hungary

After the Compromise of 1867 the level of our mathematical culture had gradually increased and slowly reached that of the international forefront partly because of practical demands and partly due to the intrinsic development of science. A whole range of external events must be referred to which either contributed to, or indicated, this development.

In 1868 Act 28 provided for compulsory schooling upon pain for children between 6—12 years of age. New and reformed curricula appeared in quick succession, each adding something new to the teaching of mathematics. The one compiled by PAULER (in use between 1871 and 1879) considerably increased the number of mathematics lessons in eight-year grammar schools and outlined a material that came to stay in the curricula up to the middle of the 20th century. The six-year polytechnical secondary schools were also converted into eight-year schools in 1875, considerably improving the level of teaching mathematics. Until then, in all schools the *formal* aspects of teaching had been predominant, but as a result of TREFORT's curriculum (in effect until 1899) there was some improvement in this regard, too, thanks to two outstanding pedagogues, MÓR KÁRMÁN[1] and GYULA KŐNIG, who took part in elaborating the teaching plan. The mathematical part of the WLASSICH curriculum of 1899—1926 stressed the teaching of the *function* concept in secondary school as a question of central importance and cut down on some of the perfectionist demands of the previous curricula (e.g., convergence criteria, cubic equations, etc.). It was another step forward when — upon MANÓ BEKE's initiative and inspired by worldwide attempts to reform the teaching of maths — the National Association of Secondary School Teachers set up the *Commission for Mathematical Reform* (1906). The Commission which included the eminent mathematicians LAJOS KOPP, IGNÁC RADOS, LÁSZLÓ RÁTZ, KÁROLY GOLDZIHER and was headed by BEKE, a man of great expertise and agility) regarded the struggle against formal education as its main task. However, the thoughtless introduction of differential and integral calculus in the secondary school curriculum contradicted this rightful purpose.[2]

---

[1] Father of the world-famous mathematician and physicist TÓDOR KÁRMÁN, who spent most of his life in the United States.

[2] The curricula mentioned and the work of the Commission of Reform are detailed in [7].

Besides the reformed curricula and some excellent textbooks following their guide-lines, there were two other factors to rouse young people's interest in mathematics. On the one hand, in 1894, DÁNIEL ARANY, a teacher at the polytechnical secondary school of Győr, started the *Középiskolai Mathematikai Lapok*. The journal, three years later taken over by LÁSZLÓ RÁTZ, soon became popular with talented young people, stimulating them to deal with maths more intensively also on their own. On the other hand, it was decided in 1894 that

"in order to make the teaching and learning of mathematics and physics more successful, the Society of Mathematics and Physics shall organize a competition in maths and physics every year in Budapest and Kolozsvár for secondary schools leavers of the year to judge the competence of the students in the above subjects." (*Mathematikai és Physikai Lapok*, Vol. 3, 1894, pp. 197—198)

The devotion and proficiency of GYULA KŐNIG, GUSZTÁV RADOS and JÓZSEF KÜRSCHÁK made it possible to lay firm foundations of a high-level competition, named after KÜRSCHÁK today. In the annual reports of the competitionsthe names of several Hungarian mathematicians, later attaining world-wide reputation, can be read.

A large number of well-trained *teacher*s were needed to increase the standards of secondary education, and highly qualified *engineer*s to solve the technical problems posed by industrial development. For the training of the latter a mathematical work-shop evolved in Budapest at the "Joseph Technical University" set up as the successor of a polytechnical school in 1871. It was a highly acknowledged institution in spite of its shortcomings as to the practical aspects of engineering. In the last quarter of the 19th century its professors of mathematics and physics included JENŐ HUNYADY, KÁLMÁN SZILY, ISTVÁN FÖLSER, GYULA KŐNIG, MÓR RÉTHY, JÓZSEF KÜRSCHÁK, GUSZTÁV RADOS and BÉLA TŐTÖSSY — a teaching staff of any contemporary technical university would have been proud of. The effect of their teaching spread beyond the university, since in the courses of the "universal class" the teacher trainees as well as those preparing for a profession in commerce or economics were also trained ac-cording to their individual aims, in certain branches of mathematics.

In 1876, after the failures of the past, a handful of zealous and ambitious professors of the Technical University decided to start a scientific but understandable periodical entirely on their own resources. In virtue of some mathematical essays and problems arousing nation-wide interest, the monthly *Műegyetemi Lapok* had a highly promising start. However, the time had not yet come for such a journal; financial difficulties compelled the editors to say farewell to their faithful but small readership after the publication of three volumes (30 issues).

"Notice. With this issue the Technical University Journal has come to the end of its career. It seems that here in Hungary, no mathematical journal can exist without financial support. Of course, if there were only half as many readers of mathematics as there are *teacher*s, things would be quite different." (*Műegyetemi Lapok*, Vol. 3, 1878, p. 316)

In addition to the technical university of Budapest, another centre for mathematics emerged at the *University of Kolozsvár* established in 1872. This university was more up-to-date than the one in Budapest concerning both its organization and legal status. While the university of humanities and sciences in our capital was closely related to the Church, the one in Kolozsvár was independent from the very beginning. Besides, the university in Kolozsvár emphasized the importance of mathematical subjects even in its structure: it had a separate faculty of mathematics and natural sciences. True, at the start this faculty belonged to the faculty of arts but it was soon granted legal independence in what was a most modern move as at that time only the University of Tübingen had a separate faculty of natural sciences.

Though the erudition of LAJOS MARTIN and SÁMUEL BRASSAI, the professors of mathematics appointed to the university of Kolozsvár at the beginning, was not sufficient to organize an active scientific life, later the students had the good fortune of being taught by such excellent specialists as GYULA FARKAS, MÓR RÉTHY, GYULA VÁLYI, LAJOS SCHLESINGER, LIPÓT FEJÉR, ALFRÉD HAAR and FRIGYES RIESZ.

Until the end of the century the mathematical life at the Budapest university of sciences — under the professorship of ÁGOSTON SCHOLTZ, OTTÓ PETZVAL and GUSZTÁV KONDOR — was far less colourful. In these years not even the lectures of the privat professors MIHÁLY DEMECZKY and JÓZSEF SUTÁK could raise the general level. The appointment of MANÓ BEKE (1900) and LIPÓT FEJÉR (1911), however, opened up the possibilities to train highly qualified scholars.

In spite of the different quality of their teaching staff, all three universities shared the view that, in addition to their educational work, they should not neglect the cultivation and development of science. SÁNDOR IMRE, the rector of the university of Kolozsvár once said:

"... the university is *not a school* but a *scientific institute* and the two should not be confused." ([3], p. 55)

In the yearbook of the Budapest university of sciences one can read the following:

"The scientific activity of teachers is not confined to lectures and practical lessons they are obliged to give on their respective departments. University professors in particular must cultivate and, if possible, promote science, namely their own field of instruction ..." ([6], pp. 60—61).

The same thought is stressed in one of KÁROLY THAN's writings:

"... instruction should be developed and constantly enriched on the basis of independent scientific research as the core and most essential source of up-to-date university education and the only guarantee for the success of autonomous teaching." (*Az MTA Értesítője*, Vol. 5, 1871, p. 24)

The demand for scholarly activity at the universities resulted in a duality which for a long time divided the lecturers' work into two main parts. On the one hand, they had to present the prescribed material considered essential for the training of secondary school teachers in lectures attended by the students compulsorily unless exempted for their excellent progress. On the other hand, they had the imperative task of training scientists.

A new regulation of 1870 calling for the establishment of *training colleges for secondary school teachers* made the first goal attainable separately. Originally, the ordinance stipulated that for *grammar* school teachers the training college should be founded as part of the faculty of arts of the university of sciences while the college for training *polytechnical* school teachers should be under the aegis of the Technical University. However, the first three years sufficed to prove that this division was unnecessary and disadvantageous, so in 1873 the two colleges were merged.

The training of creative and autonomous *scholars* was going on at the universities separately from teacher training. In the late 19th century the professors of mathematics at the Budapest Technical University and the university of sciences in Kolozsvár especially distinguished themselves in this field. We could enumerate several special courses, each embracing the essentials of a mathematical discipline of current interest and often including significant results of the lecturers. There was hardly branch of mathematics at that time deserving the epithet "modern" that was not looked into at our universities. In Budapest, preference was given to linear algebra, function theory, number theory and from the early 20th century set theory, as well as some new chapters of geometry; in Kolozsvár the main interest lay in number theory, differential equations, function theory, vector algebra and analysis, quaternions, elliptic functions and BOLYAI geometry. The efficiency and high level of this activity are attested not only by a few monographs published and many litographed lecture notes but also by the ever increasing number of papers published in journals both in Hungary and abroad.

After the Compromise of 1867 the *Hungarian Academy of Sciences* did at last take its due place: it was given absolute powers in controlling our scientific life. Although among the ordinary and corresponding members of the Department of Mathematics and Physics (MÓR RÉTHY, GYULA KŐNIG, JENŐ HUNYADY, GUSZTÁV KONDOR, OTTÓ PETZVAL, GYULA VÁLYI, LAJOS MARTIN, KÁLMÁN SZILY, LORÁND EÖTVÖS, IZIDOR FRÖHLICH, GUSZTÁV RADOS, GYULA FARKAS, JÓZSEF KÜRSCHÁK, BÉLA TŐTÖSSY) some were still coopted for their advanced age rather than for their scientific achievements, the majority were indeed scholars of world-wide fame and deserved their membership. Also the level of the meetings and the number of mathematical papers presented there increased. As to the publications of the Academy, *Értekezések a Mathematikai Tudományok Köréből* is worthy of special attention. In this series appearing from 1867 to 1894 treatises by the academicians were published in volumes comprising the material of 2—3 years (15 volumes altogether). Most of the studies in maths were written by JENŐ HUNYADY, GYULA KŐNIG and GYULA FARKAS. The *Mathematikai és Természettudományi Értesitő* (1882—1941), concerned with both mathematics and natural science, developed into a journal even of higher standard.

In both the *Értekezések* and *Math. . . . Értesitő*, the articles were written in Hungarian, not even the titles or abstracts were given in any foreign language at the beginning. Of course, due to the isolation of the Hungarian language the scientific community abroad rarely got acquainted with the new results published in them. Not only the Hungarian language but also the spelling of Hungarian names meant serious problems to foreigners.[3] Being aware of this, the academicians decided in 1883 to work out the plan of a foreign language journal which

> "shall publish all that is happening in the area of mathematics and natural science in our country; ... the Academy will be provided with as many copies of the journal as needed for the partner institutes in exchange relations with the Academy and for the foreign associates of Department III." (*Az MTA Értesítője*, 1883, Vol. 17, pp. 31—32)

Thus in 1883 the German language edition of the Review of Mathematics and Natural Science — *Mathematische und naturwissenschaftliche Berichte aus Ungarn* — appeared it ceased in 1932). Although it was not a purely mathematical journal, it played an invaluable role in our mathematical culture. It moved Hungarian mathematics out of isolation; foreign books, periodicals and review articles more and more often quoted results appearing in the "*Ungar. Berichte*".[4] Besides, in the late 19th century the number of articles by Hungarian mathematicians published in foreign journals kept increasing, which provide the rapid development of scientific research in Hungary. Below, we consider some foots concerning the last decades of the previous century in more detail.

Before the turn of the century, 20 papers were published by Hungarian mathematicians in the *Comptes Rendus* of the French Academy. GYULA FARKAS wrote a lot about the theory of sines of higher order and LAJOS SCHLESINGER published several notes on differential equations. However, the most outstanding paper was LIPÓT FEJÉR's short study in 1900 of the summation of FOURIER series. In the *Mathematische Annalen* at least 22 Hungarian writings can be found, including papers by GYULA KŐNIG on algebra and analysis, by JÓZSEF KÜRSCHÁK on the calculus of variations, and by MANÓ BEKE on differential equations. In the *Archiv für Mathematik und Physik* some 20 papers were published, most of them by GYULA VÁLYI, and some by GYULA FARKAS and LIPÓT KLUG. The *Journal für die reine und angewandte Mathematik* carried about 25 Hungarian papers, chiefly by JENŐ HUNYADY and LAJOS SCHLESINGER.

The Viennese *Monatshefte für Mathematik und Physik*, started in 1890, soon became popular with our mathematicians. Some 10 Hungarian treatises can be found in it written by GYULA KŐNIG, GYULA VÁLYI and LIPÓT KLUG.

Several Hungarian studies were published in the *Nouvelles Annales des Mathématiques*, the *Göttinger Nachrichten* and the *Acta Mathematica*.

Altogether some 120 Hungarian studies in mathematics were published in foreign journals in the last three decades of the previous century.

The launching of the first *reference* journal, *Fortschritte der Mathematik* in 1868 was of great importance for both international and Hungarian mathematical life. By

---

[3] Some examples of funny misspelling: "Vásáhely Maros" (=FARKAS BOLYAI), *Enz. d. Math. Wiss.*, III, 2.2. B, p. 2326; "Gyergioszentmiklós" D. D." (=MIHÁLY DEMECZKY), *Fortschr. d. Math.*, 11, 1879, pp. 128, 835; "Tejer L." (=LIPÓT FEJÉR), *Fortschr. d. Math.*, 31, 1900, pp. 400, 963.

[4] For a detailed history of the periodicals mentioned, see [10].

reviewing scientific results in ever thicker volumes of rising quality year by year, it greatly contributed to fostering international relations and offered a chance to specialists to cooperate and decide questions of priority. On the basis of the review articles in *Fortschritte* it can be monitored almost with mathematical precision how much the mathematical research improved in Hungary both in quality and quantity.

In such circumstances, the fate of the Hungarian mathematical results published toward the end of the last century can be summarized as follows:

1. Some were published only in Hungarian, most of them remaining unnoticed on the international stage and falling into oblivion.

2. The majority of the studies contained methodically new and didactically useful investigations. In some reviews they still bore the names of their authors, but later they were incorporated in lectures or monographs (usually without names; e.g. several results of KŐNIG on differential equations, VÁLYI's examinations of multiple perspectivity, the root approximation method of FARKAS BOLYAI, etc.).

3. There was, however, a favourable increase in the number of Hungarian results which luckily became inseparable from the names of their discoverers and still arouse the world's reverence for the creative spirit of our scientists (e.g. FARKAS BOLYAI's end-like equality of areas, the HUNYADY—SCHOLTZ theorem, KŐNIG's inequality, the KŐNIG—RADOS theorem, several fundamental results of LIPÓT FEJÉR, FRIGYES RIESZ, etc.).

After the Compromise of 1867 the foreign members of the Academy, some of whom played a very important role in Hungarian research, were selected more carefully. Foreign members coopted between 1873 and 1902 included: JÓZSEF PETZVAL, CAYLEY, HERMITE, HELMHOLTZ, KRONECKER, DU BOIS-REYMOND, LJUBOMIR KLERIČ, LAZARUS FUCHS, FELIX KLEIN, PAUL STÄCKEL, DARBOUX, MITTAG-LEFFLER — almost each name can be connected with one of the mathematical disciplines in the focus of interest in Hungary at that time.

Towards the end of the last century the idea of organizing a society to rally the mathematicians came up. Upon the initiative of LORÁND EÖTVÖS, JENŐ HUNYADY, GYULA KŐNIG, ÁGOSTON SCHOLTZ and KÁLMÁN SZILY, the mathematicians of Budapest founded a kind of private society in 1885. The meetings of the *Mathematical Society* (Mathematikai Társaság) had a programme full of educational and popularizing lectures.

"The Society had no presidents or statutes and their meetings often looked like dinner parties but for the solemn blackboard." ([8], p. 8)

Some years later the popularity of these meetings encouraged the scholars to involve physicists in the work and publish the material of the meetings in a periodical. The idea was soon realized and in June 1891 the first issue of the *Mathematikai és Physikai Lapok* came out in the edition of GÉZA BARTONIEK and GUSZTÁV RADOS. In early 1944, World War II put a stop to the publication, chiefly for want of articles.

After thorough preparations, on 5 November 1891 the predecessor of today's *János Bolyai Mathematical Society*, the *Mathematical and Physical Society* began operation upon its own statutes.[5] It was the first association to gather a large number of Hungarian mathematicians. As GYULA KŐNIG proposed, "... not everybody but all who work and teach in the fields of mathematics and physics can be chosen members". The actual demand for such a society is proven by the fact that it had 298 members right at the start and this figure rose to some 400 in only two years' time. And out of the 560 copies of the first issue of the *Mathematikai és Physikai Lapok* as many as 405 were subscribed to.

Besides the events influencing the mathematical culture of the entire country, the outstanding results of some Hungarian scholars also helped the country to gradually catch up with European standards. In the following chapters their work is reviewed.

Bibliography

[1] *A Magyar Tudományos Akadémia (MTA) Értesítője*, 1867—1889, Vols 1—23. (In Hung.)

[2] *Almanacs of the Royal University of Science of Budapest.* (In Hung.)

[3] *A history and statistics of the Royal Francis Joseph University of Science of Kolozsvár.* Kolozsvár, 1896. (In Hung.)

[4] *Almanacs of the Royal University of Science of Kolozsvár.* (In Hung.)

[5] *Programmes of the Royal Joseph Technical University.* (In Hung.)

[6] *Report on the state, work and progress of the Royal Technical University of Budapest between 1867/8 and 1874/5.* (In Hung.)

[7] FARAGÓ, LÁSZLÓ: Mathematical curricula of Hungarian secondary schools up to the turn of the century. In: *Studies in the History of Education in Hungary*, Budapest, 1957, pp. 83—116. (In Hung.)

[8] KŐNIG, DÉNES: The first 50 years of the Loránd Eötvös Mathematical and Physical Society. *Math. és Phys. Lapok*, 48, 1941, pp. 7—33. (In Hung.)

[9] OBLÁTH, RICHÁRD: On the past of the Mathematical Journal for Secondary Schools and of the mathematical competitions. *Középiskolai Matematikai Lapok*, Vol. 2, 1949, pp. 3—7. (In Hung.)

[10] SZÉNÁSSY, BARNA: Hungarian mathematical periodicals. *Matematikai Lapok*, Vol. 3, 1952, pp. 273—285. (In Hung.)

[11] — 75 years of our Society. *Matematikai Lapok*, Vol. 17, 1966, pp. 195—308. (In Hung.)

---

[5] For its history, see [11].

19. Jenő Hunyady

The majority of his essays numbering over fifty highlight geometrical subjects.[1] He had a keen eye for noticing the gaps in the geometric results multiplying rapidly as a result of the work of PONCELET, CHASLES, CAYLEY, MÖBIUS, HESSE, PLÜCKER, STEINER, STAUDT, CLEBSCH and others, for pinpointing the places where proofs could be simplified, various theorems reduced to common roots and appropriate additions, generalizations or more lucid proofs could be given. No pioneering results of theoretical significance can be associated with his name but he obtained many a minor theorem and found innumerable, methodically simpler and more elegant proofs to the results of others. All this notwithstanding, his role is noteworthy not only in the Hungarian, but also in the world history of mathematics.

In examining geometrical questions, HUNYADY almost exclusively used algebraic tools. He was the most consistent advocate of this trend which took off in the last century. He summarized his aim and method very clearly:

"... considering that analytic geometry is principally a geometrical discipline, it is the author's modest view that we have to make all efforts not to surrender the main role to algebra and analysis, confusing the means with the end, but to assign the main target — geometry — its due significance. This position is further buttressed by the argument that should we approach analytic geometry from the opposite, we might easily commit the error of reducing the analytic teaching of geometry to a compendium of exercises in algebra and analysis, which would certainly contradict the spirit of science. There is no denying, however, that proceeding in the opposite direction we are often forced to generalize the geometrical problems, and by solving these abstract problems, we obtain new viewpoints for the treatment of the very problems of geometry that have actually prompted us to launch these generalizing investigations." ([4], p. 5)

Of all the arsenal of algebra, HUNYADY chiefly used the determinants in his geometrical inquiries, and in his hands this tool became sophisticated and subtle. The

---

[1] For a complete bibliography, see the end of the study on pp. 175—202 of Volume 3 of the series *Műszaki nagyjaink* (Great engineers of Hungary). Budapest, 1968. (In Hung.)

sieve of time has gradually eliminated his less significant results, leaving some which nearly all belong to the theory of *determinants*.

Below we attempt to present his results that had stirred international repercussion, their impact, and the method of research HUNYADY applied.

*

The fact that six points are on some sort of conic section can be described through theorems formulated by various researchers including CHASLES, PASCAL, CARNOT, PAPPUS and DESARGUES. The theorem of CHASLES, for example, states that if we join two points of a conic section with four other points of it by straight lines, then the cross ratios of the four rays meeting in the first and the second point are equal. PASCAL's theorem says that the points of intersection of the opposite sides of a hexagon inscribed in a conic section are collinear. According to PAPPUS, dropping perpendiculars to the four sides of a quadrilateral inscribed in a conic section from any of its points, the ratio of the products of the pairs of segments perpendicular to opposite sides is constant. If we fix the inscribed quadrilateral and apply PAPPUS's theorem to two different points of the conic, we again arrive at the relation for six points of a conic.

In view of these theorems, several equations can be set up for the six points of a conic section. To formulate PASCAL's theorem, for instance, let us introduce the following notation: let 1, 2, 3, 4, 5 and 6 denote six points in a plane *(Figure 53)* and let x_i, y_i and z_i denote the homogeneous coordinates of the i-th point. Then the equation of the straight line connecting the points i and k is

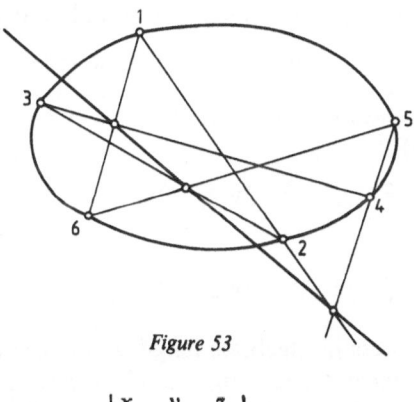

Figure 53

$$\begin{vmatrix} x & y & z \\ x_i & y_i & z_i \\ x_k & y_k & z_k \end{vmatrix} = 0.$$

The opposite sides of the hexagon defined by the given six points are

12 and 45,
34 and 61,
56 and 23.

225

Also, for brevity, we shall use the notation

$$y_i z_k - z_i y_k \equiv X_{ik},$$
$$x_k z_i - x_i z_k \equiv Y_{ik}, \tag{1}$$
$$x_i y_k - y_i x_k \equiv Z_{ik}.$$

With this notation, the condition that the points of intersection of the pairs of opposite sides are collinear is expressed by the following equation:

$$\begin{vmatrix} Y_{12}Z_{45} - Y_{45}Z_{12} & Z_{12}X_{45} - Z_{45}X_{12} & X_{12}Y_{45} - X_{45}Y_{12} \\ Y_{34}Z_{61} - Y_{61}Z_{34} & Z_{34}X_{61} - Z_{61}X_{34} & X_{34}Y_{61} - X_{61}Y_{34} \\ Y_{56}Z_{23} - Y_{23}Z_{56} & Z_{56}X_{23} - Z_{23}X_{56} & X_{56}Y_{23} - X_{23}Y_{56} \end{vmatrix} = 0. \tag{2}$$

If the opposite sides of the hexagon do not meet on an axis, then PASCAL's theorem can reversed. Therefore (2) also expresses the condition for the vertices of a hexagon to lie on a conic (if the opposite sides of a hexagon meet on an axis, then the vertices lie on either a conic section or two lines three by three).

The elements of the determinant under (2) are themselves determinants, and, in view of (1) even they are compound. A considerable part of the geometrical problems studied by HUNYADY led to compound determinants. MUIR, author of a history of determinants, noted ([15], Vol. 2, p. 202) that at least a quarter of the theorems pertaining to compound determinants were due to Hungarian mathematicians (mostly to HUNYADY, and also to ÁGOSTON SCHOLTZ, a scholar of a similar frame of mind — we may add).

Let the conic section be given by an equation in homogeneous coordinates:

$$a_{11}x^2 + a_{22}y^2 + a_{33}z^2 + 2a_{23}yz + 2a_{13}xz + 2a_{12}xy = 0$$

If this conic section passes through points 1, 2, 3, 4, 5, and 6, then the following condition must be fulfilled:

$$\begin{vmatrix} x_1^2 & y_1^2 & z_1^2 & 2y_1z_1 & 2x_1z_1 & 2x_1y_1 \\ \vdots & & & & & \vdots \\ x_6^2 & y_6^2 & z_6^2 & 2y_6z_6 & 2x_6z_6 & 2x_6y_6 \end{vmatrix} = 0 \tag{3}$$

As both (2) and (3) express in algebraic language the geometric fact that six points satisfy the equation of the conic section, it can be expected that via some witty computation they can be transformed into each other [2, 3].[2] An intermediary stage of

---

[2] HUNYADY published this statement in 1875 with the remark that "the mentioned transformations have not been fully successful yet". Joining HUNYADY's investigations, ÁGOSTON SCHOLTZ also tackled the question and, starting from determinant (3) as well as PASCAL's and CHASLES's theorems, he was the first to obtain — as a by-produce of his calculations — the final form of the determinant theorem mentioned in the sequel [17, 18]. Yet HUNYADY's priority is evident, so we have to speak of the HUNYADY—SCHOLTZ theorem rather than the SCHOLTZ theorem in contrast to the usage of some foreign mathematicians (e.g. PASCAL [16]), who know the historical backgroundless.

this transformation is the following identity:

$$
\begin{vmatrix} x_1 & y_1 & z_1 \\ x_2 & y_2 & z_2 \\ x_3 & y_3 & z_3 \end{vmatrix}^4 =
$$

$$
= \begin{vmatrix} x_1^2 & y_1^2 & z_1^2 & 2x_1y_1 & 2x_1z_1 & 2x_1y_1 \\ x_2^2 & y_2^2 & z_2^2 & 2x_2y_2 & 2x_2z_2 & 2x_2y_2 \\ x_3^2 & y_3^2 & z_3^2 & 2x_3y_3 & 2x_3z_3 & 2x_3y_3 \\ x_2x_3 & y_2y_3 & z_2z_3 & y_2z_3+z_2y_3 & x_2z_3+z_2x_3 & x_2y_3+y_2x_3 \\ x_1x_3 & y_1y_3 & z_1z_3 & y_1z_3+z_1y_3 & x_1z_3+z_1x_3 & x_1y_3+y_1x_3 \\ x_1x_2 & y_1y_2 & z_1z_2 & y_1z_2+z_1y_2 & x_1z_2+z_1x_2 & x_1y_2+y_1x_2 \end{vmatrix}
$$

With the method suggested by this identity, the determinant of order $\dfrac{n(n+1)}{2}$ can be obtained from any determinant of the n-th order. If the value of the original determinant is D and that of the higher-order one obtained from it in the above way is H, then according to the HUNYADY—SCHOLTZ *theorem*

$$H = D^{n+1}.$$

In spite of the ostensible complexity of generating the higher-order determinant, the HUNYADY—SCHOLTZ theorem proved most useful in a variety of fields. Several mathematicians have applied it, for instance, in research into conic sections and second-order surfaces in the cases $n=2$ and $n=3$.[3] Many scholars can be listed (e.g. MERTENS, PASCH, CASPARY, MÜLLER, SZABÓ and others) who put it to good use in other questions of geometry (for some of them, see: [16], pp. 103—104). Its usefulness in technical problems is else well known: let us refer to JENŐ EGERVÁRY's paper[4] in which he studies grid structures with three bars constituting a triangle or six bars constituting a tetrahedron, the structure being loaded by a system of forces in equilibrium. Several scholars (IGEL, ESCHERICH, CASPARY, PASCAL, CAZZANIGA, MUIR, HAJÓS and others[5]) dealt with generalizations of the theorem in various directions. In this connection, let us mention GYÖRGY HAJÓS's result which states that one can form a higher-order determinant from a basic one via a general procedure in such a way that the identity ob-

---

[3] HUNYADY himself used this determinant theorem to obtain, in continuation of SERRET's results, criteria [4] for 10 points to lie on a surface of the second order.

[4] Application of the HUNYADY–SCHOLTZ matrices in the theory of grid structures. *Publications of the Institute of Applied Maths. of the Hung. Acad. of Sci.*, Vol. 3, 1954, pp. 289—300. (In Hung.)

[5] Major studies with relevance here: IGEL: Zur Theorie der Determinanten. *Monatsh. f. Math.*, Vol. 3, 1892, pp. 55—67; ESCHERICH: Bestimmung einer Determinante. *Ibid.*, pp. 68—80; CAZZANIGA: Qualche complemente al teorema di Hunyady su certi determinanti. *Periodico di Mat.*, 1900, pp. 17—22; MERTENS: Sätze über Determinanten. *Journal f. reine u. angew. Math.*, Vol. 84, pp. 335—359; WHITE: Two elementary geometrical applications of determinants. *Ann. of Math.*, Vol. 1, 1900, pp. 103—107; HAJÓS, GY.: A determinant theorem. *Mat. és Term. tud. Értesítő*, Vol. 50, 1934, pp. 231—240. (In Hung.) See also: [15], Vol. 3, 11, 27, 45, 72, 222.

tained will include as specific cases not only the HUNYADY—SCHOLTZ, but also the RADOS, KRONECKER—RADOS and SCHLÄFLI—RADOS theorems.[6] Soon, the method HUNYADY suggested to decide when six points are on a conic became widely known in international literature from BALTZER's oft-published book ([1], p. 286), while the HUNYADY—SCHOLTZ theorem gained fame through E. PASCAL's work ([16], pp. 104—106).

The notion of *duplicant determinant* also originates from JENŐ HUNYADY. It has also proved useful in diverse investigations.

If A is a square matrix of order n $(n \geq 2)$ and A' is its transpose, then — in HUNYADY's terminology — the determinant

$$D = |A \pm A'|$$

is the duplicant determinant of $|A|$.

If Δ denotes the adjoint of D (with A being regular), then according to HUNYADY's attractive theorem

$$\Delta = |A|^{n-2} D. \tag{4}$$

In 1882 HUNYADY published (4) as a problem to prove[7] but the problem and with it the notion of the duplicant determinant remained unnoticed for a long time until in 1905 an anonymous mathematician proved (4), confining himself to determinants of the 3rd order.[8] This short note directed attention again to the question of duplicant determinants and, for example, THOMAS MUIR listed a long row of new theorems pertinent to them.

MUIR: The determinant of the sum of a square matrix and its conjugats. *Mess. of Math.*, Vol. 43, 1914, pp. 184—192. More recently BÉLA GYIRES has given a far-reaching generalization of HUNYADY's theorem (Verallgemeinerung eines Determinantensatzes von J. Hunyady. *Publ. Math.*, Vol. 2, 1952, pp. 290—292): if $C_k(A)$ $(1 \leq k \leq n-1)$ is the k-th derived matrix of an arbitrary square matrix A of order n, and A' is the transpose of A, then

$$|C_k(A)||C_{n-k}(A) \pm C_{n-k}(A')| = |C_{n-k}(A)||C_k(A) \pm C_k(A')|. \tag{5}$$

As in view of the well-known FRANKE theorem we have

$$|C_k(A)| = |A|^{\binom{n-1}{k-1}}; \quad |C_{n-k}(A)| = |A|^{\binom{n-1}{n-k-1}},$$

therefore, (if A is regular) it follows from (5) that

$$|C_{n-k}(A) \pm C_{n-k}(A')| = |A|^{\binom{n-1}{k}-\binom{n-1}{k-1}}|C_k(A) \pm C_k(A')|, \tag{6}$$

which is a *direct* generalization of HUNYADY's theorem.

For the detailed exposition of HUNYADY's theorems on determinants a separate monograph would be needed. By way of conclusion, let us mention one of his results

[6] In addition to the above-said, let us note that CAYLEY (1843) was the first to prove PASCAL's theorem with the help of determinants. Results similar to the HUNYADY—SCHOLTZ theorem were obtained by SCHLÄFLI in 1851 and BRILL in 1871.

[7] *Nouv. Annales de Math.*, Vol. 3, 1882, p. 384. Cf. also: [15], Vol. 2, pp. 209—210.

[8] *Nouv. Annales de Math.*, 1905, pp. 568—570.

which has proved useful in solving problems of interpolation.[9] According to HUNYADY ([5] and [15], Vol. 1, p. 104),

$$
\begin{vmatrix}
\cos\dfrac{1\cdot 1\cdot \pi}{n+1} & \cos\dfrac{2\cdot 1\cdot \pi}{n+1} & \ldots & \cos\dfrac{n\cdot 1\cdot \pi}{n+1} \\
\vdots & & & \vdots \\
\cos\dfrac{1\cdot n\cdot \pi}{n+1} & \cos\dfrac{2\cdot n\cdot \pi}{n+1} & \ldots & \cos\dfrac{n\cdot n\cdot \pi}{n+1}
\end{vmatrix}^{2}
=
$$

$$
=
\begin{cases}
\dfrac{1}{4}\left(\dfrac{n+1}{2}\right)^{n-2} & \text{if } n \text{ is even} \\[2em]
0 & \text{if } n \text{ is odd}
\end{cases}
$$

For the case of substituting sines for the cosines in this determinant MUIR[10] as well as ALLER and DATTA[11] carried out computations.

*

One more fact requires mention when we speak of the work of JENŐ HUNYADY: all his treatises reveal that he had a through knowledge of the history of the problems, he studied, and also, that he had a special affinity to the history of science. A fine example of the latter is his long commemorative speech over his great ideal PONCELET [13], an honorary member of the Hungarian Academy of Sciences. There was wide international response also to this essay of his.

It was HUNYADY's indelible merit that he directed attention to unsettled problems thus suggesting possible directions of further investigation. That is why his lectures and treatises were stimulating, why his activities influenced a lot of colleagues and students. It is largely to his credit that linear algebra has now time-tested traditions in Hungary with promising results achieved as early as the turn of the century by DÁNIEL ARANY, MANÓ BEKE, GYULA FARKAS, GYULA KŐNIG, GYULA VÁLYI, JÓZSEF KÜRSCHÁK,[12] GUSZTÁV RADOS, PÉTER SZABÓ, MIHÁLY BAUER, IZIDOR FRÖHLICH and several others.

*

[9] The problem itself was encountered in connection with the following interpolation problem. In the function $y = A_1 \cos x + A_2 \cos 2x + \ldots + A_n \cos nx$, determine the coefficients A_i in such a way that in points $x = \dfrac{\pi}{n+1}, \dfrac{2\pi}{n+1}, \ldots, \dfrac{n\pi}{n+1}$ the value of the function be $y_1, y_2, \ldots, y_n$, respectively. The problem was set by BROCARD (Nouv. Ann. Math., Vol. 11, pp. 39—41).

[10] MUIR, TH.: On a symmetric determinant connected with LAGRANGE's interpolation-problem. Proc. London Math. Soc., Vol. 13, 1882, pp. 156—161.

[11] ALLER, HANS H.—DATTA, B.: Vraagstukken 35, 36. Wiskundige Opgaven, Vol. 12, 1916, pp. 77—87; On symmetric determinants and Pfaffian. Proc. Edinburgh Math. Soc., Vol. 34, 1916, pp. 197—204.

[12] JÓZSEF KÜRSCHÁK was also concerned with the generalization of one of HUNYADY's theorems not mentioned here: On a determinant theorem of JENŐ HUNYADY. Math. és Phys. Lapok, Vol. 4, 1895, pp. 1—6. (In Hung.).

Disregarding here the method applied by HUNYADY, let us make a brief survey of some of his useful results in geometry. This sample may suffice to give the reader an inkling of the variety of problems that absorbed his attention.

Given four coplanar points A, B, C and O, the first three regarded as the points of a conic section and the fourth as its centre, the position of these points determines the type of the conic section (unless A, B and C include two points that are symmetrical to O in which case the data are insufficient). Earlier, STEINER had proposed some complicated criteria — without proof — for determining the type. HUNYADY found the following.

First we have to determine the bisecting points E, F and G of the sides of triangle ABC. If the points E, F, G and O "shut out each other", then the conic section is a hyperbola *(Figure 54)*, otherwise it is an ellipse [6, 7].

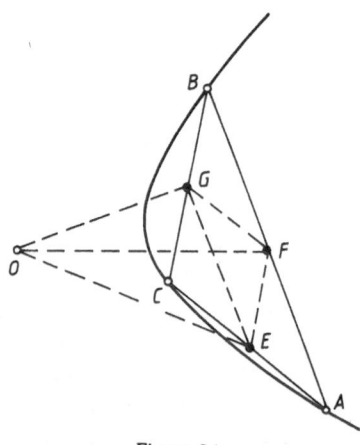

Figure 54

"Shutting out each other" means that points E, F, G and O define a convex quadrangle.

It was MÖBIUS who raised the question and partially solved it, as to when a conic section going through five given points (or: touching five given lines) is an ellipse and when a hyperbola. Here again, HUNYADY found criteria by a skilful application of determinants [8, 9].

The criteria found by HUNYADY are also noteworthy because GUSZTÁV RADOS adopted his method to find criteria by deciding the type of a second-order surface given by 9 points.[13]

It also deserves mention that HUNYADY was the first to obtain with analytic tools a complete solution of the following generalization of APOLLONIUS's problem: given the radii and centres of three circles on a sphere, to find the radius of a fourth circle tangential to the former three [10, 11].

---

[13] GUSZTÁV RADOS: On the criteria of Möbius. *Math. és Phys. Lapok*, Vol. 1, 1892, p. 113. A thorough analysis of the MÖBIUS—HUNYADY criteria was carried out by H. DURÈGE in his paper: Über die Möbius gegebenen Kriterien ... (*Wien Ber.*, Vol. 82, 1880), where some minor mistakes made by Hunyady, were also corrected.

HUNYADY's skills at systematization and his sophisticated technique of computing are most conspicuous in his great inaugural speech at the Academy published post-humously [12].

It is well known that the linear transformation

$$x' = a_{11}x_1 + a_{12}x_2 + a_{13}x_3$$

$$y' = a_{21}x_1 + a_{22}x_2 + a_{23}x_3$$

$$z' = a_{31}x_1 + a_{32}x_2 + a_{33}x_3$$

is a so-called orthogonal transformation if its coefficients satisfy the following six equations:

$$a_{k1}^2 + a_{k2}^2 + a_{k3}^2 = 1 \quad \text{and} \quad a_{k1}a_{l1} + a_{k2}a_{l2} + a_{k3}a_{l3} = 0,$$

where

$$k \neq l, \quad \text{and} \quad k = 1, 2, 3; \quad l = 1, 2, 3.$$

It is easy to prove that with the help of independent homogeneous parameters ω_1, ω_2, ω_3, ω_4 — i.e. with three independent inhomogeneous parameters — all orthogonal transformations can be represented in the following way proposed by EULER:

$$a_{11} = \rho(\omega_1^2 + \omega_2^2 - \omega_3^2 - \omega_4^2) \qquad a_{12} = \rho(2\omega_1\omega_4 + 2\omega_2\omega_3)$$

$$a_{21} = \rho(-2\omega_1\omega_4 + 2\omega_2\omega_3) \qquad a_{22} = \rho(\omega_1^2 - \omega_2^2 + \omega_3^2 - \omega_4^2)$$

$$a_{31} = \rho(2\omega_1\omega_3 + 2\omega_2\omega_4) \qquad a_{32} = \rho(-2\omega_1\omega_2 + 2\omega_3\omega_4)$$

$$a_{13} = \rho(-2\omega_1\omega_3 + 2\omega_2\omega_4)$$

$$a_{23} = \rho(2\omega_1\omega_2 + 2\omega_3\omega_4)$$

$$a_{33} = \rho(\omega_1^2 - \omega_2^2 - \omega_3^2 + \omega_4^2)$$

where

$$\rho = (\omega_1^2 + \omega_2^2 + \omega_3^2 + \omega_4^2)^{-1}.$$

Obviously, if sufficiently many data are available, the rest can be computed from the above equations. The problem, however, ramifies in different directions depending on which data are given. Partial computations were performed in this connection by many authors (EULER, MONGE, RODRIGUES, TAYLOR, HESSE and others) but HUNYADY was the one to devise an essentially complete theory of the problem by adding several results of his own.

Bibliography

[1] BALTZER, RICHARD: *Analytische Geometrie.* Leipzig, 1882.
[2] HUNYADY, JENŐ: On various forms of the condition for six points to be on a conic section. *Értekezések a Math. Tud. Köréből*, Vol. 4, 1875—1876, No. 6; Vol. 5, 1877—1878, No. 4. (In Hung.)
[3] — Über die verschiedenen Formen der Bedingungsgleichung, welche ausdrückt, dass sechs Punkte auf einem Kegelschnitte liegen. *Journal f. reine u. angew. Math.*, Vol. 84, 1877, pp. 76—85. Also: Zusatz zu meiner Abhandlung in Borchart's Journal LXXXIII, p. 76. *Ibid.*, Vol. 92, 1882, pp. 307—311.

[4] — On the determination of curves and surfaces of the second degree. (In Hung.) *Értekezések a Math. Tud. Köréből*, Vol. 7, 1879—1880. No. 18. (In Hung.) Also: Beitrag zur Theorie der Flächen zweiten Grades. *Journal f. reine u. angew. Math.*, Vol. 89, 1880, pp. 47—70.

[5] — Solution de la question 979. *Nouv. Ann. Math.*, Vol. 11, 1872, pp. 39—44.

[6] — On Steiner's criterion in the theory of conic sections. *Értekezések a Math. Tud. Köréből*, Vol. 7, 1879—1880, No. 24. (In Hung.)

[7] — Über ein Kriterium von Steiner in der Theorie der Kegelschnitte. *Journal f. reine u. angew. Math.*, Vol. 91, 1881, pp. 248—253.

[8] — On Möbius's criteria in the theory of conic sections. *Értekezések a Math. Tud. Köréből.* Vol. 7, 1880, No. 6. (In Hung.)

[9] — Über die von Möbius gegebenen Kriterien in der Theorie der Kegelschnitte. *Journal f. reine u. angew. Math.*, Vol. 89, 1880, pp. 70—79.

[10] — The problem of Apollonius on the sphere. *Értekezések a Math. Tud. Köréből.* Vol. 5, 1877—1878, No. 5. (In Hung.)

[11] — To the solution of the problem of Apollonius. *Műegyetemi Lapok*, Vol. 2, 1877, pp. 97—116. (In Hung.)

[12] — The parametric values of the coefficients of the orthogonal substitution. *Értekezések a Math. Tud. Köréből.* Vol. 14, 1889, Nos 2, 3. (In Hung.)

[13] — In memory of Jean Victor Poncelet. *Értekezések a Math. Tud. Köréből.* 1878, No. 7. (In Hung.) In German: *Literarische Berichte aus Ungarn.* Vol. 1, 1877, pp. 458—469.

[14] KŐNIG, GYULA: In memory of Jenő Hunyady. *Akadémiai Értesítő*, Vol. 2, 1891, pp. 1—9. (In Hung.)

[15] MUIR, THOMAS: *The theory of determinants in the historical order of development.* Vols 1—5, London, 1906—1930.

[16] PASCAL, E.: *Die Determinanten.* Leipzig, 1900.

[17] SCHOLTZ, ÁGOSTON: The theorem of six points on a conic section and the mysthic hexagram. *Műegyetemi Lapok*, Vol. 2, 1877, pp. 65—77. (In Hung.)

[18] — Sechs Punkte eines Kegelschnittes. *Arch. d. Math.*, Vol. 62, 1878, pp. 317—324.

20. Gyula Kőnig

His extremely varied interests kept spurring him on to explore the new, the original, the "very latest"; the second volume of several monographs he planned for two volumes fell victim to his restless mind. His scientific career can be divided into periods marked off by exact dates: each period contains some essays on a problem and a vast synthesizing work at the end.

The following discussion will touch on several of his results, theorems or fascinating proofs representing every area of his wide range of interests, yet a brief chapter like this can all but vaguely hint at the significance of his life-work.[1]

Results in set theory and mathematical logic

Some surprising results of set theory incompatible with former achievements triggered off fierce debates at the beginning and aroused scepticism in some scholars as to the value of the new discipline. GYULA KŐNIG was among the few who realized the significance of set theory and largely contributed to its consolidation.

His first result in this field — never published — concerned the power of the continuum. GEORG CANTOR[2] was the first to prove that the cardinalities of the one-dimensional and the two-dimensional continuum are equal. Other proofs were quick to follow (NETTO, PRINGSHEIM, LÜROTH, F. BERNSTEIN, JÜRGENS, THOMAE, PEANO, HILBERT, SCHOENFLIES). The essence of all these proofs is that there is a one-to-one correspondence between the points of the unit square and the points of the unit segment.

Let the points of the unit square be denoted by the ordered pairs (x, y) $(0 < x \leq 1)$; $(0 < y \leq 1)$ and those of the unit segment by $z (0 < z \leq 1)$. The first proofs of CANTOR's theorem usually expressed x and y by continued fractions and used them to produce z corresponding to (x, y) on the basis of certain prescriptions.

The employment of continued fractions suggested that the values of x, y and the corresponding z should be represented as decimal fractions, while imposing the na-

---

[1] Those interested in the details will find the book [38] illuminating. It also contains a full list of KŐNIG's treatises.

[2] CANTOR, G.: Ein Beitrag zur Mannigfaltigkeitslehre. *Journal f. reine u. angew. Math.*, 84, 1878, pp. 242—258.

tural condition that different decimal fractions should define different points and vice versa. But we immediately run into trouble here: though 0.5 and 0.49999 ... are different decimal fractions, they are represented by one and the same point of the real line. This barrier is easy to overcome: we agree that whenever we can describe a single point in two ways, with a *finite* and an *infinite* decimal fraction, we always choose the infinite one, precluding the finite decimal fractions from further considerations. Thus, e.g., the point that can also be written as 0.5 will be represented by 0.49999 .... In this way we can ensure that the fractions

$$x = 0, a_1 a_2 a_3 \ldots, \qquad y = 0, b_1 b_2 b_3 \ldots.$$

contain an infinite number of digits different from zero.

Let us now produce z from the digits of x and y in the following way:

$$z = 0, a_1 b_1 a_2 b_2 a_3 b_3 \ldots$$

This representation, however, fails to give us all the possible values of z including, e.g.

$$z = 0, c_1 c_2 0 c_3 0 c_4 \ldots.$$

This would require that x have the form $0, c_1 000 \ldots$ but under the above condition such x does not exist as the fraction does not contain an infinite number of non-zero digits.

By a witty idea of Kőnig, the Cantor theorem can still be proved with decimal fractions. All we need to do is to base the correspondence on certain collections of digits rather than — single digits in the values of x and y given by non-terminating decimals — on "molecules" rather than "atoms" as Felix Klein put it[3] — appearing in the representation. A "molecule" consists of a non-zero digit and the zeros immediately preceding it (if there are any). A collection of digits formed in this way is denoted by a letter in square brackets, as e.g. in

$$x = 0, [a_1][a_2][a_3] \ldots, \qquad y = 0, [b_1][b_2][b_3] \ldots.$$

Now, setting

$$z = 0, [a_1][b_1][a_2][b_2] \ldots,$$

it is easy to see that the proof is complete: we have succeeded in showing that to every point of the unit square there corresponds one, and only one, point of the unit segment, and vice versa.[4]

While this proof sprouted from a clever idea, in his proof of the *equivalence theorem* of set theory [2, 3] Kőnig insisted on resorting to "pure intuition", to use his words, with this intention, he neglected several tools (the concept of number, complete induc-

[3] Klein, F.: *Elementarmathematik vom höheren Standpunkte aus.* Vol. 1, Berlin, 1933, pp. 278—279.

[4] Kőnig presented this proof to a meeting of the Mathematical and Physical Society (*Math. és Phys. Lapok*, 3, 1894, p. 394) but never published it. However, it soon became known through oral communication and Schoenflies described it in his famous Festschrift (Die Entwicklung der Lehre von Punktmannigfaltigkeiten. *Jahresb. d. deutsch. Math. Ver.*, 8, 1900, p. 23). The proof was further popularized by writings of F. Klein, A. A. Fraenkel, J. Kürschák, L. Kalmár, and others.

tion) employed by others (CANTOR, BERNSTEIN, SCHRÖDER, ZERMELO, BANACH) in proving the theorem. KŐNIG's rather abstract reasoning was made more palpable and easier to teach by the graph-theoretic interpretation given by DÉNES KŐNIG and more recently by PÁL SZÁSZ.[5]

Perhaps the most significant station of GYULA KŐNIG's professional career was the international mathematical congress of 1904 in Heidelberg: the paper he read to the congress brought him both the greatest success and the bitterest disappointment. He wished to greet the little town on the Neckar that he all but idolized as the venue of his youth with an outstanding result — settling the problem of CANTOR's *continuum hypothesis*.[6] We know from the inspired recollections of JÓZSEF KÜRSCHÁK [35, 36] what a sensation the announcement of the title of KŐNIG's lecture had stirred among the participants of the congress. All section meetings were cancelled so that everyone could hear his contribution.

His paper [4] divides into two parts: in the first, KŐNIG presented the inequality named now after him, outlined the proof and drew the major conclusions from the inequality.

The definition of the *sum* and *product* of cardinals we still use today is due CANTOR. Relying on these definitions, KŐNIG's *inequality* reads follows:

Let M be any set, and let two cardinal numbers a_m and b_m belong to each $m \in M$ so that the inequality $a_m < b_m$ is fulfilled. Then:

$$\sum_{m \in M} a_m < \prod_{m \in M} b_m.$$

The proof of the theorem can be found in nearly all books on set theory. Let us now write

$$\aleph_\mu < \aleph_{\mu+1}$$
$$\aleph_{\mu+1} < \aleph_{\mu+2}$$
$$\vdots \qquad \vdots$$

and apply case KŐNIG's inequality to this special.

Using the well-known theorem that the sum of the alephs on the left-hand side is $\aleph_{\mu+\omega}$, we obtain the following *inequality*:

$$\aleph_{\mu+\omega} < \aleph_{\mu+1} \cdot \aleph_{\mu+2} \cdots \leq \aleph_{\mu+\omega}^{\aleph_0}. \tag{I}$$

At the same time, for the cardinality c of the continuum the following equality is true:

$$c = c^{\aleph_0}. \tag{II}$$

---

[5] KŐNIG, DÉNES: *Theorie der endlichen und unendlichen Graphen*. Leipzig, 1936, pp. 85—87; SZÁSZ, PÁL: On the equivalence theorem of set theory. *Matematikai Lapok*, 10, 1959, pp. 49—52. (In Hung.)

[6] The question is whether there is a cardinality greater than the countable cardinality (that of natural numbers) but less than the continuum cardinality (that of the points in the interval $\langle 0, 1 \rangle$). CANTOR conjectured that there was no set of such cardinality. The general continuum problem is whether there is another cardinality between the cardinality of an infinite set and its power set. The generalization of CANTOR's conjecture says that there is not. In 1939 GÖDEL proved that the usual axioms of set theory do not suffice to refute the generalized continuum hypothesis, and in 1963 COHEN showed that it cannot be proved, either.

Comparing (I) and (II) we obtain that the cardinality of the continuum cannot be equal to any aleph whose index is cofinal with ω.

Loosly speaking, KŐNIG's inequality does not permit us to decide which aleph is equal to the cardinality of the continuum, but with its help we can find an infinite number of alephs to which it is definitely *not equal*.

Further, let us be given a countably infinite, strictly increasing sequence of cardinal numbers $m_1 < m_2 < m_3 \ldots$. It can be written in the form

$$m_1 < m_2$$

$$m_2 < m_3 \qquad\qquad\qquad\text{(III)}$$

$$\vdots \quad \vdots$$

Applying KŐNIG's inequality to (III) and introducing the notation $\sum_{i=1}^{\infty} m_i = m$ (where, by a well-known theorem, m is greater than any m_i) we obtain the inequality:

$$m = \sum_{i=1}^{\infty} m_i < \prod_{i=2}^{\infty} m_i \leqq m_1 \prod_{i=2}^{\infty} m_i = \prod_{i=1}^{\infty} m_i \leqq mm \ldots = m^{\aleph_0}. \qquad\text{(IV)}$$

From (II) and (IV) it can be seen that *the cardinality of the continuum cannot be represented as the sum of a countably infinite (or finite) number of smaller cardinals.*

This theorem of GYULA KŐNIG, apparently the best known, was first treated monographically in HAUSDORFF's book (*Grundzüge der Mengenlehre*, Leipzig, 1914, pp. 57—58) and came to be included in nearly all subsequent monographs on set theory.

In the second part of his Heidelberg lecture KŐNIG went even further. Most regrettably, however, he based his argument on a theorem whose range he had failed to test previously. Namely, FELIX BERNSTEIN[7] in his "Habilitationsschrift" announced the following equality as generally valid without any restriction on the indices α and ν:

$$\aleph_\alpha^{\aleph_\nu} = \aleph_\alpha \cdot 2^{\aleph_\nu}.$$

Erroneously, KŐNIG applied the BERNSTEIN theorem[8] to the case $\nu = 0$, $\alpha = \mu + \omega$, assuming that c is equal to the cardinality $\aleph_\mu$ of some well-ordered set. If we also take into account that $c = 2^{\aleph_0}$, then BERNSTEIN's formula would yield the following *equality*:

$$\aleph_{\mu+\omega}^{\aleph_0} = \aleph_{\mu+\omega} \cdot 2^{\aleph_0} = \aleph_{\mu+\omega} \aleph_\mu = \aleph_{\mu+\omega}.$$

This, however, contradicts (I). Consequently, c could not be equal to any $\aleph_\mu$, hence the set of real numbers could not be well-ordered.

For KŐNIG's case, however, BERNSTEIN's theorem does not hold true. This remained unnoticed by KŐNIG, and for some time after the congress also by such outstanding mathematicians as CANTOR, HILBERT, SCHOENFLIES and KLEIN, who were quite understandably preoccupied by KŐNIG's conclusion. Even to BERNSTEIN it took some time to mask out the range of his theorem,[9] admitting that its first formulation was defective.

[7] BERNSTEIN, F.: *Untersuchungen aus der Mengenlehre*. Göttingen, 1901, p. 49.
[8] The equality is true if, and only if, $\alpha \leqq \nu + n$, where $n \geqq 0$ is a (finite) natural number.
[9] BERNSTEIN, F.: Zum Kontinuumproblem. *Math. Annalen*, 60, 1905, pp. 463—464.

KŐNIG had made a mistake by not checking the limits of the validity of BERNSTEIN's theorem, and the wrong conclusion he had drawn distressed him for years.

This fiasco and a great many modifications and refinements of his views paved GYULA KŐNIG's way to one of his most significant works, the book on the foundations of logic, arithmetic and set theory [6]. He had worked on the system laid down in it for many years until his views on the particular notions assumed their final crystalline form. The pen fell out of his hand while writing the last chapter of the book: his testament on the foundations of mathematics was posthumously published without his giving it a final touch. Only two short — though fundamental — chapters of the book appeared in Hungarian in the periodical of the Hungarian Philosophical Society [7, 8].

Let us only pick out the guiding ideas from, the work so typical of KŐNIG for its succinct but most enjoyable style and for the wealth of philosophical and psychological observations.[10] They were meant to lay the groundworks for the attainment of two — theoretically important — targets: proving the consistency of logic and various other mathematical disciplines, and the construction of set theory devoid of antinomies.

The introductory chapters of the book are preparations for a systematic exposition of synthetic logic. In them, KŐNIG sets down the rules of a discipline "preliminary to logic" in which the notions of "true" and "false" are not yet differentiated. The introductory thoughts remind us of the foundations of philosophy as presented by DESCARTES. For example, KŐNIG shares the view that consciousness and its various functions are impossible to prove but at the same time are undeniable facts. We cannot prove that there are experiences in our mind that can be different from one another, be forgotten or repeated. These and some other notions rather belonging to psychology are the "basic norms" of thinking. All that the basic norms allow for are "possible" and all that contradict them are "impossible". The aggregate of experiences in our minds constitutes the "domain of thought" (Denkbereich) which is considered to be wholly defined if we can decide clearly whether an experience belongs to the domain of thought or not.

After these preliminary considerations the axis of the book is in chapters IV and V on the formalization of true and false, on logical operations and consistency. Here, GYULA KŐNIG introduces the valuation method and gives examples of "tables of values". The following quotation may illustrate his ideas:

> "If I declare that I deem A to be true, I observe a certain state of my mind, a particular sensation which accompanies the image of A in my consciousness; and this in such a way that the connection between image and sensation is accompanied by the sense of *necessity*. I 'must' deem A to be true; that is, I evaluate A in a certain way; I attribute the value of true to A." ([7], p. 102)

---

[10] A detailed and expert description of the book was given by KŐNIG, D.: On the last work of Gyula Kőnig. *Math. és Phys. Lapok*, 23, 1914, pp. 291—302. (In Hung.) A fairly long and appreciative mention of it was made by ADOLF A. FRAENKEL: *Einleitung in die Mengenlehre*, Berlin, 1923, pp. 182—184. Some parts are analyzed from a philosophical viewpoint without a mathematical evaluation by BENCSIK, KÁLMÁN: *Paradoxes in logic and mathematics*. Budapest, 1932. (In Hung.) In any case, if is to the credit of LÁSZLÓ KALMÁR that the valid and effective ideas of the book are still alive and in the forefront.

For example:

"A mind that has declared the parallel postulate as undeniable fact deems true the theorem that the angle sum of any triangle is equal to the sum of two right angles. A mind which does not know of that postulate or does not take it for either undeniable or unacceptable will not evaluate this theorem as either true or false." *(Ibid.)*

In 1900, at the international mathematical congress of Paris, HILBERT called attention to the importance of creating methods for deciding the absolute consistency of axiom sets. In 1904, at the Heidelberg congress, he presented a method which, however, has mainly theoretical significance, for in the system he put forth as an illustrative example the axioms and methods of inference are such that only a fragmentary part of arithmetic can be built up with their help. The second step, a major one, a long this road was made by KŐNIG.

The core of KŐNIG's "valuation method" is the following, quoted here in LÁSZLÓ KALMÁR's wording:

". . . to the formulae of the given system of axioms we attach the logical values ↑ and ↓, no more than one to each formula, so that the following conditions be satisfied: (1) we attach the logical value ↑ to each axiom; (2) if we have attached the logical value ↑ to certain formulae, then we are to give the same logical value ↑ to all formulae that have been derived from these formulae with the application of certain rules of interference; (3) if we have attached to the formula f the logical value ↑, then we are to give to the formula f̄ the logical value ↓. Such an attachment is called a *valuation* of the system of axioms. If we can give a valuation to the axioms of a system of axioms, then the axiom system at issue is free from contradictions. Indeed, owing to conditions (1) and (2), we have attached the logical value ↑ to all formulae that can be proved in the given system of axioms; consequently, if f is a verifiable formula then f̄ cannot be so because due to condition (3), we have attached the logical value ↓ to f̄."[11]

Valuating a system of axioms is difficult even though we have some freedom in the process.[12]

The other major aim of GYULA KŐNIG's book was to explain in a satisfactory way the antinomies arising in set theory and to construct set theory in such a way that no antinomies should occur.

"An antinomy is born when one 'creates' from defined, well — differentiated notions — or from those believed to be such — some new concepts and, yielding to the temptation of more or less superficial analogies, one identifies the newly created notions with one or another of the 'constituting' (original) notions. The

---

[11] KALMÁR, LÁSZLÓ: *The foundations of mathematics.* Vol. 2, No. 2, p. 357. (In Hung.)
[12] An example which helps understand the method can be found in the cited book of LÁSZLÓ KALMÁR.

image of an experience A that cannot come about (cannot be lived through) without previous possession of the image of an experience B differs, for one thing, by virtue of its genesis from the latter, and we violate the basic norms of thinking if we do not regard it as different from the former." ([8], pp. 145—146)

In other words, the set of "all things" in RUSSELL's antinomy, for example, leads to a contradiction because in it the "totality of things" is conceived as something given for good and all, whereas this notion keeps expanding by the addition of newer and newer "things".[13]

In order to eschew antinomies, we have to alter CANTOR's notion of a set, without reducing its scope. The most salient feature of this modification is that in KŐNIG's view a set is not defined by its elements unambiguously as the same elements can form various sets. Notably, the definition of a set requires the specification of "how" the constitutive elements belong to it. "If, e.g., there are only red balls and black cubes in an urn, we may also speak of spherical bodies in the urn, and thereby we have created two fundamentally different «collectións»; collecting happened on the basis of completely different properties." This view does not limit the scope of set theory, as in KŐNIG's conception all notions like cardinality, equivalence, well-ordening, etc. retain sense. FRAENKEL[14] contended that KŐNIG

"gave set theory such a wide manoeuvering space that he surpassed not only WHITEHEAD and RUSSELL, but in certain respect even ZERMELO."

It is well known that some of KŐNIG's ideas were fruitfully applied by JÁNOS NEUMANN (John von Neumann) in his system of axioms for set theory.

The notion of "ordinator" introduced in the last chapter of the book could have led to the proof of the well-ordering theorem. His sudden death, however, prevented KŐNIG from crowning his life-work. Yet unfinished as it is, the book has been a major station in the development of mathematical logic and set theory.

Investigations in number theory

Up to the last quarter of the 19th century, no noteworthy results in number theory were produced in Hungary. We know from the writings of the two Bolyais that they had a high esteem for the "queen" of mathematics, yet their posthumous papers contained scarcely any ideas in this field. GYULA KŐNIG also published little ideas about the subject; more important were his special courses in number theory and the number-theoretic chapters of his monographs and textbooks. By giving a systematic treatment of the central parts of number theory, he did a pioneering work in the Hungarian literature and stimulated several researchers to independent investigations.

---

[13] By the way, KŐNIG wittily reformulated RICHARD's antinomy. See e.g. [1], pp. 252—253.
[14] FRAENKEL, ADOLF, A.: *Einleitung in die Mengenlehre*. Berlin, 1923, p. 184.

In one of his publications [10] he gave a proof of the reciprocity theorem of quadratic residues by an ingenious application of complete induction and some properties of the LEGENDRE and JACOBI symbols.

According to another — early — paper [9] of his, the following equality is true:

$$\sum a_i - \sum b_i = p\left(\sum a_i - \sum \frac{n}{b_i}\right).$$

In this formula the a_i denote the divisors of natural number n (including 1 and n), p is a prime appearing among them, and b_i are those a_i not divisible by p.

GYULA KŐNIG's best-known result in the field of number theory was put to paper by GUSZTÁV RADOS,[15] making its way into literature as the KŐNIG—RADOS theorem. The theorem says that congruence

$$f(x) \equiv A_0 + A_1 x + \ldots + A_{p-2} x^{p-2} \equiv 0 \quad (\text{mod } p)$$

$[A_0 \not\equiv 0 \,(\text{mod } p); p > 2$ a prime number; A_l a rational integer] has $p - 1 - r_p$ incongruent solutions if the rank mod p of the cyclical matrix

$$
\begin{bmatrix}
A_0 & A_1 & A_2 & \cdots & A_{p-2} \\
A_{p-2} & A_0 & A_1 & \cdots & A_{p-3} \\
A_{p-3} & \cdot & \cdot & \cdots & \cdot \\
\cdot & & & & \cdot \\
\cdot & & & & \cdot \\
A_1 & A_2 & A_3 & \cdots & A_0
\end{bmatrix}
$$

is r_p. This condition is necessary and sufficient.

The first monograph to discuss this theorem was KRONECKER's.[16] VANDIVER used it in his investigations related to FERMAT's *last theorem.*[17] RÉDEI presented and proved it generalized to finite fields.[18]

LÁSZLÓ RÉDEI and PÁL TURÁN expressed the number of incongruent solutions of the congruence above with the help of a new concept defined for cyclical matrices.[19]

[15] RAUSSNITZ (RADOS), GUSZTÁV: To the theory of higher-degree congruences. *Math. és Term. tud. Értesítő*, 1882—83, pp. 296—308. (In Hung.) Also: Zur Theorie der Congruenzen höheren Grades. *Journal f. reine u. angew. Math.*, 90, 1886, pp. 258—260. RADOS's papers reveal that during a university seminar KŐNIG answered the question as to when the above-mentioned congruence does have a solution at all. RADOS extended the investigation to find the number of different solutions.

[16] *Vorlesungen über Zahlentheorie.* Vol. 1, Leipzig, 1901, 389.

[17] Some theorems in finite field theory with applications to Fermat's last theorem. *Proc. Nat. Acad. USA*, 30, 1944, pp. 362—367; On trinomial congruences and Fermat's last theorem. *Ibid.*, pp. 367—368.

[18] *Algebra*, Vol. 1, Budapest, 1954, pp. 443—444.

[19] RÉDEI, LÁSZLÓ—TURÁN, PÁL: Zur Theorie der algebraischen Gleichungen über endlichen Körpern. *Acta Arithmetica*, 5, 1959, pp. 223—225.

Contributions to algebra

GYULA KŐNIG had diverse interests in both classical and abstract algebra, contributing considerably to the further development of mathematics. Our secondary school education benefited for decades from his textbook on algebra [12]. The author of the school-book designed in the spirit of the requirements of TREFORT's curriculum did not rest content with a formalistic presentation of the traditional chapters of álgebra but consciously strived for improving the students' *way of mathematical thinking*. KŐNIG regarded the theory of equations as particularly suitable for this. He also wished to offer a better preparation for university by treating some basic notions of calculus. A detailed discussion of infinite series, chiefly the geometrical series, led secondary schools students to acquaintance with the concept of *limit*.

The well-organized and fluently written textbook revised later by MANÓ BEKE[20] was used, in several secondary schools even at the beginning of this century.

Several of KŐNIG's independent investigations are concerned with the power series of the reciprocal

$$\frac{1}{f(x)} = c_1 + c_2 x + c_3 x^2 \ldots$$

of a polynomial

$$f(x) = a_0 x^n + a_1 x^{n-1} + \ldots + a_n = 0.$$

According to an earlier result, if the equation $f(x)=0$ with real coefficients has a positive root that is less than the absolute value of any other root, then the coefficients in the power series of $\frac{1}{f(x)}$, from a certain term onward have constant sign.

The starting point for KŐNIG's study [13] related to this result is the following. Let us be given only the *sign* of each term in the sequence

$$1, \cos \varphi, \cos 2\varphi, \cos 3\varphi, \ldots.$$

Is φ uniquely determined by this infinite sequence of *signs*, and if yes, what is its value?

The answer is affirmative, yielding

$$\varphi = \pi \lim_{n \to \infty} \frac{\varphi_n}{n}, \qquad\qquad (+)$$

where φ_n denotes the number of the changes of sign up to the n-th term of the sequence.

A nice application of this theorem [14] concerns those algebraic equations $f(x)=0$ with real coefficients whose roots of smallest absolute value are conjugate complex numbers:

$$z_1 = r(\cos \varphi + i \sin \varphi); \quad z_2 = r(\cos \varphi - i \sin \varphi) \quad (0 < \varphi < \pi).$$

---

[20] *Algebra for secondary schools*. Budapest, 1897, Revised by MANÓ BEKE.

Namely, if we form the power series of $\dfrac{1}{f(x)}$ and φ_n this time denotes the number of the changes of sign up to the n-th coefficient of this power series then according to KŐNIG's theorem, the sequence

$$\frac{\varphi_1}{1}, \frac{\varphi_2}{2}, \ldots, \frac{\varphi_n}{n}, \ldots$$

is convergent and the argument φ appearing in the two roots of smallest absolute value is also given by the relation $(+)$.

The following theorem of GYULA KŐNIG belongs to analysis but due to one of its consequences it may be classified to classical algebra (we adhere to the original and avoid function-theoretic notation.) Let $F(x)$ be regular in the neighbourhood of the point $x=0$, i.e., let it be expressed these by a power series of the form

$$F(x)=c_1+c_2x+c_3x^2+\ \ldots.$$

Further, let the singular point α of $F(x)$ nearest to $x=0$ (i.e. of least absolute value) be a pole of the first order.

Then

$$\lim_{n\to\infty}\frac{c_{n-1}}{c_n}=\alpha.$$

We know that HADAMARD was in part prompted by this result of KŐNIG to address himself in 1892 to the problem, in a sense converse to KŐNIG's, as to what conclusions for the singular points of a function could in general be drawn from the coefficients of its power series. As a discussion of the profusion of literature on this subject would carry us too far from the direction selected by KŐNIG, we mention only one consequence of his theorem.

Evidently, the function

$$F(x)=\frac{f'(x)}{f(x)}$$

belongs to the class of functions appearing in the KŐNIG theorem if $f(x)=0$ is an algebraic equation which has a (simple or multiple) root whose absolute value is less than the absolute value of the rest of the roots. As is well known,

$$\frac{f'(x)}{f(x)}=\sum_{i=1}^{k}\frac{1}{x-x_i},$$

where x_i are the roots of the equation. Simple computation yields the power series

$$F(x)=\frac{f'(x)}{f(x)}=c_1+c_2x+c_3x^2+\ \ldots,$$

in which

$$c_n=-(x_1^{-n}+x_2^{-n}+\ \ldots +x_k^{-n}). \tag{A}$$

By KŐNIG's theorem, the first-order pole of $F(x)$ having the least absolute value can be calculated from the coefficients c_n; on the other hand, it is easy to see that this

242

first-order pole of $F(x)$ is the root of least absolute value of the algebraic equation $f(x)=0$.

Of the roots $x_1, x_2, \ldots, x_k$ of the algebraic equation $f(x)=0$ let x_k have the least absolute value. Then from (A)

$$\frac{c_{n-1}}{c^n} = x_k \frac{\left(\dfrac{x_k}{x_1}\right)^{n-1} + \left(\dfrac{x_k}{x_2}\right)^{n-1} + \ldots + \left(\dfrac{x_k}{x_{k-1}}\right)^{n-1} + 1}{\left(\dfrac{x_k}{x_1}\right)^{n} + \left(\dfrac{x_k}{x_2}\right)^{n} + \ldots + \left(\dfrac{x_k}{x_{k-1}}\right)^{n} + 1}.$$

This relation immediately yields the approximation method invented by DAVID BERNOULLI (*Petrop. Comm.*, Vol. 3, 1728, p. 92): the quotient $\dfrac{c_{n-1}}{c_n}$ for sufficiently large n is a good approximation of x_k.

To sum up: from KŐNIG's theorem the root approximation on procedure offered by BERNOULLI can easily been obtained.[21]

*

GYULA KŐNIG's doctoral dissertation [19] and about ten of his papers, mainly from his youth, deal with elliptic functions applied in algebra. The results of these works together with several details obtained later on reappeared in KŐNIG's highly significant monograph on the general theory of algebraic quantities [20]. This book of KŐNIG on which he has worked around the turn of the century for years, basically follows the train of arguments of the famous *Festschrift* by KRONECKER:[22] it elaborates the material KRONECKER only outlined, adding recent results of HILBERT, HENSEL, NOETHER and first of all, KŐNIG himself. As the preface indicates, the latter amount to more than a half of the 600-page book.

The subject of the book is abstract algebra and analytical number theory. Abstract algebra taken in the present-day sense of the word: a pioneering work of international distinction by a mathematician with a penchant for abstract concepts written at a time when the method and scope of this discipline had not yet evolved. Apparently, KŐNIG had to overcome enormous difficulties to create as coherent and complete a theory as possible. In the preface he lays stress on the methodical obstacles, but we must not belittle the work he did by completing, deepening, or generalizing the results achieved by others either.

Although a remark in the preface only anticipates a knowledge of the elements of algebra and number theory for an understanding of the book, reading it needs con-

---

[21] An indication of the usefulness of KŐNIG's theorem is that many authors of the "Enzyklopädie der Mathematik" included it in their articles elaborating it from various aspects, e.g.: RUNGE (I., 1, pp. 439—440), PRINGSHEIM—FABER (II., 3, pp. 14—15), BIEBERBACH (II., 3, p. 473). Later, nice theorems about the root of least absolute value of algebraic equations were established by LIPÓT FEJÉR: *Math. és Phys. Lapok*, 17, 1908, pp. 308—324.

[22] KRONECKER, LEOPOLD: Grundzüge einer arithmetischen Theorie der algebraischen Grössen. *Journal f. reine u. angew. Math.*, 92, 1882, pp. 1—123.

centrated attention. The main difficulties arise from the subtle definition of concepts, the intricate proofs and the several philosophical remarks made by the author. Another problem is to figure out the contemporary equivalents of the technical terms KŐNIG coined and applied extensively.

The exposition of the material of the entire book rests on two algebraic structures, the *holoid and orthoid domains*, defined by KŐNIG in an axiomatic way. In today's terminology, the holoid domain is a domain of integrity of character zero with unite element; the orthoid (or *improper holoid*) domain is a field of character zero.

One could go on describing the vocabulary of KŐNIG's concepts, but the preceding examples may suffice to give an impression of the spirit of the work. The backbone of the book is a self-contained theory of the algebraic structures mentioned together with the theories of polynomials, decomposition fields of polynomials, normal field extensions, algebraically closed fields, and transcendental field extensions. GALOIS theory is briefly touched on and a nice proof of the fundamental theorem of algebra is given. In the purely algebraic proof of this theorem used nowadays (the ARTIN— SCHREIER—DÖRGE proof) the core of the procedura is KŐNIG's theorem: for every irreducible equation $f(x)=0$ of the n-th degree whose coefficients are from the field K there is a field of n-triples of elements of K which can be shown to contain all roots of the equation.

The last chapters of KŐNIG's book discuss some of the major questions of algebraic number theory.

According to JÓZSEF KÜRSCHÁK ([35], pp. 13—14), KŐNIG was contemplating the plan of writing another book on algebraic quantities different from the former in method and results. Specifically, while in the above book only *finite* sequences of operations occur, in the planned book KŐNIG wanted to settle some important open problems by means of *infinite* sequences of operations. But the work got stranded in the first phase of planning.

Investigations in analysis

GYULA KŐNIG's internationally best known results belong to set theory and mathematical logic, but it seems that the greatest impact of his work on the advancement of Hungarian mathematics came from his writings in analysis. He exerted this influence on our mathematical culture by his textbooks, rather than his research papers on analysis.

His first work of this kind [22], though featuring the word "algebra" in its title, drew equally on algebra and analysis. He meant to use this book compiled from his university lectures as a reference book for university education, "in view of our specific circumstances of education". Only the first half of the anticipated two-volume work was finished, but it is a self-contained text in itself. Though the book presumes minimal preliminary knowledge, its treatment and subject-matter are up-to-date. KŐNIG wanted to help "the reader come abreast of the level of contemporary scientific research", in which he succeeded first of all in the area of infinite series, infinite products and continued fractions.

244

An even more significant book of his is *Analysis* published in 1887 which surveys a vast material. To describe it briefly, one can say the same as of FARKAS BOLYAI's *Tentamen*: KŐNIG did not rest content with the elaboration of the traditionel material of calculus in textbook form or with the clever systematization of results achieved by others, but again and again introduced new proofs and results he had found. Our young mathematicians looked upon this book as their gospel from which they not only learned but also gleaned ideas for their independent investigations. There was a large number of investigations around the turn of the century which drew inspiration, or actual information, from this book. This is born out by many footnotes in contemporary works.

It is to be regretted that the work planned for two volumes only had its first part published and only in Hungarian at that. Had it been published in one of the world languages, it would have had a far greater impact.

"Analysis" is again to be interpreted in the broad sense, typically of GYULA KŐNIG, including in addition to the regular material, problems of real functions and analytic number theory. It is particularly remarkable that just over a decade after the creation of set theory KŐNIG deemed several of its parts ripe for elaboration in a textbook — in a country, to boot, where for example not long ago JÁNOS ÁRMIN VÉSZ had thought that the continuity and differentiability of a function were equivalent properties.[23] In the course of introducing the real numbers and in the discussion of the derivative several considerations related to the cardinality of sets were incorporated and in the meantime, KŐNIG created the Hungarian technical terms for several concepts of set theory (e.g. countableness, element, cardinality) in general use today. It applies to the whole work what we can read in a footnote ([23], p. 147):

> "A systematic treatment of analysis must not use geometrical arguments since their ultimate bases — whether taken as axioms or hypotheses — are always connected to an external system of thought of which all mathematical truths are totally independent. This, of course, does not preclude either the geometric illustration of the abstract relations of numbers or the demonstration of the scientific value and significance of the methods developed by *applying analysis* in geometry (and other natural sciences)."

Instead of giving a detailed account of the contents of the book, let me pick an example which shows that if contains useful new results. In the discussion of infinite *products*, KŐNIG presents his result which in the literature is known to be the first example of a *conditionally convergent* infinite product ([23], p. 257).

Let

$$u_n = (-1)^n \frac{1}{n+1}$$

in the product $\prod_{0}^{\infty} (1 + u_n)$.

---

[23] VÉSZ, JÁNOS ÁRMIN: *Outlines of Advanced Mathematics. I—II.*, Pest, 1861—1862. (In Hung.)

Then

$$\prod_0^\infty (1+u_n) = (1+1)\left(1-\frac{1}{2}\right)\left(1+\frac{1}{3}\right)\left(1-\frac{1}{4}\right)\ldots =$$

$$= 2 \cdot \frac{1}{2} \cdot \frac{4}{3} \cdot \frac{3}{4} \cdots \frac{2m}{2m-1} \cdot \frac{2m-1}{2m} \cdots = 1.$$

By contrast, in the rearranged product

$$(1+1)\left(1+\frac{1}{3}\right)\left(1-\frac{1}{2}\right)\left(1+\frac{1}{5}\right)\left(1+\frac{1}{7}\right)\left(1-\frac{1}{4}\right)\ldots$$

$$\ldots \left(1+\frac{1}{4k+1}\right)\left(1+\frac{1}{4k+3}\right)\left(1-\frac{1}{2k+2}\right)\ldots$$

taking the factors by threes, the value of the product will be a number greater than $\frac{4}{3}$, for the value of the first "triad", is $\frac{4}{3}$, and that of the others is greater than 1.[24]

It should be noted that this example prompted MIHÁLY BAUER to probe into the problem of the convergence of infinite products, the fruit of which was his fundamental study on the subject.[25]

*

Some of GYULA KŐNIG's essays on problems of analysis are useful mainly for didactical purposes. In one of them [24], e.g., he derives the theory of gamma functions from the integral given by LEGENDRE in 1811,

$$\int_0^\infty t^{x-1} e^{-t} dt$$

but with the application of the inequality

$$0 \le e^{-nx} - (1-x)^n < \frac{1}{n}$$

$(0 \le x \le 1; n$ a natural number) he can avoid some cumbersome integrals, and therefore his treatment is simpler than the usual one.[26] Another of his papers [25] was also

---

[24] It can be proved that the rearranged product is greater than $\frac{4}{3}$ even if we do not group the terms by threes.

[25] BAUER, MIHÁLY: Contributions to the theory of infinite products. *Math. és Phys. Lapok*, 7, 1898, pp. 19—26. (In Hung.)

[26] For the treatment given by König the less restrictive inequality

$$0 \le e^{-nx} - (1-x)^n < \frac{e}{n}$$

is also sufficient. The validity of the latter was proved by KÜRSCHÁK: On an inequality. *Math. és Phys. Lapok*, 17, 1908, pp. 305—307. (In Hung.)

written for didactic purposes but has theoretical importance as well. In it he showed how the second mean value theorem of integral calculus can be derived from the first. The chief asset of the paper is that it contains the definition of the STIELTJES integral. It is well known that physicists had used this kind of integral in the seventies of the last century, but had not realized its theoretical significance. Relying on ideas of HERMITE, in 1894, STIELTJES published his results on the new concept of integral.[27] So, STIELTJES has the priority of publication, but KÖNIG had introduced relevant ideas in his lectures much earlier. In the essay in question KÖNIG remarks that the second mean value theorem of integral calculus can be derived in an original way, "if we generalize the notion of definite integral in a new direction — a fairly important subject in itself". Then he adds:

"The generalization of the concept of definite integral mentioned is given by the following definition:

$$\int_a^b \varphi(x)\,d\psi(x) = \lim_{x_r - x_{r-1} = 0} \sum_{r=1}^n \varphi(\xi_r)\,[\psi(x_r) - \psi(x_{r-1})],$$

where $x_0 = a$, $x_n = b$, $x_r - x_{r-1}$ are numerical values of the same sign tending to 0, further $\xi_r = x_{r-1} + \vartheta_r(x_r - x_{r-1})$, $0 \leq \vartheta_r \leq 1$, finally $\varphi(x)$ and $\psi(x)$, for the time being are to be taken as finite and single-valued functions in the interval ab only. Obviously, the expression on the right side will not always have a limit independent of the selected x's and ξ's; only if we use the symbol appearing on the left-hand side, and then we will say that φ is integrable with respect to ψ over the interval ab. We call attention to the fact that here $d\psi(x)$ does not mean $\psi'(x)\,dx$ in general, for we do not assume $\psi(x)$ to be differentiable."

In contrast to the above-mentioned paper, another of KÖNIG's writings [26] on analysis incurred wide international response. The results it contains fit organically into the line of investigations into real functions initiated by WEIERSTRASS, DINI, DARBOUX, HANKEL, LÜROTH and others in the second half of the last century.

Let $f(x)$ be a continuous function of the real variable x on some interval such that in any neighbourhood of an arbitrary point x_0 there exist values ξ, x_1 and x_2 ($x_1 < \xi < x_2$) with both $f(x_1)$ and $f(x_2)$ being either greater or smaller than $f(\xi)$. Then this requirement is equivalent with the following two:

1. In an arbitrarily small interval there are two (in fact, infinitely many) different points at which the values of the function are equal.
2. In any neighbourhood of any point, an extremum of the function can be found.

Finally, if a function meets one of these equivalent requirements, then in any neighbourhood of any point there is a point ξ to which one can find a sequence $h_n \to 0$ such

[27] STIELTJES, T. J.: Recherches sur les fractions continues. *Annales de Toulouse*, 8, 1894, pp. 1—122; 9, 1895, pp. 1—47. This definition was soon followed by various generalizations (LEBESGUE, YOUNG, FRÉCHET, RADON, DANIELL, and others).

that

$$\lim_{h_n \to 0} \frac{f(\xi + h_n) - f(\xi)}{h_n} \to c,$$

where c is an arbitrary constant given in advance ($c = 0$ being also admitted). It easily follows from the above-said that there is no function with derivative infinite everywhere.

From a Hungarian viewpoint, the greatest merit of KŐNIG's paper is that it turned the attention of ZOÁRD GEŐCZE towards problems of real functions, the direct outcome of which was his example of an everywhere continuous and nowhere differentiable function (see Chapter 23), and an indirect outcome was GEŐCZE's entire mathematical life-work. I depend for this statement on GEŐCZE's posthumous manuscripts that were unfortunately destroyed in the war.[28]

*

The commemorative writings on KŐNIG emphasize that he was averse to a forced application of mathematical results; "he was not glad to see men of practice use complicated geometrical theorems for the derivation of results that might be more apparent through simple mechanical considerations." (KÜRSCHÁK)

Many quotations could be presented to testify the correctness of this statement, but probably his treatises on differential equations provide the best evidence. As for their starting point, they are expressly applicational (problems of dynamics, of the best air-vane and ship-screw, etc.) but one can hardly if at all realize this when reading the essays: KŐNIG quickly removes himself from the practical problem that prompted the research and seeks to establish an abstract, possibly general theory.

It is well known that in mechanics and in problems of the calculus of variations and geometry the following first-order partial differential equation, the so-called HAMILTON–JACOBI equation, plays an important role:

$$\frac{\partial V}{\partial t} + H\ (t, x_1, \ldots, x_n; p_1, \ldots, p_n) = 0. \tag{I}$$

Here V is the unknown function, H is HAMILTON's function, $t, x_1, \ldots, x_n$ are the independent variables, and $p_i = \dfrac{\partial V}{\partial x_i}$. JACOBI proved that the integration of (I) is equivalent to the canonical system

$$\frac{dx_i}{dt} = \frac{\partial H}{\partial p_i}; \quad \frac{dp_i}{dt} = -\frac{\partial H}{\partial x_i} \quad (i = 1, 2, \ldots, n) \tag{II}$$

consisting of $2n$ ordinary differential equations. Previous to KŐNIG, however, almost all examinations started from equation (I), the most important theorem being due to

[28] Undoubtedly, KŐNIG's results cited above had an influence on A. ROSENTHAL who in his dissertation (*Über die Singularitäten der reellen ebenen Kurven.* Munich, 1912) studied the exceptional points (peak, turning-point, multiple point, point of discontinuity, etc.) of real plane curves, and established theorems on their cardinality under restrictions on the curves similar to those imposed by KŐNIG.

JACOBI. The latter says that if

$$V = V(t, x_1, \ldots, x_n; \ a_1, \ldots, a_n) + a_{n+1}$$

is a complete integral of (I)[29] then it provides $2n$ integrals of the canonical system (II) by simple differentiation.[30]

GYULA KŐNIG took a different course [27, 28]: disregarding the above-mentioned relation between (I), and (II), he separately examined the problem of integration for the system (II), using only a few of JACOBI's, LIE's and MAYER's results. It would take up much space to describe his way to the solution of system (II), so we confine ourselves to his most important theorem: according to it, if a single *first* integral of the system (II) is known,[31] then the canonical system can be reduced to $2(n-1)$ equations, and more importantly, the system thus obtained is of the same type as (II). Consequently, the solution of the system, originally consisting of $2n$ equations requires the determination of a *first* integral for each of several systems consisting of

$$2n, \ 2n-2, \ 2n-4, \ \ldots, 4, \ 2$$

equations, respectively.

By perfect analogy, when k first integrals are known, the number of equations can be reduced to $2(n-k)$. This reduction is possible not only with type (II), but with a wider class of differential systems.

The significance of KŐNIG's discoveries is evident if we consider that the integration of first-order partial differential equations can be reduced to the solution of systems of ordinary differential equations. It must have been the importance of this article that induced TOEPLITZ to review it at length and in highly appreciative terms.[32]

TOEPLITZ reviewed this treatise at unusual length — on five pages — in the *Fortschritte*, 16, 1884, pp. 309—314. Several result of the treatise are quoted in the Enzyklopädie article by WEBER: *Partielle Differentialgleichungen. II.*, 1.1, pp. 294—399. PASCAL refers to it in: *Repertorium der höheren Analysis.* Leipzig–Berlin, 1927, Vol. I., 2, pp. 569—570. SPECKMANN also makes mention of KŐNIG's treatise: De Darboux'sche methode ter integratie der nietlineaire partieele differentiaalvergelijkingen van de tweede ordre (*Amst. Versl. en Meded.*, 9, 1892, pp. 441—497); and RIQUIER in his essay: Sur une question fondamentale du calcul integral. (*Acta Math.*, 23, 1900, pp. 203—331). It is cited in the bibliography of the German edition of DARBOUX's fundamental work *Über die partiellen Differentialgleichungen zweiter Ordnung* (originally in French).

Of all writings of KŐNIG on the theory of differential equations, the treatise — of impressive length — on second-order partial differential equations with two independent variables [30, 31] earned the greatest international acclaim. Work on this class of equations had got stuck after the fundamental but obscure studies of AMPÈRE and some more systematic essays by BOUR, BOOLE, IMSCHENETZKY and others. Then DARBOUX (1870) staked out a completely new direction for research by asking the conditions under which a second-order partial differential equation with two independent variables can be reduced to a system of ordinary differential equations.

[29] A *complete* integral of a first-order partial differential equation is a solution containing as many independent parameters as the number of independent variables.

[30] See, e.g., STEPANOV, V. V.: *A textbook of differential equations.* Budapest, 1952, pp. 360—361. (In Hung.)

[31] I.e., a solution containing the unknown functions and *one* arbitrary constant.

[32] *Fortschritte d. Math.*, 16, 1884, pp. 306—308.

Without loss of generality, equations of this type can be written in the form

$$r + F(x, y, z, p, q, s, t) = 0 \qquad (1)$$

where, in the usual *Monge* notation for the function $z = f(x, y)$,

$$p \equiv \frac{\partial z}{\partial x}; \quad q \equiv \frac{\partial z}{\partial y}; \quad r \equiv \frac{\partial^2 z}{\partial x^2}; \quad s \equiv \frac{\partial^2 z}{\partial x \partial y}; \quad t \equiv \frac{\partial^2 z}{\partial y^2}.$$

If we can find equations

$$u(x, y, z, p, q, s, t) = a_1 \qquad (2)$$

and

$$v(x, y, z, p, q, s, t) = a_2, \qquad (3)$$

(where a_1 and a_2 are arbitrary constants) such that the function z of the two variables satisfies each of (1), (2) and (3), and if we resolve equations (1)—(3) for r, s, and t then, according to MAYER's investigations, the general solution to (1) is obtainable from the resulting system by solving a system of ordinary differential equations. Now, KŐNIG proved that, in order to meet the compatibility requirement, the functions u and v must satisfy two partial but *first-order* differential equations (since these equations as well as KŐNIG's criteria for systems of first-order partial differential equations — to be discussed later — have a complicated form, we do not dwell on them).

In addition to the result just mentioned, KŐNIG succeeded in characterizing all second-order partial differential equations whose integration can be reduced to the solution of systems of ordinary differential equations.

These results of KŐNIG and the subsequent — and, in some cases, related — investigations of SONIN, BÄCKLUND, BEUDON and HAMBURGER soon became integrated in the general theory of partial differential equations and made their way to the monographs via GOURSAT's works.

We touch on one more problem treated by KŐNIG. Let us be given a system of first-order partial differential equations with independent variables

$$x_1, x_2, \ldots, x_n$$

and unknown functions

$$z_1, z_2, \ldots, z_n.$$

MAYER studied the integrability of such systems consisting of mn equations; LIE answered the question in the special case $m = 1$. KŐNIG, largely generalizing the results of MAYER and LIE, obtained integrability criteria for systems in which the number of equations is different from mn. His investigations furnished new and simple proofs to MAYER's and LIE's theorems.[33]

---

[33] KŐNIG's relevant criteria are also applied in some investigations of modern differential geometry. See, e.g., KNEBELMANN, M. S.: Collineations of projectively related affine connections. *Ann. of Math.*, 29, 1928, pp. 389—394; SOÓS, GYULA: *Continuous transformation groups of line element manifolds*. Budapest, 1955 (dissertation for candidate's degree). (In Hung.) The *Encyklopädie* also refers to KŐNIG several times: II., 1.1, pp. 305, 355, 376, 380, 395, 396.

Regarding its starting point, KŐNIG's treatise in which he derived the laws of motion from some well-known principles of mechanics, belongs to theoretical physics but, considering its mathematical tools, it is a work on differential equations [32, 33].[34]

In this study, KŐNIG carried on previous investigations learning on the foundations of mechanics as given by HERTZ, but his results are more general in imposing no restriction on the forces and constraints. We leave the detailed analysis of his results to physicists, but give a brief survey of the method he used and some notions he defined.

In the introduction of GYULA KŐNIG's treatise we read the following ([32], p. 3):

"The *state of motion* of a system of n mass points can be described by specifying the coordinates (x_i, y_i, z_i), the velocity components $(\dot{x}_i, \dot{y}_i, \dot{z}_i)$ and masses (m_i) of the points for a given value of the time coordinate (t). If the masses are constant, then including them among the state variables is unnecessary; although there has been no need for such assumption in the study of phenomena observed so far, *the possibility is not excluded that masses change with time or, more generally speaking, with the state of motion of the system in a well-defined manner.* Let me therefore note here once and for all, though I lay no special stress on it, that the following discussions also apply under all such more general assumptions concerning the masses."

As KŐNIG's paper was published in 1887, we see that the part of this quotation in italics may provide interesting information on the history of rocket theory.

In addition to some new results in physics, a remarkable feature of the treatise is that by defining some new notions KŐNIG was able to state some well-known theorems more concisely. He introduced the terms *velocity virial, acceleration virial* and *energema,*[35] together with their definition, into mechanics. CLAUSIUS had defined the *force virial* by the formula

$$\sum_{i=1}^{n} (X_i x_i + Y_i y_i + Z_i z_i),$$

where X_i, Y_i and Z_i are the components of the force acting upon the point (x_i, x_i, z_i). By analogy, KŐNIG's *velocity virial* and *acceleration virial* are

$$\sum_{i=1}^{n} (X_i \dot{x}_i + Y_i \dot{y}_i + Z_i \dot{z}_i)$$

and $\sum_{i=1}^{n} (X_i \ddot{x}_i + Y_i \ddot{y}_i + Z_i \ddot{z}_i)$, respectively. Further, energema is given by the expression

$$\sum_{i=1}^{n} \frac{X_i^2 + Y_i^2 + Z_i^2}{m_i}.$$

---

[34] For a detailed review, see: *Fortschritte d. Math.*, 20, 1888, pp. 928—930.
[35] The word *energema* was coined by KÁLMÁN SZILY SR.

Kőnig's model for non-Euclidean geometries

The least known work of GYULA KŐNIG is his early paper on geometry [34] which was conspiciously ignored by commentaries and reviews: in it he presented his thoughts on how to obtain an intuitive description of three-dimensional non-Euclidean geometries. GUSZTÁV RADOS ([37], pp. 20—21) made a passing remark about it, but there is no trace of reference to KŐNIG's investigations in the writings of the most authentic person, FELIX KLEIN. This is doubly conspicuous as KŐNIG's note immediately followed the publication of the CAYLEY–KLEIN model in the *Göttinger Nachrichten* journal well and supported by KLEIN. We do not know the reason of KLEIN's silence, but the question unwittingly arises whether professional partially might have had a share in it.[36]

However, to do justice to everymore, it should be admitted that this neglect may have been caused by objective reasons as well. KŐNIG's study is very sketchy and his statement are difficult to check for lack of proofs. On the other hand, the mere objective of the paper lends it significance: it is one more evidence of KŐNIG's inclination to modern problems; in fact, as early as 1872 he realized the usefulness of intuitive descriptions or models as tools for the understanding of non-Euclidean geometries and — what he did not state explicitly — as a help in proving their consistency.

In the Hungarian literature LÁSZLÓ KALMÁR[37] emphasized the significance of KŐNIG's way of description in proving the relative consistency of non-Euclidean geometries.

KŐNIG's model is rather an idea than an elaborated model, so it is not of equal value with BELTRAMI's or KLEIN's work. The starting point of the paper is the fact recognized by RIEMANN and HELMHOLTZ that the analytic description of the intrinsic structure of non-Euclidean spaces requires not only the coordinates of points but also the *curvature* of the space at each point. In the case of *two-dimensional* non-Euclidean figures we may consider curvature to be a third coordinate and regard the resulting ordered *triple* as a point in three-dimensional Euclidean space. This method, however, cannot be used for representing three-dimensional non-Euclidean spaces, as that would only be possible in four-dimensional Euclidean spaces which are not accessible for perception.

According to KŐNIG, we can also represent three-dimensional curved spaces, at least in theory, if we establish a one-to-one correspondence between the *points of a non-Euclidean space* and suitably chosen *lines* of the three-dimensional *Euclidean space*. In this way we can avoid representing a curved space in a Euclidean space of dimension higher by 1.

This mapping is easy to understand if we follow KŐNIG and apply it first to a space of zero curvature (Euclidean space). Take a *base plane* and a *base line* outside of, but

---

[36] Another example can be adduced to show that this is probably not an unwarranted assumption. When reviewing the important treatise "Die Fundamentalgleichungen der nichteuklidischen Geometrie auf elementarem Wege abgeleitet" (*Arch. f. Math. u. Phys.*, 58, pp. 416—423) by MÓR RÉTHY, KLEIN confines himself to a mention by title (*Fortschritte d. Math.*, 8, 1876, p. 314).

[37] *A matematika alapjai* (The foundations of mathematics). Budapest, 1959, Vol. 2, No. 2, p. 344. (in Hung.)

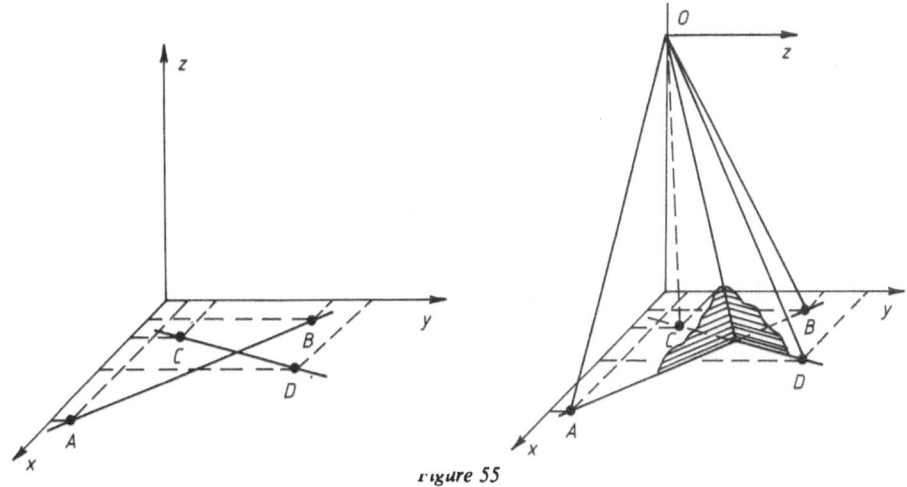

Figure 55

parallel to it. Fix a rectangular system of coordinates in the base plane and a starting point on the base line. If with point $P(x, y, z)$ of the Euclidean space we associate the line that connects point (x, y) of the system of coordinates in the base plane with point z of the base line, then the figures of the Euclidean space transform into certain complexes of lines. E.g., the *lines* of the base plane turn into planes *(Figure 55)*, the image of a *helical arc* will be a *ruled surface (Figure 56)*. This mapping can be performed for any geometrical figure, but usually the image is hard to describe. At any rate, such geometrical notions as "incidence", "betweenness", "ordering", etc. do have a counterpart under the mapping; it is also possible to introduce angles, since, e.g., the image of two straight lines subtending an angle consists of *two* intersecting families of lines *(Fig. 55)*.

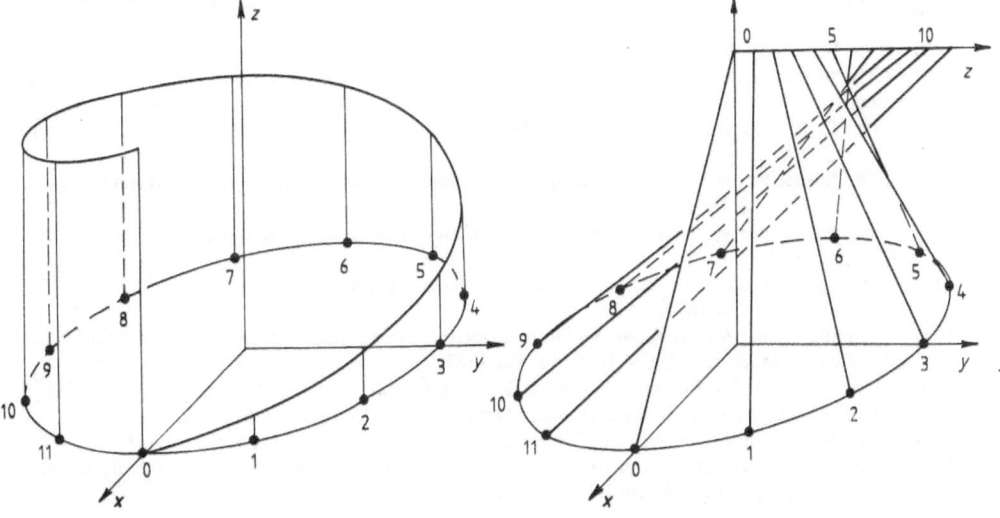

Figure 56

Now, KŐNIG contended that this method might be used for mapping non-Euclidean spaces into families of lines if we assure that the image has something corresponding to the curvature of the space. To achieve this, instead of the above-mentioned "base plane" and "base line", we choose for base objects a *curved surface* and a *curved line*, their curvature coinciding with the curvature of the original non-Euclidean space. Otherwise the process is the same as before: we fix a superficial system of coordinates on the "base surface" and a starting point on the "base curve". With the point $P(x, y, z)$ of the non-Euclidean space we associate the line that connects points (x, y) of the superficial coordinate system with point z of the base curve. Thus, the "points", "lines" and "planes" of the non-Euclidean space will be represented by "lines", "ruled surfaces" and "families of lines", respectively, while the curvature of the non-Euclidean space corresponds to the "curvature" of the base objects. If we replace the base surface and the base curve by the tangent plane and tangent line at points where these are parallel, we "in small" obtain the model we have used for representing Euclidean space. This corresponds to the theorem which says that in suitably small parts of curved spaces the theorems of Euclidean geometry are valid.

Let us stop here our brief account of KŐNIG's considerations. The foregoing may suffice to indicate his idea and his awareness of the significance of modelling.

Bibliography

[1] KALMÁR, LÁSZLÓ: *The foundations of mathematics*. Vol. 1, 1953. Felsőoktatási Jegyzetellátó Vállalat. (In Hung.)

[2] KŐNIG, GYULA: To the theory of sets. *Math. és Phys. Lapok*, 15, 1906, pp. 253—255. (In Hung.)

[3] — Sur la théorie des ensembles. *Comptes Rendus*, 143, 1906, pp. 110—112.

[4] — Zum Kontinuumproblem. *Verh. des III. Internationalen Mathematiker-Kongresses in Heidelberg*, 1905, pp. 144—147.

[5] — Zum Kontinuum-Problem. *Math. Annalen*, 60, 1905, pp. 177—180.

[6] — *Neue Grundlagen der Logik, Arithmetik und Mengenlehre*. Leipzig, 1914.

[7] — Formalizing the concept of true in synthetic logic. *A Magyar Filozófiai Társaság Közleményei*, 1910, pp. 98—106. (In Hung.)

[8] — The basic facts. *Ibid.*, 1913, pp. 133—148. (In Hung.)

[9] — The theorem concerning the divisors of integers. *Műegyetemi Lapok*, 1, 1876, pp. 186—187. (In Hung.)

[10] — Das Reciprocitätsgesetz in der Theorie der quadratischen Reste. *Acta Mathematica*, 22, 1898, pp. 181—192.

[11] — The general theory of first-order congruence systems in several unknowns. *Műegyetemi Lapok*, 2, 1877, pp. 113—121. (In Hung.)

[12] — *Algebra for the upper classes of secondary schools*. Budapest, 1879, 1880, 1881, 1882 (in four booklets). (In Hung.)

[13] — On a general solution to the algebraic equations of the *n*-th degree. *Műegyetemi Lapok*, 1, 1876, pp. 20—28, 112—116. (In Hung.)

[14] — Ein allgemeiner Ausdruck für die ihrem absoluten Betrage nach kleinste Wurzel der Gleichung *n*-ten Grades. *Math. Annalen*, 9, 1876, pp. 530—540.

[15] — On a property of power series. *Math. és Term. tud. Értesítő*, 1, 1882—1883, pp. 60—62. (In Hung.)

[16] — Über eine Eigenschaft der Potenzreihen. *Math. Annalen*, 23, 1884, pp. 447—449.

[17] — Darstellung von Functionen durch unendliche Reihen. *Math. Annalen*, 5, 1872, pp. 310—340.

[18] — Die Factorenzerlegung ganzer Functionen und damit zusammenhängende Eliminationsprobleme. *Math. Annalen*, 15, 1879, pp. 161—173.

[19] — *Zur Theorie der Modulargleichungen der elliptischen Functionen.* Inaugural-Dissertation. Heidelberg, 1871.

[20] — *Outlines of the general theory of algebraic quantities.* Budapest, 1903. (In Hung.)

[21] — *Einleitung in die allgemeine Theorie der algebraischen Grössen.* Leipzig, 1903.

[22] — *An introduction to higher algebra.* Budapest, 1876. (In Hung.)

[23] — *Analysis. An introduction to the system of mathematics.* Budapest, 1887. (In Hung.)

[24] — An elementary discussion of gamma functions. *Math. és Phys. Lapok,* 1, 1892, pp. 5—16. (In Hung.)

[25] — On the theory of definite integrals. *Math. és Term. tud. Értesítő,* 15, 1897, pp. 380—394. (In Hung.)

[26] — Über stetige Functionen, die innerhalb jedes Intervalles extreme Werte besitzen. *Monatshefte f. Math. u. Phys.,* 1, 1890, pp. 7—12.

[27] — The Hamilton systems and the general theory of partial differential equations of the first order. *Értekezések a Math. Tud. Köréből,* 8, 1881, No. 10. (In Hung.)

[28] — Über die Integration der Hamiltonschen Systeme und der partiellen Differentialgleichung erster Ordnung. *Math. Annalen,* 23, 1884, pp. 504—519.

[29] — Über die Integration simultaner Systeme partieller Differentialgleichungen mit mehreren unbekannten Functionen. *Math. Annalen,* 23, 1884, pp. 520—526.

[30] — *The theory of second-order partial differential equations with two independent variables.* Budapest, 1885. (In Hung.)

[31] — Theorie der partiellen Differentialgleichungen zweiter Ordnung mit zwei unabhängigen Variabeln. *Math. Annalen,* 24, 1884, pp. 465—536.

[32] — On the meaning of the fundamental equations of dynamics. *Értekezések a Math. Tud. Köréből,* 14, 1887, No. 1. (In Hung.)

[33] — Über eine neue Interpretation der Fundamentalgleichungen der Dynamik. *Math. Annalen,* 31, 1888, pp. 1—42.

[34] — Über eine reale Abbildung der s. g. Nicht-Euclidischen Geometrie. *Gött. Nachrichten,* 1872, pp. 157—164.

[35] KÜRSCHÁK, JÓZSEF: Gyula Kőnig. *Magyar Mérnök- és Építész-Egylet Közlönye,* 15, 1914. (In Hung.)

[36] — GYULA KŐNIG: *Mat. és Fiz. Lapok,* 40, 1933, pp. 1—23. (In Hung.)

[37] RADOS, GUSZTÁV: In memory of Gyula Kőnig. *Az Akadémia elhunyt tagjai felett tartott emlékbeszédek,* 17, 1915, No. 3. (In Hung.)

[38] SZÉNÁSSY, BARNA: *Gyula Kőnig 1849—1913.* Budapest, 1965. (In Hung.)

21. The emergence of the Bolyai cult

We have often remarked that initially the mathematical achievements of the BOLYAIS stirred no significant repercussion either at home or abroad. Both died without having earned recognition or success, painfully lacking understanding. It is far less known who and when began to recognize the significance of their work, and what the actual course was along which they eventually arrived at their deserved places in the history of mathematics.

In seeking an answer to this question the first thing we have to make clear is that the pioneers of discovering the BOLYAIS were not Hungarians; our scientific community needed foreign initiative to come alive to the fact that they had a debt in this regard. So, our scholars only continued rather than started what was their duty.[1]

The first writing that introduced the name BOLYAI to a relatively wide public was a study devoted to the memory of GAUSS in 1856. Having learnt of the correspondence between FARKAS BOLYAI and GAUSS, SARTORIUS VON WALTERSHAUSEN, the administrator of GAUSS's estate, asked FARKAS BOLYAI for the letters received from GAUSS. Making use of this material in his essay [26], he often mentions FARKAS BOLYAI's name.

The first discussion of some *mathematical results* of the two BOLYAIS came from RICHARD BALTZER.[2] This endeavour of BALTZER's is especially important as his work *Die Elemente der Mathematik* (I—II, 1860, 1862), which saw several editions, was one of the most popular textbooks in the last century.

BALTZER's book contains only a small sample of the two BOLYAIS results. It presents, e.g., FARKAS BOLYAI's solution of the trinomial equation by iteration, complemented with a few considerations on the convergence of the algorithm ([1], I, pp. 254—255);

---

[1] It is unfortunately for beyond the scope of this work, to discuss the commentaries on the *Appendix* published at the turn of the century. Nevertheless, we try to shed some light on the less known investigations connected with other results of the two BOLYAIS and on the historical events.

[2] HEINRICH RICHARD BALTZER (Meissen, 1818—1887, Giessen) was a student of archaeology, mathematics and philosophy at the University of Leipzig. He received the title of doctor of philosophy and mathematics in 1841. He taught first in Chemnitz (today Karl-Marx-Stadt) and later in Dresden. His *Theorie und Anwendung der Determinanten* was first published in 1857, and largely contributed to enhancing the appeal of the discipline (Hungarian translation: Genova, 1877). From 1869 to his death he was professor of mathematics at the University of Giessen. He devoted much energy to the publication of the collected works of MÖBIUS and JACOBI.

it describes JÁNOS BOLYAI's definition of parallelism (*ibid.*, II, pp. 12—13) and his theorem that parallelism is a transitive relation even in this case (*ibid.*, II, p. 144). These were the very first of the BOLYAIS results to reach the mathematicians of the world.

It was a milestone in the history of the BOLYAI geometry that in 1867 the *Appendix* appeared in French and in 1868 in Italian [5, 6]. The careful translations were done by HOÜEL and BATTAGLINI.[3] In this way the scientific world could get easier access to the *Appendix* than through clumsy Latin of the original text. Around this time the *Bulletin* of the Hungarian Academy of Sciences carried the following comment:

> "Corresponding member JENŐ HUNYADY reported on the treatise published by the Academy of Bordeaux with the title: »La Science absolue de l'espace par J. Bolyai, précédée d'une notice sur la vie et les travaux de W. et de J. Bolyai, par F. Schmidt, architecte à Temesvár.« He also proposed that the writings of the Hungarian mathematicians FARKAS and JÁNOS BOLYAI who had received European renown should be asked for use by the Hung. Acad. of Sci. from the Reformed College of Marosvásárhely where the mentioned documents are preserved" (*MTA Értesítője*, 2, 1868, p. 224).[4]

Returning to the French edition of the *Appendix*, let us recall a few data. HOÜEL learned about the activities of JÁNOS BOLYAI and LOBACHEVSKY via BALTZER. Realizing the importance of the subject with surprising acumen, he first published LOBACHEVSKY's *Geometrische Untersuchungen*, and then the *Appendix* in the same year. It belongs to the precedents of the French (and Italian) publication of the *Appendix* that, independently of the BOLYAI case, HOÜEL had been in correspondence with an architect of Temesvár, FERENC SCHMIDT, at the beginning only to keep each other informed of high-standard books in technical and mathematical literature published in their respective countries. But as time passed, SCHMIDT became one of the most devoted BOLYAI researchers, and also the correspondence between HOÜEL and SCHMIDT assumed its real importance in the BOLYAI research. SCHMIDT's father had known JÁNOS BOLYAI personally and he urged his son to start collecting biographical data about the two BOLYAIS. HOÜEL got to know this from the correspondence and realizing the importance of the matter, asked FERENC SCHMIDT to compile the biographies of the two BOLYAIS. SCHMIDT's short reply soon appeared in GRUNERT's *Archiv* [27], becoming the first writing to carry biographical information on the two BOLYAIS in a major language. The reprints of the French translation of the *Appendix* were also complemented with SCHMIDT's writing, which was soon published by BONCOMPAGNI

---

[3] GUILLAUME-JULES HOÜEL (Thaon, 1823—1886, Periers) was a professor of Bordeaux University. He excelled in translating original works on non-Euclidean geometry into French. GIUSEPPE BATTAGLINI (1826—1894) was a professor of advanced geometry in Naples, Rome and again in Naples. An ardent advocate of non-Euclidean geometries, he also wrote a book on LOBACHEVSKY.

[4] Probably unaware of the Hungarian initiative, BONCOMPAGNI (Rome, 1821—1894, Rome), the noted Italian historian of mathematics asked the Hungarian authorities in 1869 (cf. Some writings of LORÁND EÖTVÖS, the scholar and culture politician, Budapest, 1964. 48—49 and through TIVADAR PAULER, minister of public education, the Academy in 1871 (*Az MTA Értesítője*, 5, 1871, p. 233) to conduct an investigation of the manuscripts of the two BOLYAIS. Then, however, this work had been going on.

in Italian, in the journal of mathematical history *Bollettino di Bibliografia e Storia delle Scienze.*[5]

Although all these writings arouse interest in JÁNOS BOLYAI's work, it took absolute geometry a long time to strike root in mathematics. LOBACHEVSKY's life-work fared similarly, and we know that RIEMANN's famous inaugural dissertation (1854) was not immediately crowned with success either: it only appeared in 1867.

At any rate, a great change was brought about in the indifferent, even hostile reception of non-Euclidean geometries by the publication of BELTRAMI's treatise in 1868,[6] in which he proved that the theorems of non-Euclidean geometry are locally valid on the pseudosphere. So, he has been the first to reduce the problem of consistency of non-Euclidean geometries to the problem of consistency of Euclidean geometry.

But staunch opposition continued even after BELTRAMI, although the opponents of non-Euclidean geometries were not so much mathematicians as philosophers. The struggle was spearheaded by RUDOLF HERMANN LOTZE (1817—1881), a philosopher of great authority in Göttingen who deemed non-Euclidean geometries nonsense. His arguments included some non-Euclidean geometrical theorems that were inaccessible to the Euclidean approach as well as the anti-Kantian philosophical consequences of the new concept of space.

As late as the sixties and seventies of the last century, writings kept appearing which condemned the non-Euclidean geometries as unacceptable. To illustrate this, let us quote but one passage with special relevance to Hungary. It should be noted in advance that in the academic year 1871/72 the eminent Austrian mathematician J. FRISCHAUF[7] already gave a course on non-Euclidean geometries at the university of Graz, drawing chiefly on the *Appendix*. That was the first detailed exposition of the *Appendix*, the importance of which was enhanced by the fact that in 1872 FRISCHAUF published the subject-matter of his lectures in book from [15]. FRISCHAUF incurred severe criticism from the Austrian educational authorities for his courageous move.[8] A typical document of the controversy can be read in the preface to FRISCHAUF's book. The author first notes that reading his book requires little preliminary knowledge of mathematics and then adds:

"I have thought I should emphasize this circumstance as the imperial and royal Ministry of Education reproached me for my lecture »Pangeometry and projectiv-

[5] FORTI, ANGELO: Nota intorno alla vita ed agli scritti di Wolfgangi e Giovanni Bolyai di Bolya, matematici ungheresi. *Boll.*, 1, pp. 277—299.

[6] BELTRAMI, EUGENIO: Saggio di interpretazione della geometria non euclidea. *Giorn. Mat.*, 6, 1868, pp. 285—315.

[7] JOHANNES, FRISCHAUF (Vienna, 1837—1924, Graz) studied at the university of science and technology in Vienna. From 1860 he worked at the observatory, then accepted the inivitation of the university of Graz to the chair of mathematics in 1866. He published several papers in the *Acta of the Vienna Academy*, the *Zeitschrift f. Math. u. Physik*, and the *Journal f. reine u. angew. Mathematik*. His didactically well-organized books on arithmetic, geometry, physics and astronomy-geodesy incorporating the most recent results greatly improved the standard of university education in Austria.

[8] The fierce dispute in which absolute geometry was defended by FRISCHAUF and KILLING took place in Volumes 7—8 (1876—77) of *Zeitschrift f. math. u. naturw. Unterricht.*

ity« held in the winter term of the 1871/72 academic year being too difficult to understand, although there is supposed to be freedom of teaching and learning at our universities. As a matter of course, I rejected this objection based on simply incredible ignorance. The gentleman in charge may find evidence of the rightfulness of my previous reply on the basis of pangeometry developed in this book, and as for projective geometry, he is kindly advised to take a book on elementary geometry in his hands, or one of the more detailed textbooks of new geometry. I refrain, here from expounding the opinion we must have of the Austrian educational reforms if the imp. and roy. Ministry have officials in charge of university lectures who have hardly seen secondary-school text adopting up-to-date viewpoints."[9]

Let us remain with FRISCHAUF's mentioned book for a little while. The book was the first and for a long time the only treatise which, relying on the *Appendix*, con-structed absolute geometry in an elementary synthetic way. The preface also discloses that FRISCHAUF's original plan was to publish the *Appendix* with commentarias, but he gave up this project when he learned that GYULA KŐNIG was preparing a work of this kind. In the introduction, the author gives a brief summary of the contents of the *Appendix* section by section, but later he does not insist on this sequence, making changes and regroupings for methodological purposes. His discussion does not touch on the problems of construction JÁNOS BOLYAI solved in hyperbolic space.

Similar is the content of FRISCHAUF's *Elemente der absoluten Geometrie* (Leipzig, 1876), a more detailed and comprehensive book. The *Appendix* became known both at home and abroad mainly through these two books as only few had read the ori-ginal: moreover, FRISCHAUF's work lent great assistance to further elaborations. This must be stressed, for some of the later commentators did not even mention FRISCHAUF's name.

In making non-Euclidean geometries recognized and proving relative consistency, an outstanding role was played by FELIX KLEIN. At one place ([18], 1, pp. 152—153) he tells the story of the genesis of his fundamental papers. We know from here that he got acquainted with the theories of LOBACHEVSKY and JÁNOS BOLYAI as late as 1869 — and only in the second-hand presentation of STOLZ. The next year KLEIN held a lecture on CAYLEY's measure in the WEIERSTRASS seminar in Berlin where he raised the question whether this concept of measure could be applied to the BOLYAI–LOBACHEVSKY geometry or not. In endless debates with STOLZ — at a time when he had admittedly never read a single word of LOBACHEVSKY or BOLYAI in the original — he arrived at the conviction that non-Euclidean geometries could be treated as special chapters of projective geometry. He summarized his ideas in a short essay [19] that contained the construction called the CAYLEY–KLEIN *model* ever since.

All these examinations gave momentum to the processing of the BOLYAIS *posthumous papers*. The "Bolyai Committee" set up in 1871 (including JÁNOS ÁRMIN VÉSZ, GYULA KŐNIG, JENŐ HUNYADY and FERENC SCHMIDT) made a report, finding some 10—15 sheets of the two BOLYAIS writings worthy of publication. They commissioned GYULA KŐNIG to edit the works and advised Department III of the Academy to put aside a

[9] [15], p. VII.

certain amount annually for these expenses. But it took a very long time to have the plan completed.

Yet the years of protraction were not lost: in the course of elaborating the posthumous papers newer and newer results of the BOLYAIS were discovered that had to be completed or made generalizations and further investigations possible. *This is how the scientific literature relying on the achievements and writings of the two Bolyais evolved in Hungary in the last two decades of the 19th century.*

The prelude was MÓR RÉTHY's lecture delivered at Kolozsvár University in 1874, which soon appeared in print [20]. This paper throws light on RÉTHY's efforts to introduce and popularize absolute geometry. His aim was to kindle interest in reading the *Appendix* itself, and with this in mind, he gave reformulations of several basic definitions and concepts (BOLYAI's definition of parallelism, the paracycle, parasphere, hypercycle, hypersphere) that are easier to follow than their originals in the *Appendix*. And he went a step further: starting from the fact that in absolute geometry in infinetly small segments of space the theorems of Euclidean geometry hold true, and relying on the recognition that spheric trigonometry is independent of EUCLID's 5th postulate, RÉTHY built up BOLYAI trigonometry independently. Thus, he set the trend aiming at the simplest possible exploration of absolute trigonometry relying on the fewest possible axioms. Of the wealth of literature in this vein, let us only mention the articles by PÁL SZÁSZ.

Another merit of RÉTHY is that he was the first to review and elaborate the hyperbolic constructions of the Appendix in detail.

PAUL STÄCKEL also mentions MÓR RÉTHY's lecture of 1874 at one place ([29], 1, p. 202). The data he presented can be found in several subsequent sources but dressed in a wording that misleads the reader into thinking RÉTHY had held a whole *course* of lectures on the *Appendix*. We have not found any trace of that anywhere. It is, however, indubitable that RÉTHY did very much to have the two BOLYAIS recognized and advocated the position that mathematical research in Hungary must start from their results.

Making a brief detour, let us correct a *mistake* by STÄCKEL at this point: he claims that GYULA VÁLYI regularly held special seminars on the material of the *Appendix* from 1887 ([29], 1, p. 203). The correct starting date is the second term of the academic year 1891/2: that was the first time, according to the time-table of Kolozsvár University, that GYULA VÁLYI announced a special seminar on "János Bolyai's Appendix".

The publication and detailed elaboration of some significant results of FARKAS BOLYAI began in the seventies. In an early study of his [13], GYULA FARKAS enlarged upon the root approximation method for trinomial equations discussed briefly in the *Tentamen*, paying special attention to the questions of convergence of the algorithm.

Through the good offices of BALTZER's above-mentioned book and GYULA FARKAS's study, the BOLYAI algorithm became well known and inspired several Hungarian and foreign mathematicians to tackle the problems of its generalization, application and convergence up to this day.

Of the earlier writings, without aiming at completeness, we mention the following. ÅSTRAND: Neue einfache Transformation und Auflösung der Gleichungen von der Form $x^n - ax \pm b = 0$. *Astr. Nachr.*, 89, 1877, pp. 347—350; *idem*: Om en ny Methode for Lösning af trinomiske Ligninger af n-te Grad. *Arch. för*

Math. og Naturvindenskab., 6, 1880, pp. 448—449; DICKSON: Continued roots. *Analyst*, 5, 1878, pp. 20—21; GÜNTHER: Eine didaktisch wichtige Lösung trinomischer Gleichung. *Zeitschrift f. math. u. naturw. Unterricht*, 11, 1880, pp. 68—72; SANIO: Bemerkungen über Gleichungsauflösung. *Arch. d. Math. u. Phys.*, 2, 1885, pp. 332—336; NETTO: Über einen Algorithmus zur Auflösung numerischer algebraischer Gleichungen. *Math. Ann.*, 29, 1886, pp. 141—147; *Idem: Vorlesungen über Algebra*, Vol. I, Leipzig, 1896, p. 301 (without mention of the name); ISENKRAHE: Über die Anwendung iterirter Functionen zur Darstellung der Wurzeln algebraischer und transcendenter Gleichungen. *Math. Annalen*, 31, 1888, pp. 309—317; *Idem: Das Verfahren der Functionswiederholung, seine geometrische Veranschaulichung und algebraische Anwendung*, Leipzig, 1897 (this voluminous reprint is the then fullest summary of iterative methods); ARISTOV: Über Iteration der Functionen. *Nachr. Univ. Kasan*, 10, 1900, pp. 15—49, 85—131.

More recent publications treating or applying the BOLYAI algorithm are the following. ONOFRI-MAMBRIANI: Su algoritmi infiniti generati da certe equazioni ricorrenti. *Boll. Un. Mat. Ital.*, 14, 1934, pp. 71—78; VERESS, P.: The Bolyai algorithm and the problem of the interest rate in computing annuities. *Mennyiségtani és Természettudományi Didaktikai Lapok.* 1, 1943, pp. 57—62 (in Hung.); EVERETT: Representation for real numbers. *Bull. of the Amer. Math. Soc.*, 52, 1946, pp. 861—869; RÉNYI, A.: On algorithms for the production of real numbers. *Az MTA Mat. és Fiz. Tud. Oszt. Közl.*, 7, 1957, pp. 265—293 (in Hung.); MYRBERG: Iteration von Quadratwurzeloperationen. Suomalaisen Tiedeakatemian Toimituksia. *Annales Acad. Scientiarum Fennicae*, 1958, No. 259; and several other papers by Myrberg.

But there is more to GYULA FARKAS's credit than the few results listed above, as he addressed the question of convergence of *more general* iterative procedures as well [14]: given the function $f(x)$, let us substitute $f(x)$ for x again and again, indefinitely. This yields an infinite sequence of functions.

Now, according to GYULA FARKAS:

1. If $f(z)$ is an analytic function of the complex variable z, and if the condition is satisfied that while z changes over a domain T, the corresponding values $f(z)$ are also within T (even if T shrinks to a point), then the iteration sequence is convergent.

The analogue of this theorem in the real case is:

2. Let x and $f(x)$ be real, and let $f(x)$ increase together with x. If the inequality

$$p < f(p) < f(x) < f(q) < q$$

is fulfilled for $p < x < q$, then the mentioned iteration process is convergent.

SÁMUEL BRASSAI delivered his memorial oration over FARKAS BOLYAI [11] in 1886, that is, with a delay of 30 years. The essay, useful from the viewpoint of biographical data but sub-standard and downright denunciatory regarding the presentation of the mathematical results, seems to have had the deliberate aim of discrediting the name BOLYAI. It is most regrettable that the highly prestigious professor of the university of Kolozsvár undertook that ignoble task but it is also strange that as late as 1898 the Academy thought it fit to publish BRASSAI's posthumous essay on the *XIth axiom* [12]. The study which was behind the times by a good 50 years professionally, contains sentences like this: ". . . we should not raise the theorem of 'non-Euclidean' geometry to the rank of mathematical truth." It is no excuse for BRASSAI that his mathematical knowledge was too defective to understand the *Tentamen* and the *Appendix*, and that he, as the first translator of Euclid, was unable to depart from the realm of thought described in the *Elements*.

Undoubtedly, it can be attributed to BRASSAI that the enthusiastic upswing of the seventies in the fight for the cause of the two BOLYAIS slightly slackened over the eighties. The planned publication of their bequest was also dragging on. Only some superficial signs of esteem can be cited from these years. Directly or indirectly, all this was presented by the indefatiguable FERENC SCHMIDT. Upon his initiative the board of the Marosvásárhely college erected a tombstone over the unmarked grave of FARKAS BOLYAI in 1884; in 1893 FERENC SCHMIDT tracked down JÁNOS BOLYAI's grave with the help of the faithful servant JULIA SZŐTS. The Mathematical and Physical Society erected a monument there in 1894. In 1911 the mortal remains of the BOLYAI's were disinterred and laid to rest in one grave.

As has been mentioned, FARKAS BOLYAI presented three theorems on the "end-like equality of areas" in the *Tentamen* (cf. p. 155), but only proved the first. This shortcoming was discovered during the preparation of the second edition of the book. This triggered off MÓR RÉTHY's interest in the problem in the nineties. First of all he proved that FARKAS BOLYAI's third theorem follows from the second. To verify the second theorem, he presented a possible way of decomposition. This can be sketched as follows.

Let us call two plane figures congruent "in equal sense" if one can be brought to overlap with the other by rotating it in the plane by φ. Now, let A and B be simply connected domains congruent in equal sense with intersection K and centre of rotation O exterior to both of them *(Figure 57)*. According to RÉTHY, we can dissect the non-overlapping parts of A and B by rotating the intersection K around centre O at

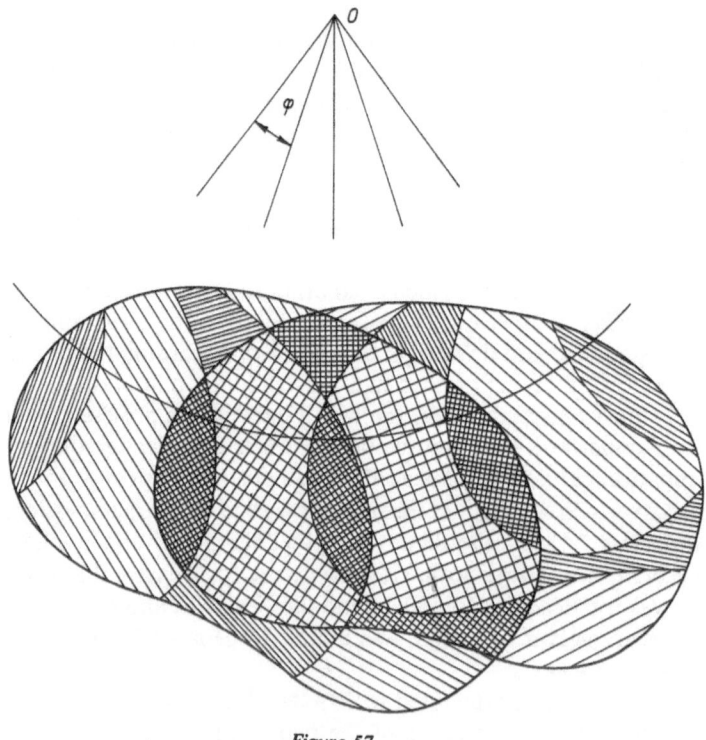

Figure 57

angles φ, 2φ, 3φ, ..., then $-\varphi$, -2φ, -3φ, ... the necessary number of times and mask the contour of K every time.

Improving on FARKAS BOLYAI's results, MÓR RÉTHY found also the following. For the end-like equality of two plane figures of equal area it is necessary and sufficient that their non-straight peripherical arcs partly consist of pairwise congruent pieces whose sense of curvature relative to the interior of the respective figure coincides, and partly of pieces which occur on the periphery of one and the same plane figure the same number of times in the positive and negative sense. This criterion carries over to figures on spheres of equal radii.

RÉTHY's papers [22—25] on this subject were soon followed by several writings[10] and today there is a vast number of investigations applying the concept of the end-like equality of areas in some way or other. Around the turn of the century, those mathematicians engaged in working out the strict foundations of geometry (DEHN, KAGAN, KILLING, SATUNOVSKY, SCHUR, STOLZ, ZOLT, and others) regarded this concept as a suitable starting point for the introduction of area measure.

DAVID HILBERT's by now classic definitions and the results formulated with their help mark in a certain sense the completion of the investigations began by FARKAS BOLYAI. Following HILBERT we say that two polygons are *equal for dissection* (zerlegungsgleich) if they can be decomposed into a finite number of pairwise congruent triangles (= they are end-like equal). Two polygons have *equal areas* (inhaltsgleich) or are *of equal area* (P. von gleichem Inhalte) if we can attach end-like equal polygons to them so that also the resulting composite polygons are end-like equal.

According to HILBERT's famous theorem, the end-like equality of two polygons of equal area cannot be proved without using the EUDOXUS—ARCHIMEDES axiom of continuity (or an equivalent axiom) ([16], pp. 53—63).[11]

Besides GYULA VÁLYI's university course to be discussed later (Chapter 22), another significant event of the 1890s was the involvement of HALSTED,[12] professor of math-

---

[10] SPIEGL, ZSIGMOND: Contribution to the theory of end-like equal areas. *Math. és Phys. Lapok*, 2, 1893, pp. 17—20 (in Hung.); DOBRINER, HERMANN: Bemerkungen zu der Abhandlung des Herrn M. RÉTHY über "Endlichgleiche Flächen". *Math. Ann.*, 42, 1893, pp. 275—284; *idem*: Der Satz "Congruentes von Congruentem giebt Gleiches" in seiner Anwendung auf ebene Flächen. *Math. Ann.*, 42, 1893, pp. 285—296; RAUSENBERGER, OTTO: Das Grundproblem der Flächen- und Rauminhaltslehre. *Math. Ann.*, 43, 1893, pp. 601—604.

[11] From the profuse literature on the subject, let us mention two further investigations by Hungarian authors. TAMÁS VARGA reformulated FARKAS BOLYAI's proof of theorem 1 (cf. p. 156) so that it can be taught in secondary schools (Farkas Bolyai's theorem of dissection. *Matematikai Lapok*, 5, 1954, pp. 101—114 (in Hung.). PÁL SZÁSZ also gave a proof of theorem 1; he used EUDOXUS' axiom and the axioms of incidence, ordering and congruence, but not the area concept or HILBERT's *segment calculus* (Farkas Bolyai's theorem on the dissection of polygons (*Matematikai Lapok*, 7, 1956, pp. 230—237; in Hung.).

[12] GEORGE BRUCE HALSTED (Newark, 1853—1922, New York) studied at Princeton University. He was a professor at the universities of Austin, Ohio and finally Colorado. His main fields of interest were the philosophical problems of mathematics, mathematical logic, and non-Euclidean geometries. In the latter field he did immense work by editing works on the subject in America and also writing books on his own. He was greatly attracted to the work of SACCHERI, JÁNOS BOLYAI, LOBACHEVSKY and POINCARÉ. He was a devout follower of Darwinism and as a proof of the non-Euclidean nature of the universe, proposed the interesting argument that Darwinism was only compatible with non-Euclidean geometries (Darwinism and non-Euclidean geometry. *Nachr. Univ. Kazan*, 6, 1897, pp. 22—25).

ematics at the University of Austin, Texas, in the research into absolute and hyperbolic geometries. First he published several treatises by LOBACHEVSKY in his translation, shortly followed by the English translation of the *Appendix* — published four
times within three years [7]. HALSTED's activities are especially important for Hungary
because, as an unbiased person deeply committed to the cause, he thought it his duty
to earn world-wide appreciation for BOLYAI and LOBACHEVSKY on grounds of parity.
Prior to his efforts, LOBACHEVSKY had been commented, and his writings published
more often, partly became the Russian had done more to have their outstanding
mathematician acknowledged. For example, a statue was erected in honour of
LOBACHEVSKY in Kazan, and a prize of substantial value was offered in his memory.
Commenting on the latter event, HALSTED wrote in the December 1895 issue of the
magazine *Science* of New York:

"And in the country for which the genius of JÁNOS BOLYAI had brought just as
much glory by creating non-Euclidean geometry — in Hungary — they rest content
with the single move of the Mathematical and Physical Society of Budapest erect-
ing a commemorative stone over his collapsed grave at Marosvásárhely."

That HALSTED's respect for JÁNOS BOLYAI was no lip-service was proved by his
pilgrimage to the grave at Marosvásárhely in the summer of 1896. On that occasion
he encouraged a professor of the college there, JÁNOS BEDŐHÁZI, to write a detailed
biography of the two BOLYAIS — as a footnote to BEDŐHÁZI's book reveals ([2], pp.
452—453). It is only too typical that a foreigner had to bring his Hungarian colleague
alive to the importance of writing a book of this kind.

In 1894 the Hungarian scientists achieved a great success in the fight to have
LOBACHEVSKY's and JÁNOS BOLYAI's names mentioned aequo loco. In that year the
"Congrès international de bibliographie des sciences mathématiques" were preparing a major bibliographic publication under the chairmanship of POINCARÉ, one chapter
of which was planned to be "Géométrie de Lobatchewsky". The committee set up to
compile the relevant Hungarian writings (GUSZTÁV RADOS, BÉLA TŐTÖSSY, JÓZSEF
KÜRSCHÁK and LAJOS KOPP) intervened to have the title complemented with JÁNOS
BOLYAI's name. That marks the date since when the two names have jointly hallmarked
the first non-Euclidean geometries.

After long preparatory work dragging on for decades, in 1897 GYULA KŐNIG at last
presented the first volume of the second edition of the *Tentamon* to a session of the
Academy. The hard work of editing was undertaken by KŐNIG and MÓR RÉTHY. As
the mathematician editors did not feel competent enough in the Latin language, they
turned for help to the outstanding philologist HENRIK FINÁLY. FINÁLY's mastery of
Latin and his intelligence in mathematics greatly improved the quality of the publication. But the second volume, of Tentamen treating geometry (and the supplemen-
tary volume containing the figures) appeared much later: MÓR RÉTHY presented it
at the end of 1904 (*Akadémiai Értesítő*, 16, 1905, pp. 49—52).

The delay was caused by personal and financial reasons. After the appearance of
the first volume GYULA KŐNIG resigned from further work as he felt incompetent in
geometry. He was succeeded in this job by JÓZSEF KÜRSCHÁK and the figures were

drawn by BÉLA TŐTŐSSY. Due to the death of HENRIK FINÁLY his son GÁBOR helped to revise the Latin text.

Our country redeemed its long-standing debt to JÁNOS BOLYAI in 1897 when the *Appendix* appeared in Hungarian at once in two translations. One was made by IGNÁC RADOS [8], the other — and better — by JÓZSEF SUTÁK [9]. The latter came out as a separate volume with SUTÁK's detailed but often objectionable comments added.

Then at the turn of the century there was a massive upswing in research into the life and work of the two BOLYAIS. FERENC SCHMIDT's decades-long endeavour was at last crowned with success: the correspondence of FARKAS BOLYAI and GAUSS appeared [4]. As was mentioned earlier, FARKAS BOLYAI sent the letters received from GAUSS to Göttingen where they were attached to the collection of the Gauss manuscripts. After many abortive attempts, in 1896 SCHMIDT obtained from Göttingen the accurate copies of the letters, followed by the permission for publication, through the good offices of the secretary of the local mathematical society WILAMOWITZ-MOELLENDORFF, and a young colleague of his, PAUL STÄCKEL.[13] The volume was prepared by SCHMIDT, NÁNDOR GRUBER and STÄCKEL. STÄCKEL followed the example of HALSTED and visited Marosvásárhely to get a deep insight into the BOLYAI question. Not even the difficulties of the language could prevent him from thoroughly studying and processing the writings left by the BOLYAIS. He published his findings in a series of articles which revealed several significant ideas of JÁNOS BOLYAI not included in the *Appendix*. Despite its minor inaccuracies and errors, STÄCKEL's two-volume work [29] comprising the above findings as well as ample biographic data and excerpts from the *Tentamen* and *Appendix* has remained a standard work which provides the fullest picture of the activities of the two BOLYAIS. It was a very modest way of thanking STÄCKEL for his invaluable services to coopt him among the foreign members of the Academy in 1900.

Then came the centenary of JÁNOS BOLYAI's birth which was celebrated at Kolozsvár, on 15 January 1903. The university published a representative volume of studies in Latin for this occasion, which — together with the commemorative addresses delivered during the festivities — soon appeared also in Hungarian ([17], [10]). It was a sign of increasing interest and appreciation that apart from the papers by STÄCKEL, SCHLESINGER and BONOLA published in the commemorative volume, there were several high-standard writings on JÁNOS BOLYAI in the 1903 issues of the *Mathematikai és Physikai Lapok*.[14] They clearly show that the memory of the BOLYAIS was cherished not only in Kolozsvár but also in Budapest. IZIDOR FRÖHLICH had all justification to say in his memorial address that "... the bright ideas of JÁNOS BOLYAI's mind

---

[13] SAMUEL GUSTAV PAUL STÄCKEL (Berlin, 1862—1919, Heidelberg) finished his studies in his native town in 1884. During his adventurous life he taught at the universities of Halle, Königsberg, Kiel, Hannover, Karlsruhe and finally Heidelberg. He was chiefly interested in problems of algebra, complex functions and theoretical mechanics. His original contributions published mostly in *Math. Annalen* and CRELLE's *Journal* were influenced by questions raised by EULER, LAGRANGE, GAUSS, ABEL and JACOBI.

[14] BEKE, MANÓ: The Bolyai trigonometry. *Math. és Phys. Lapok*, 12, 1903, pp. 30—49 (In Hung.); KÜRSCHÁK, JÓZSEF: On the angle of parallelism. *Ibid.*, pp. 50—52 (in Hung.); RÉTHY, MÓR: A survey of János Bolyai's "New different world". *Ibid.*, 1—29, pp. 303—320 (in Hung.); SCHLESINGER, LAJOS: János Bolyai. *Ibid.*, pp. 57—88 (in Hung.); SZABÓ, PÉTER: On a basic theorem of absolute geometry. *Ibid.*, pp. 321—326 (in Hung.)

are being disseminated by the departments of our university" ([10], p. 57). Regrettably enough, we cannot refuse the statement in FELIX KLEIN's telegramme of felicitation saying that the importance of the *Appendix* was first recognized by a foreign mathematician, BALTZER, and not by the Hungarians ([10], p. 166).[15]

On the occasion of the centenary, a plaque was placed on the native house of JÁNOS BOLYAI in Kolozsvár up to now a memento of the great mathematician. The Academy of Sciences established a "Bolyai prize" on the model of the "Lobachevsky prize". The charter of the prize proclaimed that the prize was to be awarded first in 1905 and then in every fifth year "for the best mathematical investigation published anywhere in any language". The prize was 10,000 Hungarian crowns and a gold medal worth 600 crowns. It was also provided for that a four-member panel decide upon the awarding of the prize, two of them Hungarians and two foreigners.

Unfortunately, the prize could not be awarded more than twice: in 1905 the committee (KŐNIG, RADOS, DARBOUX, KLEIN) honoured POINCARÉ, in 1910 they (KŐNIG, RADOS, MITTAG-LEFFLER, POINCARÉ) nominated HILBERT. During the First World War international scientific contacts were interrupted, Hungarian currency was inflated and the fine initiative broke off.

Parallel to the establishment of the *Bolyai prize*, the Academy announced a composition for writing a monograph on absolute geometry. To our knowledge, LAJOS SCHLESINGER, who held a jubilee course at the university of Kolozsvár under the title *The absolutely true science of space* encompassing a vast material, began to draw up the work but it was never published.

*

That marks the end of the first great period of "discovering" the two BOLYAI's. Over the subsequent decades there were always some, though not many, who cherished the memory of the two mathematicians. This work gathered some new momentum both at home and abroad from the recent BOLYAI anniversaries.

Bibliography

[1] BALTZER, RICHARD: *Die Elemente der Mathematik.* 5th edition. Leipzig, 1875.
[2] BEDŐHÁZI, JÁNOS: *The two Bolyais.* Marosvásárhely, 1897. (In Hung.)
[3] BOLYAI, FARKAS: *Tentamen. I—II.* 2nd ed. Budapest, 1897, 1904.
[4] *Briefwechsel zwischen Carl Friedrich Gauss und Wolfgang Bolyai.* Leipzig, 1899.
[5] BOLYAI, JÁNOS: La science absolue de l'espace. *Bordeaux Mém.,* Vol. 5, 1867, pp. 207—248.
[6] — Sulla scienza dello spazio assolutamente vera. *Giorn. Mat.,* Vol. 6, 1868, pp. 97—116.
[7] — *The science absolute of space.* Austin 1891, Tokyo 1891, Austin 1894, Austin 1896.
[8] — The absolutely true science of space. Transl. by Ignácz Rados. *Math. és Phys. Lapok,* 6, 1897, pp. 145—192. (In Hung.)
[9] — *The absolutely true science of space.* Intr. and annot. by JÓZSEF SUTÁK. Biography of János Bolyai by FERENC SCHMIDT. Budapest, 1897. (In Hung.)
[10] *Bolyai, János . . . Jubilee.* Kolozsvár, 1903. (In Hung.)

---

[15] FELIX KLEIN was only mistaken in forgetting about FARKAS BOLYAI.

[11] BRASSAI, SÁMUEL: Commemorative oration over Farkas Bolyai. *Az Erdélyi Múzeum-Egylet Kiadványai*, 3, Kolozsvár, 1886, pp. 209—247. (In Hung.)

[12] — The XIth axiom. *Akadémiai Értesítő*, 9, 1898, pp. 415—427. (In Hung.)

[13] FARKAS, GYULA: The Bolyai algorithm. *Értekezések a Math. Tud. Köréből*, Vol. 8, No. 3, 1881. (In Hung.)

[14] — Sur les fonctions itératives. *Journal de Math.*, 10, 1884, pp. 101—108.

[15] FRISCHAUF, JOHANN: *Absolute Geometrie nach Johann Bolyai.* Leipzig, 1872.

[16] HILBERT, DAVID: *Grundlagen der Geometrie.* 5th ed. 1922.

[17] *Ioannis Bolyai in memoriam.* Claudiopolis (Kolozsvár), 1902.

[18] KLEIN, FELIX: *Vorlesungen über die Entwicklung der Mathematik im 19. Jahrhundert.* Berlin, 1926.

[19] — Über die sogenannte Nicht-Euklidische Geometrie. *Gött. Nachr.*, 1871, pp. 419—433; *Math. Annalen*, 4, 1871, pp. 573—625.

[20] RÉTHY, MÓR: The so-called non-Euclidean plane trigonometry of the three-dimensional homogeneous space. *Értekezések a Math. Tud. Köréből*, Vol. 6, No. 7, 1875. (In Hung.); Die Fundamentalgleichungen der nicht-euklidischen Geometrie auf elementarem Wege abgeleitet. *Arch. f. Math. u. Phys.*, 58, 1875, pp. 416—423.

[21] — A survey of János Bolyai's "new, different world". *Math. és Phys. Lapok*, 12, 1903, pp. 1—29; 303—320. (In Hung.)

[22] — End-like equal areas. *Math. és Term. tud. Értesítő*, 8., 1890, pp. 176—202; idem., *Math. és Phys. Lapok*, 2, 1893, pp. 1—16; 118—129; 241—253. (In Hung.)

[23] — Endlich gleiche Flächen. *Math. Annalen*, 38, 1891, pp. 405—428.

[24] — Über endlich-gleiche Flächen. *Math. Annalen*, 42, 1893, pp. 297—307.

[25] — Zum Beweise des Hauptsatzes über die Endlichkeit zweier ebenen Systeme. *Math. Annalen*, 44, 1894, pp. 471—472.

[26] SARTORIUS VON WALTERSHAUSEN: *Gauss zum Gedächtnis.* Leipzig, 1856.

[27] SCHMIDT, FERENC: Aus dem Leben zweier ungarischen Mathematiker, Wolfgang und Johann Bolyai von Bolya. *Archiv. f. Math. u. Phys.*, 48, 1867, pp. 217—228. The same in French: *Bordeaux Mém.*, 5, 1867, pp. 191—205.

[28] SPIEGL, ZSIGMOND: A contribution to the theory of end-like equal areas. *Math. és Phys. Lapok*, 2, 1893, pp. 17—20. (In Hung.)

[29] STÄCKEL, PAUL: *Wolfgang und Johann Bolyai. Geometrische Untersuchungen. I—II.* Leipzig and Berlin, 1913. The same in Hungarian: Budapest, 1914.

[30] SZÉNÁSSY, BARNA: Some data from the history of how the two Bolyais were discovered. *Matematikai Lapok*, 29, Budapest, 1977—1981, pp. 71—95. (In Hung.)

[31] VÁLYI, GYULA: *On János Bolyai's Appendix.* Litographed after the lectures of Dr. Gyula Vályi. Kolozsvár, 1904. (In Hung.)

22. Gyula Vályi

He was not a mathematician to publish profusely. His extreme self-criticism and modesty kept him back from publicity. He was especially shy of coming out with "semi-finished goods"; what he published was always aimed at illumining a so-far unsettled problem of mathematics from as many aspects as possible.

A part of his ideas have been preserved not by his publications but by lecture notes taken down more or less accurately by his students and circulated in duplication. They testify to the high standards of VÁLYI's activities pursues for many years at the university of Kolozsvár in teachers and researchers.

Internationally, his most widely noted work was his doctoral dissertation [4]. The roots of the mathematical investigation in it go back to the problem of the best marine screw, to which MÓR RÉTHY had drawn his attention. The dissertation is closely connected to papers on this subject (see Chapter 17).

The dynamic core of the problem is the following. Let physical action be composed of actions exerted on the elements of a surface; we want to find the surface $z = f(x, y)$ upon which the action is maximal. Generally speaking, the elementary actions may depend on the position and orientation of the surface element df. If, however, we assume a dependence of this kind, the mathematics of the problem is too complicated. It became simpler under the restriction that the action V depends on the orientation of the surface element but not on its position (which is not the case e.g. with the steam turbine). In this case we have to solve the variational problem

$$\iint V(p, q)\,dx\,dy$$

in two variables.[1] [In this chapter, as well, the letters p, q, r, s, t denote the first and second partial derivatives of the function $z = f(x, y)$.] This variational problem leads to the following partial differential equation of the second order:

$$\frac{\partial^2 V}{\partial p^2}r + 2\frac{\partial^2 V}{\partial p \partial q}s + \frac{\partial^2 V}{\partial q^2}t \equiv 0. \tag{I}$$

In his doctoral dissertation, GYULA VÁLYI first of all examined necessary and sufficient conditions in order that this second-order partial differential equation for z be

---

[1] For the details, see TIBOR WESZELY's book [3], where GYULA VÁLYI's work is surveyed at length.

reducible to a first-order equation provided that V is known. According to the important criterion given by VÁLYI the following condition should be satisfied:

$$2D\frac{\partial^2 D}{\partial q^2} - 2D^3\frac{\partial^2 D}{\partial P^2} - 3\left(\frac{\partial D}{\partial q}\right)^2 - D^2\left(\frac{\partial D}{\partial P}\right)^2 \equiv 0; \quad \left(P = \frac{\partial V}{\partial q}\right). \tag{II}$$

Here

$$D = \left[\left(\frac{\partial^2 V}{\partial p \partial q}\right)^2 - \frac{\partial^2 V}{\partial p^2}\frac{\partial^2 V}{\partial q^2}\right]^{\frac{1}{2}}. \tag{III}$$

So, if (II) is fulfilled then the partial differential equation (I), and hence the variational problem, can be regarded as solved, for we have exact methods to integrate first-order partial differential equations. The significance of VÁLYI's criterion is that differential equations of type (I) often arise in practice.

Further, VÁLYI also posed the question. For which action functions V can the variational problem be solved by the above procedure? To get an answer, we have to integrate the partial differential equation (II). In this part of his dissertation, however, VÁLYI committed an error that was overlooked for quite a long time. In 1905 KAPTEYN checked the computations and found the correct solution to (II).[2]

JÓZSEF KÜRSCHÁK carried on the investigations begun by VÁLYI in several directions. In one of his treatises[3] he gave simpler proofs to VÁLYI's results. He also examined the case where the variational integral depends not only on the orientation of the surface element but also on its position, i.e., he considered the variation of

$$\iint V(x, y, z, p, q)\,dx\,dy. \tag{IV}$$

According to JÓZSEF KÜRSCHÁK's theorem,[4] the corresponding partial differential equation can be reduced to an equation of the first order, if and only if Eq. (III) is identically zero. Finally, KÜRSCHÁK also arrived at notable results for the variational problem with $n > 2$ variables, obtained by generalization of Eq. (IV).[5]

Owing to the large monograph by FORSYTH,[6] another of VÁLYI's contributions to the theorem of partial differential equations became well known, namely his criterion for the integrability of the system of simultaneous second-order partial differential equations

$$F_1(x, y, z, p, q, r, s, t) = 0$$

$$F_2(x, y, z, p, q, r, s, t) = 0.$$

VÁLYI argued as follows. Let us find an equation

$$F(x, y, z, p, q, r, s, t) = 0$$

[2] KAPTEYN, W.: Sur l'équation différentielle de Monge. *Archiv. f. Math. u. Phys.*, (3) 9, 1905, pp. 313—329.
[3] KÜRSCHÁK, JÓZSEF: Über die partielle Differentialgleichung des Problems $\delta \iint V(p, q)\,dx\,dy = 0$. *Math. Annalen*, 44, 1894, pp. 9—16.
[4] KÜRSCHÁK, JÓZSEF: Über partielle Differentialgleichungen zweiter Ordnung mit gleichen Characteristiken. *Math. Annalen*, 37, 1890, pp. 317—320
[5] KÜRSCHÁK, JÓZSEF: Über eine Klasse der partiellen Differentialgleichungen zweiter Ordnung. *Math. u. naturw. Berichte aus Ungarn*, 14, 1895—96, pp. 285—318.
[6] FORSYTH: *Theorie der Differentialgleichungen*. Leipzig, 1893. 358.

such that r, s, t (as functions of x, y, z, p, q) calculated from the system of equations

$$F_1=0, \qquad F_2=0, \qquad F=0$$

supposed to be competible make the system of total differential equations

$$dz = p\,dx + q\,dy$$

$$dp = r\,dx + s\,dy$$

$$dq = s\,dx + t\,dy$$

integrable. This requirement yields for F two simultaneous linear first-order partial differential equations [5], and by solving them, in theory we have solved the original problem.[7]

Also interesting are VÁLYI's examples [6] showing how differentiation, a well-known method of solving ordinary differential equations, may be useful with certain partial differential equations. Again by a clever treatment of partial differential equations, he found a simpler way to MONGE's and ENNEPER's results on the determination of those surfaces all normals of which are tangent to a given sphere [7].

Another less known merit of GYULA VÁLYI's was the recruitment of enthusiastic explorers of *number theory*, a mathematical discipline previously neglected in Hungary, by his courses held for many years. His efforts to this effect in Kolozsvár were similar to GYULA KŐNIG's in Budapest, though their approaches were apparently different: while KŐNIG was attracted to all that was modern and complex, VÁLYI was first of all engaged by classical problems possibly treatable by elementary means. In these he was perfectly at home with his sharp logic. Also in his geometrical investigations, to be discussed later, he often applied number-theoretic methods. His attitude and whole personality is clearly shown by the fascinating elementary problem quoted below almost word for word from his lectures on number theory ([8], pp. 163—169; [9]).

The problem of ancient origin is the following: find all triangles whose sides are integers and whose area and perimeter are measured by the same number. Using HERON's formula, the condition can be written as

$$s(s-a)(s-b)(s-c) = (a+b+c)^2, \tag{A}$$

where s is the half of the perimeter. Another form of (A) reads

$$\frac{(a+b+c)(b+c-a)(a+c-b)(a+b-c)}{16} = (a+b+c)^2$$

This shows that $a+b+c=2s$ must be an even number, or else the left side could not be divisible by 16. Hence s is a natural number. Simplifying Eq. (A) we obtain:

$$(s-a)(s-b)(s-c) = 4s \tag{B}$$

---

[7] In the theory of simultaneous differential equations studied by VÁLYI, also applying the above argument BIANCHI soon arrived far-reaching results: Sulle soluzioni comuni a due equazioni a derivate parziali del 2^0 ordine con due variabili. *Rom. Acc. L. Rend.*, [4], 2, 1886, pp. 218—223; 237—241; 307—310.

Let us introduce the notations

$$s-a=\alpha,$$
$$s-b=\beta, \qquad \text{(C)}$$
$$s-c=\gamma.$$

Obviously, α, β and γ are also integers.

Summing the equations (C), we get

$$s=\alpha+\beta+\gamma. \qquad \text{(D)}$$

On the other hand, multiplying the equations (C) yields

$$\alpha\beta\gamma=4(\alpha+\beta+\gamma). \qquad \text{(E)}$$

We first prove that α, β and γ are different numbers. For let, e.g., $\beta=\alpha$. Then from (E)

$$\alpha\beta^2=4\alpha+8\beta,$$
$$\alpha\beta^2-8\beta=4\alpha.$$

Multiplying the last equation by α, and adding 16 to both sides, we get:

$$\alpha^2\beta^2-8\alpha\beta+16=4\alpha^2+16,$$
$$(\alpha\beta-4)^2=4(\alpha^2+4).$$

Thus the assumption that $\beta=\gamma$ has led to a contradiction as the left side of the last equation is a perfect square while the right side is not a perfect square for any positive integer α.

It means no restriction if we assume that

$$\alpha>\beta>\gamma. \qquad \text{(F)}$$

We show that $\gamma>2$ is impossible. For, if we assume that $\gamma\geq3$, then from (E):

$$3\alpha\beta\leq4(\alpha+\beta+\gamma)<4\alpha+8\beta<12\alpha.$$

This means that, contrarily to (F), β would be less than 4. Hence γ can only be 1 or 2.

Let $\gamma=1$, then from (E)

$$\alpha\beta=4(\alpha+\beta+1).$$

This can also be written as

$$(\alpha-4)(\beta-4)=20.$$

The last equation and (F) are satisfied by the values

$$\alpha=24;\ 14;\ 9;$$
$$\beta=\ 5;\ \ 6;\ 8.$$

For $\gamma = 2$ the same reasoning gives

$$\alpha = 10; \ 6;$$

$$\beta = \ 3; \ 4.$$

When the corresponding values of α, β and γ are known, with the help of equations (D) and (E) we arrive at the side lengths of the five triangles that satisfy the condition.[8]

Besides his courses in number theory, two other series of lectures enhanced GYULA VÁLYI's fame. One was on function theory — repeated as was his usual practice — at certain intervals, constantly improving and polishing the material presented. According to ALFRÉD HAAR, perhaps nowhere in the world was the theory of complex functions set forth at a higher level at that time ([1], p. 62). The other course of lectures, a landmark in the development of Hungarian mathematical culture, dealt with JÁNOS BOLYAI's *Appendix*. It was first given in the second term of the academic year 1891/92 and repeated nearly unchanged every four years.

The 102 pages of lithographed notes taken of the lectures is a fascinating and entertaining reading [10]. It reveals that some one-third of his course was devoted to the presentation of the historical precedents, followed by a commentation of the *Appendix* in the order of the sections. VÁLYI made additions to the proofs and resolved the extraordinarily succinct wording by inserting some explanatory notes. Here and there he also borrowed some results of LOBACHEVSKY in order to compare absolute and hyperbolic geometry.

It is undoubtedly the result of VÁLYI's activity that despite all fault-finding by SÁMUEL BRASSAI, Kolozsvár grew into a citadel of the BOLYAI cult, and many of VÁLYI's colleagues and students took a great share in popularizing the work of the two BOLYAIS.[9]

Regarding the applied methods, the majority of VÁLYI's geometrical investigations are similar to those initiated by JENŐ HUNYADY in Hungary: they found answers to geometrical questions by analyzing certain determinants. For instance, the decision of the problem whether a surface given by point coordinates is a surface of revolution is equivalent to the question on the existence of a multiple root of the LAPLACE equation

$$\Delta \lambda = \begin{vmatrix} a_{11} - \lambda & a_{12} & a_{13} \\ a_{21} & a_{22} - \lambda & a_{23} \\ a_{31} & a_{32} & a_{33} - \lambda \end{vmatrix} = 0 \qquad (a_{ik} = a_{ki})$$

belonging to the central equation of the surface. In a paper of his, VÁLYI put this question under scrutiny [12]. He found that λ is a triple root (the case of the sphere) if and only if

$$a_{23} = a_{31} = a_{12} = 0.$$

[8] $a = 6 \quad 7 \quad 9 \quad 5 \quad 6$
$b = 25 \quad 15 \quad 10 \quad 12 \quad 8$
$c = 29 \quad 20 \quad 17 \quad 13 \quad 10$
[9] We are referring to the fundamental treatises by LAJOS SCHLESINGER, an important paper by HENRIK KIRÁLY (On the geometry valid on surfaces of constant curvature. *Math. és Phys. Lapok*, 10, 1901, pp. 111—144 (in Hung.) and subsequent books by LAJOS DÁVID.

272

Then

$$\lambda = a_{11} = a_{22} = a_{33}.$$

Further, λ is a double root if the following is satisfied:

$$\frac{a_{11}a_{23} - a_{12}a_{13}}{a_{23}} = \frac{a_{22}a_{13} - a_{12}a_{23}}{a_{13}} = \frac{a_{33}a_{12} - a_{13}a_{23}}{a_{12}} \neq 0$$

$(a_{23}, a_{13}, a_{12} \neq 0)$. These cases provide the necessary and sufficient conditions for a second-order surface to be a surface of revolution.[10]

Easy to handle, hence useful also for didactic purposes is the table in which VÁLYI exhibits how the type of a second-order curve given by homogeneous coordinates can be obtained from the symmetric determinant containing the coefficients [11].

Significant interest was aroused abroad by GYULA VÁLYI's investigations into certain properties of triangles and tetrahedra. The lines connecting the feet of the altitudes of a triangle determine its first pedal triangle. The pedal triangle of the latter is called the second pedal triangle of the original triangle, etc. If the angles of the first triangle are α, β and γ, and all of them are smaller than $\frac{\pi}{2}$, then the angles of the first pedal triangle are:

$$\alpha_1 = \pi - 2\alpha, \quad \beta_1 = \pi - 2\beta, \quad \gamma_1 = \pi - 2\gamma.$$

If one of the angles (e.g. α) of the first triangle is obtuse, then the angles of the first pedal triangle are:

$$\alpha_1 = 2\alpha - \pi, \quad \beta_1 = 2\beta, \quad \gamma_1 = 2\gamma.$$

The angles of the second, third, etc. pedal triangles of the original triangle can be expressed in a likewise simple manner.

GYULA VÁLYI posed the following problem [13]. Find all triangles similar to their n-th pedal triangle.

The answer is given by the solutions of a system of elementary congruences, set up from the values α_n, β_n and γ_n. For the cases $n = 1, 2, 3$ VÁLYI determined the angles of the triangles satisfying the condition by calculation as well. For example, two triangles are similar to their first pedal triangle, namely those with angles

$$\frac{\pi}{3}, \frac{\pi}{3}, \frac{\pi}{3}, \quad \text{or} \quad \frac{\pi}{7}, \frac{2\pi}{7}, \frac{4\pi}{7}.$$

Denoting by $\Phi(n)$ the number of triangles similar to their n-th pedal triangle, but

---

[10] However, surprising it may sound, the first to systematize these criteria well-known today, was GYULA VÁLYI. On pp. 242—246 of his book *Vorlesungen über analytische Geometrie des Raumes* (Leipzig, 1953), KOMMERELL discusses the very cases that VÁLYI elaborated, using his method as well. These criteria seem to have found their way into standard textbooks starting from VÁLYI through the analytic geometries of STAUDE (1910) and SCHÖNFLIES (1925).

not similar to any of their lower-order pedal triangles, we have:

$$\Phi(1) = 2 \qquad \Phi(4) = 228$$
$$\Phi(2) = 10 \qquad \Phi(5) = 990$$
$$\Phi(3) = 54 \qquad \Phi(6) = 3966$$

Several of VÁLYI's studies belong to projective geometry and discuss *multiple* perspectivity. DESARGUES's theorem says that if the lines joining the corresponding vertices of two triangles are concurrent, then the intersections of corresponding sides are collinear, and conversely. In this case the two triangles are said to be in perspective.

The question can be raised at this point whether there are triangles of multiple perspectivity, i.e. such that they are in perspective for several ways of correspondence, between their vertices. Denote by A, B, C the vertices of one triangle, and by 1, 2, 3 those of the other. Assume both triangles are in perspective if joining A with 1, B with 2, and C with 3. Denote this perspectivity by A_1, B_2, C_3.

VÁLYI asked how many of the perspectivities

$$A_1B_2C_3, \quad A_1B_3C_2, \quad A_2B_1C_3, \quad A_2B_3C_1, \quad A_3B_1C_2,$$

and

$$A_3B_2C_1$$

were possible simultaneously for two triangles. Depending on whether 1, 2, 3, ... 6 are realized *simultaneously*, we speak of simple, double, ..., sextuple perspectivity of two triangles.

ROSANES and SCHRÖTER were the first to tackle this problem in *geometric* terms.[11] In their wake, VÁLYI presented a full theory of the problem in *analytic* terms.[12]

VÁLYI found [14, 15] that there is a maximum *four-fold* perspectivity between two real triangles and a maximum sixfold perspectivity between non-real triangles. For instance, two concentric, equilateral triangles are triply perspective, two concentric equilateral triangles with parallel sides are quadruply perspective. VÁLYI arrived at his conclusions through relatively elementary arguments after having aptly chosen the notations and the projective systems of coordinates. This might be the main reason why papers on this subject profusely cite his works.[13]

---

[11] ROSANES: Über Dreiecke in perspectivischer Lage. *Math. Annalen*, 2, 1870, pp. 541—552; SCHRÖTER: Über perspectivisch liegende Dreiecke. *Ibid.*, pp. 553—562.

[12] I deem it important to emphasize this because in an article (Bemerkungen über den Aufsatz von VÁLYI und dessen Vorgänger. *Archiv d. Math. u. Phys.*, 70, 1884, pp. 334—335) HOPPE accused VÁLYI of cribbing from ROSANES and SCHRÖTER. Actually, VÁLYI never claimed priority of posing the problem, and his method of investigation differed from that of his predecessors, as was admitted by SCHRÖTER, one of the authors concerned. In a publication in Hungarian (*Math. és Phys. Lapok*, 7, 1898, p. 105) VÁLYI mentioned the names of ROSANES and SCHRÖTER. In his long treatise (Beiträge zur Theorie der mehrfach perspectiven Dreiecke und Tetraeder. *Math. Annalen*, 28, 1886, pp. 107—200), E. HESS relied chiefly on the method and results of VÁLYI. Another Hungarian mathematician engaged in the study of multiple perspectivity was LIPÓT KLUG: Über mehrfach perspective Tetraeder. *Archiv d. Math. u. Phys.*, (2) 6, 1887, pp. 93—104.

[13] See, e.g., the Encyklopädie entries by A. SCHOENFLIES and BERKHAM-MEYER: III, 1.1, p. 428, and III, 1.2, p. 1221.

The concept of multiple perspectivity can also be transferred to tetrahedra, but analytic considerations are more difficult here. VÁLYI's results showed [17] that only single, double and quadruple perspectivity were possible here. A very nice theorem of his states that in case of quadruple perspectivity the four centres of perspectivity define a tetrahedron that is in quadruple perspectivity with both of the original tetrahedra.[14]

We will not dwell on papers of VÁLYI stating theorems on the altitudes of tetrahedra. We only call attention to the result which says that he number of points from which all sides of a triangle are seen at right angle is 0, 1 or 2 depending on whether the triangle is obtuse, right-angled or acute [17]. The latter case, as is well known, is important in axonometry.

VÁLYI also did research into polar reciprocity. According to a theorem known previously, if the triangles ABC and 1 2 3 are in perspective, then there always exists a conic section with respect to which the two triangles are reciprocal polar. This means that the sides 23, 31 and 12 are the polars of the vertices A, B and C, and the sides $\overline{BC}$, $\overline{CA}$ and $\overline{AB}$ are the polars of the vertices 1, 2 and 3.

Now, VÁLYI studied the question: what happens if two triangles are n-fold perspective ($n = 1, 2, 3, 4, 5, 6$). In this case there are namely n conic sections for which we have polar reciprocity. VÁLYI found interesting relations between these conic sections, e.g., in the case of double perspectivity the conic sections are doubly tangent to each other [16].

GYULA VÁLYI used the far from elementary tools of elliptic functions in investigating third-order curves of the first kind and fourth-order space curves [18, 19]. The crux of the method is that the homogeneous coordinates of the points of these curves can be represented by various elliptic functions of *one* parameter.[15] A common feature of these representations is that the sum of the parameters of the points of intersection of the plane curve and a straight line, or the space curve and an algebraic surface is the period of the elliptic function. This relation makes it possible to reduce the investigation of point configurations of these curves to the examination of number-theoretic congruences, which is much simpler conclusions than purely geometric reasoning. We mention only one theorem of VÁLYI's achieved with this method: on the third-order plane curve (of the sixth class) without multiple points there are 3-fold perspective triangles and r-fold perspective r-gons.[16]

---

[14] As a corollary of his investigations, VÁLYI found a new proof of DESARGUES's theorem. This was later simplified by JÓZSEF KÜRSCHÁK: Desargues's Theorem. *Math. és Phys. Lapok*, 15, 1906, pp. 201—202. (in Hung.)

[15] It should be noted that this highly efficient method — the use of elliptic functions in the investigation of the above-mentioned curves — originated from ARONHOLD (1861). Many scholars including CLEBSCH, HARNACK, HUMBERT, O. SCHLESINGER, HALPHEN and others exploited this possibility.

[16] The latter investigations of VÁLYI influenced the subject selected by FRIGYES RIESZ and GYULA SZŐKEFALVI-NAGY for their doctoral dissertations. See FRIGYES RIESZ: The position-geometric treatment of point configurations on fourth-order space curves of the first kind. *Math. és Phys. Lapok*, 11, 1902, pp. 293—309; 346—360; 13, 1904, pp. 191—204. (in Hung.); SZŐKEFALVI-NAGY, GYULA: On the arithmetic properties of algebraic curves. *Math. és Phys. Lapok*, 18, 1909, pp. 331—348; 21, 1912, pp. 58—66.

Bibliography

[1] OBLÁTH, RICHÁRD: Gyula Vályi. *Matematikai Lapok*, 7, 1956, pp. 61—70. (In Hung.)

[2] RÉTHY, MÓR: In memory of corr. member Gyula Vályi. *Commemorative orations over the late members of the Hungarian Academy of Sciences*, 17, 1915, No. 5. (In Hung.)

[3] WESZELY, TIBOR: *Gyula Vályi*. Bucharest, 1983. (In Hung.)

[4] VÁLYI, GYULA: *Contribution to the theory of partial differential equations of the second order.* Kolozsvár, 1880. (In Hung.); 2nd ed.: *Math. és Phys. Lapok*, 15, 1906, pp. 256—269. (In Hung.); Zur Theorie der partiellen Differentialgleichungen zweiter Ordnung. *Arch. d. Math. u. Phys.*, (3), 15, 1910, pp. 294—304.

[5] — On the integration of second-order simultaneous partial differential equations with two independent variables. *Math. és Term. tud. Értesítő*, 1, 1882, pp. 309—312 (In Hung.); Über die Integration simultaner partiellen Differentialgleichungen zweiter Ordnung mit zwei unabhängigen Variabeln. *Journal f. reine u. angew. Math.*, 95, 1883, pp. 99—101.

[6] — Integration einiger partiellen Differentialgleichungen zweiter Ordnung. *Arch. d. Math. u. Phys.*, 70, 1883, pp. 219—233; Zusatz zum Aufsatze: Integration ... *Arch. d. Math. u. Phys.*, (2), 1, 1884, pp. 109—110.

[7] — Die Flächen, deren sämtliche Normalen eine Kugelfläche berühren. *Arch. d. Math. u. Phys.*, 68, 1882, pp. 217—219.

[8] — *Number theory*. Kolozsvár, 1898. Duplicated lecture notes. (In Hung.)

[9] — A problem of number theory in geometry. *Math. és Phys. Lapok*, 1, 1892, pp. 56—57. (In Hung.)

[10] — *On János Bolyai's "Appendix"*. Kolozsvár, 1904. Duplicated lecture notes. (In Hung.)

[11] — The classification of second-order surfaces. *Math. és Term. tud. Értesítő*, 8, 1889—90, pp. 218—219 (In Hung.); On the classification of surfaces of the second order. *Math. és Phys. Lapok*, 1, 1892, pp. 341—346. (In Hung.); Klassification der Flächen zweiter Ordnung. *Arch. d. Math. u. Phys.*, (2), 9, 1892, pp. 223—224.

[12] — On second-order surfaces of revolution. *Math. és Phys. Lapok*, 3, 1894, pp. 1—4. (In Hung.)

[13] — On pedal triangles. *Math. és Phys. Lapok*, 10, 1901, pp. 309—321. (In Hung.)

[14] — Mehrfache Collineation von zwei Dreiecken. *Arch. d. Math. u. Phys.*, 70, 1882, pp. 105—110.

[15] — Multiply perspective triangles in the plane. *Math. és Phys. Lapok*, 7, 1898, pp. 105—114. (In Hung.)

[16] — Multiply collinear triangles and conic sections. *Math. és Term. tud. Értesítő*, 2, 1883—84, pp. 170—174. (In Hung.); Mehrfach collineare Dreiecke bei Kegelschnitte. *Arch. d. Math. u. Phys.*, (2), 2, 1885, pp. 320—324.

[17] — On the theory of perspective tetrahedra. *Math. és Term. tud. Értesítő*, 4, 1885, pp. 55—56. (In Hung.); Multiply perspective tetrahedra. *Math. és Term. tud. Értesítő*, 4, 1885, pp. 6—8. (In Hung.); Zur Lehre vom perspectiven Tetraeder. *Arch. d. Math. u. Phys.*, (2), 3, 1886, pp. 441—445.

[18] — On the theory of third-order curves. *Math. és Term. tud. Értesítő*, 8, 1890, pp. 23—28.; 9, 1891, pp. 18—25; 10, 1892, pp. 2—13. (In Hung.)

[19] — On fourth-order space curves of the first kind. *Math. és Term. tud. Értesítő*, 10, 1892, pp. 244—251. (In Hung.)

23. Zoárd Geőcze

All his mathematical activity focussed on problems that are solvable by the theory of real functions. In his master's thesis and his first printed paper [1] he gave a geometrical construction of an everywhere continuous curve which, however, has infinite are length in an arbitrarily small interval. His example (in the original wording) is the following:

"Let $z = f_1$ be linear and increasing with x, and let its length in $[0, a]$ be G. Let $z = f'_2$ also be like that, but its length be $2G$. It is easy to see that if δ is given, there is a function $z = f_2$ whose geometric image *(Figure 58)* is a broken line consisting

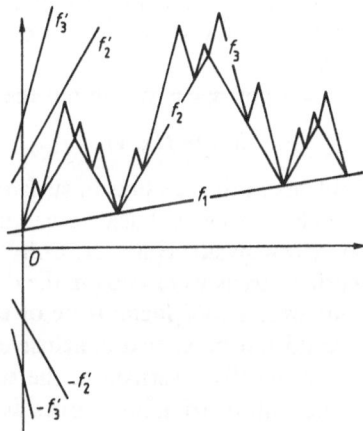

Figure 58

of an even number of segments parallel to the lines $z = f'_2$ and $z = -f'_2$ alternately, and every second vertex of the broken line counted from $x = 0$ as the first towards $x = a$ is on the line f_1. Let the abscissas of the vertices be $0 = x_0 < x_1 < \ldots < x_{2r} = a$. And we can achieve that $0 = \leq f_2 - f_1 < \delta$ by choosing r to be sufficiently large and (x_i, x_{i+1}) sufficiently small. Just as f_2 for f_1 in $[0, a]$, there is an f_3 in every (x_i, x_{i+1}) whose segments are alternately parallel to the rising and falling lines $z = f'_3$ and

277

$z = -f_3'$ of length $4G$ in $[0, a]$, and $0 \leq f_3 - f_2 < \dfrac{\delta}{2}$ throughout $[0, a]$. One can form f_4, etc., in a similar way."

GEŐCZE proved that the function

$$f = f_1 + \sum_{i=1}^{\infty} (f_{i+1} - f_i)$$

is continuous everywhere but even in an arbitrarily small interval its arc length is infinite; moreover, as was proved later,[1] it is nowhere differentiable.

The rest of GEŐCZE's studies is concerned with the measurement of surface area. For a better understanding of his investigations, it is expedient to recount a few notions and historical facts.

To build up the calculus of surface areas, one can start from two, theoretically different, definitions of surface. Common sense prefers to define a surface patch as the image of the unit square under an ore-to-ore continuous mapping of bounded stretching. Let the surface thus defined be called *simple* (or *proper*, to use GEŐCZE's term). But surface can also be defined as follows: in the parameter plane (u, v) we fix a closed JORDAN curve, and on the parameter domain bounded by this curve we consider three single-valued continuous functions $x(u, v)$, $y(u, v)$ and $z(u, v)$. We call this triple of functions a continuous surface. The image of the function

$$F : x = x(u, v); \quad y = y(u, v); \quad z = z(u, v) \tag{1}$$

in a rectangular system of coordinates is a closed set. If the latter definition is chosen, then — essentially — the calculus of surface areas becomes part of the analysis of functions of several variables.

In the special case when $x = u$ and $y = v$, we obtain the more feasible function

$$S : z = f(x, y). \tag{2}$$

There were unsettled problems in the calculus of surface areas as late as the beginning of our century. Although — due to basic research done by EULER, MONGE, RODRIGUES, GAUSS, PEANO, MINKOWSKI, HERMITE, SCHWARZ and others — several definitions of surface and surface areas were known, the difficulties constantly arising showed that these definitions were either incomplete or erroneous and the problem of measuring surface area could not be settled conclusively with their help.

Earlier, plane triangles were usually inscribed in the surface and the surface area was defined as the limit of the sum of triangular areas as the side lengths of the triangles tended to zero. This approach is analogous to defining *arc length* through approximation by chords. In a letter to GENOCCHI, however, H. A. SCHWARZ showed in 1880 that when the passage to the limit is appropriately chosen, the area of a network of triangular polyhedra inscribed in a straight cylinder does not tend to the area of the cylinder.

---

[1] KÁNTOR, SÁNDOR: Zoárd Geőcze's function is everywhere continuous but nowhere differentiable. *Matematikai Lapok*, 8, 1957, pp. 264—267. (In Hung.); CSÁSZÁR, ÁKOS: A comment on Zoárd Geőcze's function. *Ibid.*, pp. 268—271. (In Hung.)

From the end of the last century, SCHWARZ's example — included in textbooks today — urged many a mathematician to examine the definition of surface area more thoroughly. Some imposed restrictions on the inscribed triangles,[2] others gave entirely new definitions of surface measure. For instance, PEANO suggested the following procedure.[3]

Divide the surface into a finite number of pieces in some way and, separating these pieces from one another, them without deformation in to some end-position. Project each of these pieces perpendicularly to a given plane, and let the total area of the projections be T. The *supremum* of the values T taken for all possible dissections of the given surface patch and all possible end-positions of the pieces is the PEANO surface. Denote it by $P(F)$.

When defining surface measure, MINKOWSKI started from the recognition that computing volumes is less problematic, so the computation of surface areas should be derived from cubature. Let us therefore draw spheres of radius r from the points of the surface as centres. The points of space that are on or inside at least one of these spheres form a solid. Let its volume be $V(r)$. If there exists $\lim\limits_{r \to 0} \dfrac{V(r)}{2r}$, then it is the MINKOWSKI *surface*.[4]

MINKOWSKI's surface measure is comparatively easy to apply to closed, convex surfaces. There are, however, examples where the area calculated with MINKOWSKI's method is smaller than the measures yielded by other definitions ([10], pp. 258—259).

Modern investigations usually take LEBESGUE's definition of surface area for starting point. For *simple* surfaces it can be described as follows. Let $A(P)$ denote the area of the simple (i.e. composed of plane triangles) polyhedral surface P in the elementary sense (i.e., the sum of the areas of the triangles). Also, let S be any simple surface and P_n be a sequence of simple polyhedral surfaces converging to S. In general, the numerical sequence $A(P_n)$ has no limit, but $\underline{\lim}\, A(P_n)$ always exists, its value depending on S and the selection of the sequence $\overline{P_n}$. The greatest lower bound of the set of numbers $\underline{\lim}\, A(P_n)$ corresponding to all sequences P converging to S is the LEBESGUE *area of the surface* S. We denote by $L(S)$.

Obviously, the functional $L(S)$ depends on the surface S only. There is no room here to enumerate all the major theorems for $L(S)$ or to give a precise definition of the *convergence* concept (due to FRÉCHET) involved. We note, however, that LEBESGUE's definition can be carried over, mutatis mutandis, to *continuous* surfaces, and so the functional $L(F)$ (for F, see p. 280) can be defined.[5]

---

[2] E.g., according to O. HÖLDER (*Beiträge zur Potentialtheorie*. Dissertation. Stuttgart, 1882, p. 29) all angles of the inscribed triangles must remain between certain bounds. RADEMACHER (Über partielle und totale Differenzierbarkeit. *Math. Annalen*, 79, 1919, pp. 340—359) required that, when passing to the limit, the plane of the inscribed triangles should tend to the tangent plane at the respective point of the surface. For the Schwarz example it is easy to prove that this condition is sufficient.

[3] PEANO, GIUSEPPE: *Applicazioni geometriche del calcolo infinitesimale*. Torino, 1887, p. 164.

[4] MINKOWSKI, HERMANN: Über die Begriffe Länge, Oberfläche und Volumen. *Jahresbericht d. Deutschen Math. Ver.*, 9, 1901, pp. 115—12i.

[5] For the definition of surface area the treatise of LEBESGUE: Intégrale, Longueur, Aire (*Annali di Mat.*, III, 7, 1902, p. 315) is the usual reference. Let us mention for historical truth that LEBESGUE published his definition earlier. Cf. Sur la définition de l'aire d'une surface. *Comptes Rendus*, 129, 1899, pp. 870—883.

According to LEBESGUE's definition the polyhedral sequence converging to the surface only provides an upper bound for the area measured. ZOÁRD GEŐCZE used this feature of the definition for giving an example [12, 15] — frequently quoted in the literature — which exhibits, so to say, the worst extreme.

Define a surface by the continuous transformations (1) where the parameters u, v range over the unit square Q. Divide Q with lines parallel to the u and v axes into q^2 smaller congruent squares, then divide every small square with a diameter into two triangles. This yields $2q^2$ congruent triangles. If the vertices of such a triangle are A, B, C, let the points of the surface corresponding to them be A^0, B^0, C^0.

Let the sum of the areas of the plane triangles with vertices A^0, B^0, C^0 be T_q, then

$$L(F) \leqq \lim_{q \to \infty} T_q.$$

Now, if the surface is the function of only one parameter, say u, that is

$$F : x = x(u); \quad y = y(u); \quad z = z(u), \tag{3}$$

then the area of each plane triangle with vertices A^0, B^0, C^0 is zero as, due to the construction, two of its three vertices coincide. In this way, however, $\lim_{q \to \infty} T_q = 0$ and hence $L(F) = 0$. When (3) defines a space curve that fills a cube — PEANO proved the existence of a space curve of this kind[6] — then we obtain a surface that fills a cube but has surface measure zero.

In the case of a *simple* surface, however, the LEBESGUE area is always greater than zero, the first exact proof of which was given by GEŐCZE [14].

*

A scholarship report of ZOÁRD GEŐCZE informs us that he got acquainted with LEBESGUE's surface area definition only during his first stay in Paris in 1908. Prior to that, in 1906 he had given an independent definition of surface area [2] that proved to be very useful.

Divide the portion of surface F into pieces F_1, F_2 ..., F_n and project these pieces perpendicularly to the coordinate planes xy, yz, zx. Let the area in the elementary sense of the three projections of F_i be a_i, b_i, c_i and consider the following sum:

$$\sum_{i=1}^{n} (a_i^2 + b_i^2 + c_i^2)^{\frac{1}{2}}.$$

The supremum of this sum for all possible divisions of F is the GEŐCZE *area*.

It is logical to expect that, provided that certain conditions for the surface are satisfied, LEBESGUE's and GEŐCZE's surface measures coincide; one of the very aims of relevant research was to find these conditions. But a whole row of other questions arise: how are the various definitions of surface area interrelated; what is the scope

---

[6] PEANO, GIUSEPPE: Sur une courbe qui remplit toute une aire plane. *Math. Annalen*, 36, 1890, pp. 157—160.

of the classic double integral in the calculation of surface area;[7] how can the results be generalized to hypersurfaces, etc. All these problems can be raised in connection with surfaces defined by (2) which are easy to treat, and with surfaces defined in the parametric form (1) where the circumstances are far more involved.

In a part of these problems ZOÁRD GEŐCZE found conclusive results while in others he initiated a new chapter of modern research into analysis.

Mention should be made of the great merits of TIBOR RADÓ (1895—1967) in making GEŐCZE's overcomplicated investigations better known and more easily understandable. He was encouraged by FRIGYES RIESZ in the mid-twenties to look into GEŐCZE's treatises and the question of surface quadrature. We owe it largely to RADÓ that today we have thick monographs about the problems of surface measurement[8] which synthesize the most outstanding results of research papers estimated at several hundreds, published over the past 3—4 decades.

The majority of GEŐCZE's investigations deal with surfaces given by equation (2). His definition for these surfaces is as follows. Divide the unit square Q into a finite number of smaller squares. Let the vertices of fixed elementary square q be (x_1, y_1); (x_2, y_1); (x_2, y_2); (x_1, y_2). Thus it is true for any point (x, y) of the square q that $x_1 \leq x \leq x_2$; $y_1 \leq y \leq y_2$.

Introduce the following notations:

$$a(z, q) = \int_{x_1}^{x_2} |f(x, y_2) - f(x, y_1)|\, dx,$$

$$b(z, q) = \int_{y_1}^{y_2} |f(x_2, y) - f(x_1, y)|\, dy,$$

$$c(z, q) = (x_2 - x_1)(y_2 - y_1).$$

Obviously, $c(z, q)$ is the area of the square q, while $a(z, q)$ is the area of the figure in the xz plane bounded by the curves $z = f(x, y_1)$ and $z = f(x, y_2)$, together with the lines $x = x_1$ and $x = x_2$ of the same plane; the meaning of $b(z, q)$ is similar in the yz plane. Now, GEŐCZE's *basic quantity* can be obtained by the spatial Pythagorean theorem:

$$g(z, q) = (a^2 + b^2 + c^2)^{\frac{1}{2}}.$$

---

[7] The classical double integral is

$$I(F) = \iint_Q (EG - F^2)^{\frac{1}{2}}\, du\, dv \qquad \text{for} \quad (1)$$

and

$$I(f) = \iint_Q \left[\left(\frac{\partial f}{\partial x}\right)^2 + \left(\frac{\partial f}{\partial y}\right)^2 + 1 \right]^{\frac{1}{2}} dx\, dy \qquad \text{for} \quad (2)$$

The meaning of the symbols appearing here is well known from analysis.

[8] RADÓ, TIBOR: Length and area. *Amer. Math. Soc. Coll. Publ.*, 30, New York, 1948; LAMBERTO CESARI: *Surface area.* Princeton, 1956. Their ample bibliographies exempt us from listing the research papers.

Let D denote the decomposition of the unit square, and let us introduce the following function:

$$G(z, D) = \sum_{1}^{n} g(z, q).$$

GEŐCZE proved, among other things, the following fundamental theorems for the function $G(z, D)$:

If D^+ is a subdivision of D, then $G(z, D) \leq G(z, D^+)$;
if $z_n(x, y) \to z(x, y)$, then $G(z_n, D) \to G(z, D)$.

It can also be proved that, if the surface is *continuous*, then for all possible divisions of the square Q there exists

$$\lim \sup G(z, D) = \Gamma(z),$$

and is equal to the GEŐCZE *area of the surface patch.*

In general,

$$\Gamma(z) \leq \frac{L(z)}{I(z)}. \tag{4}$$

One of the most significant results of GEŐCZE's mathematical work is that, in the case of rectifiable surfaces[9] in (4) the equality sign is valid. In 1927 TIBOR RADÓ proved that the statement holds for *continuous* surfaces as well.

It is to be noted that when proving the above theorem, the well-established apparatus of real functions was not yet at the disposal of ZOÁRD GEŐCZE and in many cases he was compelled to create the mathematical tools for his investigations himself.

As a consequence, he achieved useful results also in the theory of semi-continuous functions [9] and tangent planes [17].

Let the function $f(x, y)$ of two variables be defined over the unit square, and let $V_y(x)$ denote the total variation of the function of one variable obtained from $f(x, y)$ by fixing the value of x in the interval $(0, 1)$ and letting y vary in the interval $(0, 1)$; the total variation $V_x(y)$ is defined in a similar way. With the help of GEŐCZE's basic quantity it can be proved that

$$L(z) \leq \int_0^1 V_y(x)\,dx + \int_0^1 V_x(y)\,dy + 1.$$

Consequently, if there is a positive number K such that

$$\int_0^1 V_y(x)\,dx + \int_0^1 V_x(y)\,dy \leq K. \tag{5}$$

---

[9] The surface $z = f(x, y)$ can be rectified if it satisfies a LIPSCHITZ condition, i.e. for $(x_1, y_1) \neq (x_2, y_2)$ there is a positive constant K such that

$$f(x_1, y_1) - f(x_2, y_2) \leq K |[(x_1 - x_2)^2 + (y_1 - y_2)^2]^{\frac{1}{2}}|.$$

then the LEBESGUE area of the surface $z = f(x, y)$ is finite. This criterion is not only sufficient but also necessary for the finiteness of the surface area [18].[10]

ZOÁRD GEŐCZE established some results also for non-parametrically defined *hypersurfaces*. This generalization entails no special difficulties. E.g., similarly to (5), the hyperarea of the hypersurface $u = f(x, y, z)$ is finite if and only if the sum

$$\int_0^1 \int_0^1 V_z(x, y)\,dx\,dy + \int_0^1 \int_0^1 V_y(x, z)\,dx\,dz + \int_0^1 \int_0^1 V_x(y, z)\,dy\,dz$$

is bounded from above. Here $V_z = (x, y)$ is the total variation of the function obtained by fixing x and y and letting z vary ([13], [18], pp. 80—81).

While the question of the area of surfaces given by equation (2) was settled relatively early thanks to the achievements of GEŐCZE, LAMPARIELLO, TONELLI, RADÓ and others, the solution of the area problem for surfaces defined in terms of parameters was hindered by certain topological difficulties until the recent years when TIBOR RADÓ and LAMBERTO CESARI obtained definitive results. GEŐCZE began to work out this circle of problems but he had no time left to carry out this impressive programme. His last, mostly posthumous, papers drew up the outlines of the work to be done. Let us pick out two of his results in this field. One is the criterion which says that the area of a surface defined with parameters is finite if and only if for $v_1 \neq v_2$ there exists positive constant K such that

$$|x(u, v_1) - x(u, v_2)| + |y(u, v_1) - y(u, v_2)| + |z(u, v_1) - z(u, v_2)| \leq K|v_1 - v_2|$$

[11]. According to another of his results [17, 19] the equalities

$$L(F) = G(F) = P(F) = I(F)$$

hold also for *rectifiable* surfaces defined parametrically.

Incidentally, the question of surface quadrature has a feature is common with some problems of number theory: the problem and the result are easy to state but the way leading to the result is extremely hard. Nevertheless, TIBOR RADÓ was able to simplify the proofs of ZOÁRD GEŐCZE considerably so that he replaced GEŐCZE's projections by a new concept, the *projection kernel*, which did not mean a substantial change but compiled better satisfy with the requirements of the theory of real functions.[11]

*

More recent results achieved by other authors may interest specialists only. Even in our days there are publications dealing with unsettled problems of quadrature; in these, GEŐCZE is mentioned as a pioneer of the field. Thus his life-work is that of the "founder of a school": he had created the overall strategic plan but, due to his untimely death, the elaboration of the details was left to posterity.

---

[10] Later the same criterion was given by TONELLI: Sulla quadratura delle superficie. *Rend. della Ac. dei Lincei*, 1926, pp. 357—362.

[11] For this, see e.g. RADÓ, TIBOR: To the theory of surface measurement. *Math. és Term. tud. Értesítő*, 45, 1928, pp. 225—244. (in Hung.)

Bibliography

[1] GEŐCZE, ZOÁRD: On the arc length of curves constituting a continuous system. In the *Az ungvári reáliskola 1904/5. tanévi Értesítője*, 32 pp. (In Hungarian)

[2] — The quadrature of the surface of revolution. In the *Az ungvári reáliskola 1905/6. tanévi Értesítője*, 12 pp. (In Hung.)

[3] — *The quadrature of the surface z = ƒ(x, y)*. Mimeographed study. (In Hung.)

[4] — Quadrature des surfaces courbes. *Comptes Rendus*, 144, 1907, pp. 253—256.

[5] — Contributions to the quadrature of the surface z = ƒ(x, y). *Math. és Term. tud. Értesítő*, 26, 1908, pp. 475—512. (In Hung.)

[6] — Quadrature des surfaces courbes. Paris, 1908. The same: *Math. u. naturw. Berichte aus Ungarn*, 26, 1910, pp. 1—88.

[7] — Recherches générales sur la quadrature des surfaces courbes. *Math. u. naturw. Berichte aus Ungarn*, 27, 1911, pp. 1—21, 131—163; 30, 1914, pp. 1—29.

[8] — Contribution à la quadrature des surfaces courbes. *Comptes Rendus*, 152, 1911, pp. 678—679.

[9] — Sur la fonction semi-continue. *Bulletin de la Soc. Math. de France*, 39, 1911, pp. 256—295.

[10] — On the measurement of area. *Math. és Phys. Lapok*, 20, 1911, pp. 255—301; 21, 1912, pp. 24—57. (In Hung.)

[11] — Sur la quadrature des surfaces courbes. *Comptes Rendus*, 154, 1912, pp. 1211—1213.

[12] — Sur l'exemple d'une surface dont l'aire est égale à zéro et qui remplit un cube. *Bulletin de la Soc. Math. de France*, 41, 1913, pp. 29—31.

[13] — Sur la quadrature des variétés. *Comptes Rendus*, 157, 1913, pp. 910—912.

[14] — On the theory of surface measurement. *Math. és Term. tud. Értesítő*, 31, 1913, pp. 306—318. (In Hung.)

[15] — Example of a surface of zero area filling a cube. *Math. és Phys. Lapok*, 23, 1914, pp. 115—117. (In Hung.)

[16] — On surfaces of zero area. *Math. és Term. tud. Értesítő*, 33, 1915, pp. 730—748. (In Hung.)

[17] — On rectifiable surfaces. *Math. és Term. tud. Értesítő*, 34, 1916, pp. 337—354 and 587. (In Hung.)

[18] — On the necessary and sufficient condition for a surface patch to have a finite measure. *Math. és Phys. Lapok*, 25, 1916, pp. 61—81. (In Hung.)

[19] — On Peano's definition of the area of a surface patch. *Math. és Term. tud. Értesítő*, 35, 1917, pp. 325—358. (In Hung.)

[20] — On the general surface. *Math. és Term. tud. Értesítő*, 35, 1917, pp. 359—360. (In Hung.)

[21] SZÉNÁSSY, BARNA: *Commemorative speech over Zoárd Geőcze*. Budapest, 1941. (In Hung.)

[22] — Zoárd Geőcze's mathematical life-work and recent results of surface measurement. *A Szent István Akadémia Értesítője*, 1943, pp. 118—142. (In Hung.)

24. At the outset of the 20th century

This book is concerned with the history of Hungarian mathematics up to the early 20th century. If a subsequent volume is written it will require the collaboration of a team of authors.

This chapter is meant to supplement the previous ones, listing older Hungarian results that have not been touched upon yet. Inserting it here is justified by our ambition to present as complete a picture as possible. It also includes the outlines of certain results that were achieved at the very beginning of our century but the investigations related to them began towards the end of the last century and their influence on later research was also significant. In this way it is nevitable that the present volume and the continuation that will hopefully be completed some time, will overlap at certain points — for which we herewith apologize to the Reader. Let it be put forth in my defence that the time of the completion and publication of the next volume cannot be foreseen.

It is to be noted that in arranging the material of the chapter, we kept to the order used in those issues of *Fortschritte der Mathematik* having appeared around the turn of the century.

*

History of mathematics. The *universal* history of mathematics remained outside the interest of earlier Hungarian researchers, whereas recently noteworthy findings have been published, especially about Greek mathematics.

Certain events of the past of *Hungarian* mathematics, however, intrigued our researchers in the 19th century as well. The largest material accumulated around the end of the last century and it was concerned with the two Bolyais. Several significant data were discovered and published. But the writings of this time were not free from unwarranted romantic stories surrounding the figures of the two scholars.

For some time KÁLMÁN SZILY SR. conscientiously collected memorabilia related to old Hungarian arithmetics and generally to the past of the natural sciences in Hungary. His publications are still informative although the reader should check them for inaccuracies and errors. Especially his view of history is objectionable. For example, he looks at the old arithmetics from the vantage point of the late 19th century, wrested from their contemporaneous economic and cultural background. Criticism from his stance is inevitably condemning. Several of SZILY's works on the history of science are predominated by this view, leading to most offensive conclusions in his articles about APÁCZAI.

From the viewpoint of science history, somewhat more useful are the commemorations published initially in *Értekezések a Mathematikai Tudományok Köréből* (Studies in the Mathematical Sciences) and from 1882 in *Emlékbeszédek az Akadémia elhunyt tagjai felett* (Commemorative orations over the late members of the Hungarian Academy of Sciences). The shortcoming of these obituaries was that they were almost exclusively confined to biographic data and personal reminiscences, leaving the presentation of mathematical results and the evaluation of their impact unsaid; even in the best cases no more than a few words were spared for that. Several outstanding scholars of Hungary did not fare even that well, and with the passing of time, it is ever harder to recall their memory.

Algebra. A method of construction worked out for the solution of systems of linear equations by VILMOS CSILLAG proved most practicable [10]. The backbone of the method is to interpret the products $a_{ik}x_k$ in the system as torques, i.e. we take one factor to be the force and the other the arm of the force; using the so-called force polygon theorem of graphic statics as well as DESARGUES's theorem, a path can be given along which we can arrive at the roots after a certain number of consecutive constructions.

The most significant results of KÁLMÁN SZILY's mathematical investigations are concerned with the sum of the squares of binomial coefficients [56]. Particularly useful are the following two formulae, also recently applied at several places:

$$\sum_{k=0}^{a} \binom{a}{k}^2 = \sum_{k=0}^{m} 2^{a-2k} \binom{a}{k}\binom{a-k}{k}$$

$\left(a \text{ an even number, } m = \dfrac{a}{2} \right)$, and

$$\sum_{k=0}^{a} \binom{a}{k}^2 = (-1)^a \sum_{k=0}^{2a} (-1)^k \binom{2a}{k}^2.$$

Three noted experts of algebra, GUSZTÁV RADOS (1862—1942), JÓZSEF KÜRSCHÁK (1864—1933) and MIHÁLY BAUER (1874—1945) began their investigations in the late 19th century but their lifework extended well into the 20th century, so a detailed exposition of their results belongs elsewhere.

RADOS's first results were mainly concerned with bilinear and quadratic forms, and the roots of the characteristic equations of induced and adjoint substitutions. His theorems belonging to the theory of determinants and matrices are extremely diverse. We mention only an early result of his because of its important applications in algebra and number theory. KRONECKER considered such kind of matrix multiplication that the entires of the product matrix are all two-factor products of entires of the factor matrices. KRONECKER stated without proof that the determinant of the matrix obtained from the n-th order quadratic matrix A and the m-th order B quadratic matrix by this "direct multiplication" is equal to the product of the m-th power of the determinant of A and the n-th power of the determinant of B.

RADOS was the first to prove this theorem [43]. Later he returned to the problem several times and generalized the theorem substantially.

Relying on an article by LÁSZLÓ RÉDEI, we can summarize the scientific activity of MIHÁLY BAUER as follows: "In algebraic number theory: classic *(Dedekind)* ideal theory including various questions of factorization into prime ideals the problem of composite fields, in essential divisors of the discriminant, cyclotomic field, HENSEL's field of p-adic numbers. In algebra (partly in connection with algebraic number theory): linear substitutions, characteristic polynomial, number of subgroups of finite groups, simplicity of the alternating groups, finite groups, equations without affect, irreducibility theorems, root determination by iteration."

In his first investigations, BAUER received problems from, and was encouraged by GUSZTÁV RADOS. For example, in his very first paper [1] relying on the result of RADOS, BAUER proved that the non-zero eigenvalues of a real skew-symmetric matrix are purely imaginary and appear in conjugate pairs. In another article of his [2] belonging to RADOS's circle of interest, he determined in a rational way the polynomial whose zeros are the products $a_i b_k$ formed from the roots $a_1, \ldots, a_m$, and $b_1, \ldots, b_n$ of $f(x)$ and $g(x)$, respectively.

After several interesting investigations, JÓZSEF KÜRSCHÁK arrived at really outstanding algebraic results in the 1910s. KÜRSCHÁK's undoubtedly classic *valuation theory* [39] which followed STEINITZ's famous paper within a few years showed that the idea of algebraic closure could be applied also to abstract fields by suitably generating the concepts of absolute value, convergence and limit. Moreover, as OSTROWSKI soon proved, KÜRSCHÁK's is the only possible valuation which, in case of the algebraic extension of the perfect field, does not modify the valuation of the original elements.

Number theory. Despite all the indefatigable efforts of GYULA KŐNIG and GYULA VÁLYI, research in number theory was slow to start in Hungary. GYŐZŐ ZEMPLÉN's short paper [57] in which he presented a witty proof of the fact that every natural number can be written in one and only one way as the algebraic sum of different powers of 3 is at the elementary level and has didactical value only. MIHÁLY DEMECZKY [13] gave new proofs of mostly well-known results, furnishing criteria for the number of different solutions of a system of linear congruences.

One of the pioneers of serious number-theoretical research in Hungary was MIHÁLY BAUER. He was engaged in several problems of elementary and algebraic number theory. In his first papers he settled those special cases of the DIRICHLET theorem on arithmetic progressions where the general term of the progression can be written in the form $ax + 1$ or $ax - 1$ [3] (the latter only for a being a power of an odd prime). In another of his early papers [4] he considered the quantities $kx + 1$ $(x = 1, 2, \ldots, m)$, and determined how many of them are prime to m. MIHÁLY BAUER began to examine the so-called identical congruences in our century: his results on this subject aroused wide international interest and became the source of many new investigations.

The special course held by LIPÓT FEJÉR (1880—1959) first at the university of Kolozsvár (second term of the academic year 1908/9) with the title "The application of analysis in number theory" (three hours a week) and later at the university of Budapest greatly contributed to the intensification of number-theoretic research in Hungary. His lectures played a pioneering role in popularizing analytic number theory as LANDAU's fundamental two-volume work on prime numbers appeared only in 1909.

Analysis. The profusion of new results, some of them essential, achieved by Hungarian researchers in this discipline significantly influenced the development of mathematics.

As has been seen in the last century our scientists obtained particularly nice results in the theory of differential equations. Besides the persons dealt with in the previous chapters, special mention must be made of JÓZSEF PETZVAL who, after a whole range of articles on remarkable results of his own, around the middle of the last century published a two-volume monograph on differential equations that was the only guide in the field for a long time [41]. Decades later it was still referred to as a major source. Among his original results, the most important ones concern the singular solutions to linear differential equations with a complex variable. He went deeper in this subject than RIEMANN and was ahead of POINCARÉ in several respects. The nowadays fairly complete theory of RICCATI's differential equations owes a lot to him.

JÓZSEF PETZVAL was the founder of the Vienna school of differential equations — its well-known members were WINCKLER and SPITZER — which had substantial achievements but he also influenced another Hungarian mathematician, LAJOS SCHLESINGER (1864—1933).

In solving certain linear differential equations, GYULA FARKAS made a successful use of the "higher order sine"[1] [14] defined in the early 19th century by WRONSKI. In another paper he studied this function separately [15].

MANÓ BEKE's (1862—1946) efforts bore fruit mostly in the problem of the reducibility of differential equations. His results have been extensively quoted and improved on. Let the coefficients of a linear differential equation belong to a well-defined class of functions; we call the equation reducible if it shares a root with a differential equation of lower order having coefficients in the same class of functions. It was FROBENIUS who introduced the notion of reducibility in 1873 and studied the question for the case where the coefficients were single-valued analytic functions. MANÓ BEKE launched the investigation of the case in which the coefficients were entire rational functions. He recognized that in this case reducibility could be decided with a finite number of steps, namely by exploring the properties of certain algebraic groups [7]. Later, in line with BEKE's researches, FABRY, BENDIXSON, LANDAU, LOEWY and chiefly SCHLESINGER addressed themselves to the question of reducibility LOEWY and SCHLESINGER also extended their investigations to systems of linear differential equations.

Incidentally, it was BEKE's favourite procedure to interpret an algebraic equation as the characteristic equation of a differential equation [8], and this interpretation enabled him to discover splendid algebraic results.

The sine of higher order is defined by the following infinite series:

$$\Phi_{m-1}(x) = \frac{x^{m-1}}{1 \cdot 2 \ldots (m-1)} - \frac{x^{2m-1}}{1 \cdot 2 \ldots (2m-1)} + \frac{x^{3m-1}}{1 \cdot 2 \ldots (3m-1)} \pm \ldots$$

The order is given by the value $m-1$. (See e.g. *Fortschritte der Math.*, 10, 1878, p. 302).

Probably the most prominent scholar to deal with differential equations on the basis of function theory was SCHLESINGER. There are few subjects in this field to which he did not contribute something essential. Starting from GAUSS's theory of the arithmetico-geometric mean and largely refining the investigations of POINCARÉ, PICARD and LAZARUS FUCHS, SCHLESINGER had ample results at the turn of the century to write systematic monographs [50, 51] on the theory of ordinary differential equations built on the study of complex functions. Little could be added to his books by similar monographs written since then. A pupil of SCHLESINGER, LAJOS DÁVID (1881—1962) further elaborated the theory of the arithmetico-geometric mean in several of his early treatises [11]. Later he was preoccupied with questions of the history of mathematics. Posterity remembers him as one of the most eminent BOLYAI researchers.

LAJOS SCHLESINGER's work catalyzed both Hungarian and world mathematics. Without going into detail, let us only mention the little known fact that upon his direct influence did LIPÓT FEJÉR turn to the theory of differential equations, substantially simplifying previous proofs for the existence theorem of ordinary differential equations (with a complex variable) [23].

At the outset of this century another outstanding Hungarian scientist, TÓDOR KÁRMÁN (1881—1963), who spent a considerable part of his life in the United States, began his extensive research work on the numerical integration of partial differential equations encountered in physical and technical problems.

Another line of research launched at the beginning of our century was LIPÓT FEJÉR's that led to a significant enrichment of our knowledge of classic polynomials. Taking the LEGENDRE polynomials for starting point, he produced a new, perhaps the simplest, proof of WEIERSTRASS's polynomial theorem as early as 1902. A little later he started studying the CHEBYSHEV polynomials, and in this field he was soon joined by JENŐ EGERVÁRY, GÁBOR SZEGŐ, MIHÁLY FEKETE, JÁNOS NEUMANN and others.

A modest but, in Hungary, pioneering paper on the theory of interpolation, a favourite terrain of Hungarian mathematical research today, was due to GYŐZŐ ZEMPLÉN. In an early paper [58], following MARKOV's ideas, ZEMPLÉN stated the problem of interpolation in a highly general way and summarized the existing results very lucidly, obtaining at the same time a new method for decomposing rational functions into partial fractions.

A new direction of interpolation theory was designated by LIPÓT FEJÉR whose works on LAGRANGE interpolation became the source of a vast literature. With the passage of time, his interest extended to all the areas involved in the problem: he tried to find connections between various interpolations, and possibilities for generalization. In the literature one often comes across FEJÉR's interpolation polynomial; this is the polynomial of no more than $(2n-1)$-th degree whose values in n basic points coincide with the values of the given function and whose derivative in the same points is equal to zero.

The history of 20th century Hungarian mathematics will have a separate chapter on various investigations and results about series. A trail-blazing work in this field was FEJÉR's doctoral dissertation published in an abridged form in 1900 in the French *Comptes Rendus* [19] and in detail in *Mathematikai és Physikai Lapok* [20]. The method of summation FEJÉR published in them can be rightly regarded as the start of the renaissance of FOURIER series. Over the 19th century the theory of FOURIER series, so important in physics and many other areas, was enriched by substantial new results, but then a standstill ensued for lack of a method to clarify convincingly the problems of convergence and divergence. FEJÉR's very first results gave sudden impetus to research; he suggested that the unsettled problems should be approached not through the sequence of partial sums of a FOURIER series, but through the arithmetic means of the terms of this sequence. The importance of averaging as a method soon outgrew the original aim and became a strong weapon in various examinations of convergence. FEJÉR himself proved a whole range of significant theorems, and FEJÉR's "summation", "kernel function", etc. are still among the basic notions of the theory of FOURIER series and orthogonal series in general. There are hundreds of papers based on his ideas. Following his first publications, almost no year passed without the major international mathematical periodicals carrying weighty findings by FEJÉR himself or by members of the first generation of his pupils. Let it suffice to mention the names of MARCELL RIESZ, FRIGYES RIESZ, MIHÁLY FEKETE, OTTÓ SZÁSZ, FERENC LUKÁCS and PÁL DIENES. It says much of the prestige of Hungarian mathematics in this field that the two lengthy articles[2] on infinite series in the *Encyklopädie* were written by Hungarians — MARCELL RIESZ and OTTÓ SZÁSZ —, and as a footnote indicates, the advisors also included LIPÓT FEJÉR, FRIGYES RIESZ and LAJOS SCHLESINGER!

This field of problems partly overlaps with Hungarian research in the theory of real functions. The results of FRIGYES RIESZ, MARCELL RIESZ, ALFRÉD HAAR, SIMON SIDON, MIHÁLY FEKETE, and DÉNES KŐNIG who started research at a very young age, together with the investigations of ZOÁRD GEŐCZE discussed elsewhere in this book, belong to the most radiant pages of the history of Hungarian mathematics.

In 1905, FRIGYES RIESZ (1880—1956) achieved significant results in generalizing certain facts of linear set theory to an arbitrary n-dimensional space. He was one of the first to realize the great theoretical importance of the LEBESGUE integral (several monographs introduce this integral in the way suggested by RIESZ), and relying on this, he preceded Germany's E. FISCHER[3] by two months in publishing one of the fundamental theorems of the theory of real functions, the RIESZ—FISCHER theorem [48]. It says that a sequence of element f_n of the space L^2 is converge at, in square integral, to an element f of the space L^2 if and only if $\|f_m - f_n\| \to 0$ as $m, n \to \infty$.

Here $\|f\|$ denotes the square root of the square integral of f.

Translating the RIESZ—FISCHER theorem into the language of orthogonal series:

$$a_0 + (a_1 \cos x + b_1 \sin x) + \ldots + (a_\nu \cos \nu x + b_\nu \sin \nu x) + \ldots$$

is the FOURIER series of a function with its square integrable in the sense of LEBESGUE if and only if the series

$$\sum_{\nu=1}^{\infty} (a_\nu^2 + b_\nu^2)$$

is convergent.

---

[2] HILB, E., RIESZ, M.: Neuere Untersuchungen über trigonometrische Reihen. *Enz. d. Math. Wiss.*, II, 3.2, pp. 1189—1228; HILB, E.—SZÁSZ, O.: Allgemeine Reihenentwicklungen. *Ibid.*, pp. 1229—1276.
[3] FISCHER, E.: *Comptes Rendus*, 144, 1907, pp. 734—736.

Having examined the function spaces L^2, FRIGYES RIESZ went on to elaborate the theory of the more general function spaces L^p. He was joined in these researches by MIHÁLY FEKETE, SIMON SIDON and several world-famous mathematicians somewhat later.

ALFRÉD HAAR (1885—1933) began his scientific career with a noteworthy result: in his doctoral dissertation [26] he presented his famous example of an orthogonal function system.

The inexhaustible material on real functions is now well known from various text-books, monographs, and commemorative writings about LIPÓT FEJÉR and FRIGYES RIESZ. Let us finally mention RIESZ's paper read to the Rome Congress in 1908 in which he presented the system of axioms of proximity spaces ignored at first but highly appreciated later [49].

The theory of functions of a complex variable aroused the attention of Hungarian researchers as early as the 1870s. Independent research into this field was initiated, among others, by GYULA FARKAS, whose investigations in connection with PICARD's theorem and doubly periodic functions deserve special mention. Following a couple of essays on the subject by JÓZSEF SUTÁK (properties of analytic functions), the most prominent scholar to continue this line of research was again LAJOS SCHLESINGER who, during studying of the posthumous papers of GAUSS, found several topics for further elaboration. LIPÓT FEJÉR's inquiries (partly together with CARATHÉODORY) into the interrelation between the points of extreme and the coefficients of a harmonic function, or the behaviour of these functions on the boundary of their domain of convergence already fall to the 20th century. These were the main subjects that such outstanding scholars of the theory of complex functions as PÁL DIENES, GÁBOR SZEGŐ, FRIGYES RIESZ, OTTÓ SZÁSZ, GYÖRGY PÓLYA, FERENC LUKÁCS and others were attracted to at the beginning. Of the many results, we refer to the proof of the fundamental theorem of conformal mapping given by FEJÉR and RIESZ, which from that time on is considered by BÉLA SZŐKEFALVI-NAGY "the" proof of the theorem.

Both research and application of vector algebra and vector analysis were rare around the turn of the century. Although the theoretical physicist IZIDOR FRÖHLICH was not unfamiliar with this field, he and many of his pupils avoided this discipline, then unusual and even slightly despised in Hungary. The pioneers in this area were MÓR RÉTHY, GYŐZŐ ZEMPLÉN and, first of all, GYULA FARKAS. In the study of physics, FARKAS was indined to use the theory of vectors, and he wrote a thick essay to popularize the discipline [18].

Probability theory. At the end of the last century this discipline was badly neglected in Hungary. All that was practically done was a few university lectures (at Budapest's Technical University and in Kolozsvár) and brief chapters in some textbooks introducing the subject. Everyday life, however, posed a lot of problems that would have required competence in probability theory and statistics. To adduce an example, let us mention a study by VILMOS FEST [24] in which he tried to decide on the basis of rather defective statistical data whether the waterways, railways or paved roads were the most economical in our country. His calculations favoured the railroad. Trying to estimate the expected income of the railways SÁNDOR KISFALUDI LIPTHAY computed some coefficients from the data at his disposal with the help of which he could predict the expected revenues quite precisely [28]. He based his reasoning on the following — rather natural — assumption: if the number of passangers and freight tons in railway traffic is compared to the number of inhabitants living along the railway line, the values obtained for similar lines will fluctuate within a narrow margin.

Research of scholarly merit in probability theory began in Hungary with the work of KÁROLY JORDAN (1871—1959) and GYÖRGY PÓLYA (1887—1985) in the 1910s only. Statistical data collection in Hungary was institutionalized relatively early.

A statistical group was set up in 1867 and converted into a separate office in 1871. The first directors of the "Hungarian Bureau of Statistics" (KÁROLY KELETI, JÓZSEF KŐRŐSY) organized the collection of data with expertise and, as a result, Hungarian statistical publications were held in high esteem all over the world. The data on the working classes, however, were rather defective in the early publications since at that time this social class was still weak.

With a view to training experts in commerce, the *Budapest Academy of Commerce* was set up in 1857. Towards the end of the century its professors also included noted mathematicians (GÉZA GHICZY, IGNÁC RADOS, MANÓ BEKE). It was a consequence of the work of the Academy of Commerce that *commercial* and *political arithmetics* came to be separated (the former dealing with problems of commercial practice, the latter with problems of accounts, annuities, insurance policies, etc.). In commercial arithmetic, operations of limited accuracy also got a role.

In line with the targets of the discipline, the bulk of the literature in financial mathematics consisted of various tables making calculations easier, and textbooks for secondary and college education. In respect of the former, the most important step was that, at the beginning of this century, GYULA ALTENBURGER and KÁROLY GOLDZIHER started to compile the first Hungarian mortality tables. The best-known and highest-level book was the *Political arithmetic* in two volumes by K. BEIN–S. BOGYÓ–M. HAVAS (Budapest, 1907).

Two individual results deserve special mention. In the 1890s GYULA ALTENBURGER worked out a method for the evaluation of group reserve funds. This method allows the insurants of the same age to be grouped together and their reserve funds computed at once irrespective of how old they were they took out the policy. This method spread all over the world and is briefly called the "Hungarian" or "Altenburger" method.

As for the other result, GÉZA GHYCZY found an iterative process for the solution of the trinomial equation appearing in the rate-of-interest problem of the calculus of annuities to which no tables are needed [25]. However, the method is slow to produce a root so it did not prove practicable.[4]

Geometry. As was seen in the previous chapters, the geometric investigations carried out towards the end of the last century mostly relied on the heritage of the two BOLYAIS and dealt with the completion, generalization and explanations of their results. Apart from these, separate mention must be made of some nice results of JÓZSEF KÜRSCHÁK. His series of articles on cyclometry [36] amounting to a whole book can serve as a

---

[4] See e.g. BEIN, K.—BOGYÓ, S.—HAVAS, M.: *Political arithmetic.* Vol. 1, p. 75. (In Hung.)

model example of a fine and understandable presentation of results with a wide appeal — the subject being the quadrature of the circle. Methodically rather simple is this famous essay containing an exact proof of the theorem which says that among the *n*-gons inscribed in the circle the regular one has the largest area while among the circumscribed *n*-gons the regular one has the smallest area [35]. The flair for the elementary on the part of a mathematician devoted to deep abstraction is evidenced by a short note in which he proved intuitively that the area of the regular 12-gon inscribed in the circle is three times the area of the square constructed with the radius [37].

One of the best known Hungarian results in geometry around the turn of the century was connected with the theory of constructions. Setting up the five groups of axioms for geometry, DAVID HILBERT made a scruting of what constructions were possible within each group and with which tools. He found that for the (so-called discrete) constructions based on the first four groups of axioms no circle had to be drawn, so the role of the compasses was reduced to transferring distances. GYULA KŐNIG called KÜRSCHÁK's attention to the problem whether, relying on this fact, one could say something more.

In his elaborated paper [38] JÓZSEF KÜRSCHÁK gave a construction by which *any* *segment* can be transferred through the use of a *gauge* (Eichmass), i.e. an instrument with a fixed opening suitable for the transfer of a distance fixed once for all. This result belongs to the field of problems intensely studied since MOHR and MASCHERONI, and leading HILBERT, HJELMSLEV, VAHLEN and others to notable results.

It shows the significance of KÜRSCHÁK's theorem that, from the second edition onward, one chapter of HILBERT's *Grundlagen der Geometrie* bore the title *Die geometrische Konstruktionen mittels Lineals und Eichmasses*.

As the length of the segment fixed in the gauge is indifferent in this construction, the question arises whether the gauge could be replaced by some distance, or if that were impossible, by some plane figure bounded by arbitrary straight lines. According to MIHÁLY BAUER, the answer is negative: a *movable* gauge cannot be replaced by any polygon [5].

KÁROLY JORDAN (1871—1959), who was originally engaged in chemical problems presented his first major contribution to mathematics in the field of integral geometry. After a series of minor essays published with his colleague RAYMOND FIEDLER who died early, they summarized their pioneering researches in a monograph in 1912 [27]. Let us quote BÉLA GYIRES to sum up their relevant achievements:

"Calling the set of boundary points of a bounded convex domains as a convex curve, the equation $p(\alpha)$ of his curve in tangential polar coordinates is such an everywhere continuous function of period 2π and expandable into a uniformly convergent FOURIER series which has a left-hand and a right-hand derivative at every point, while $p'(\alpha)$ is not necessarily continuous but also expandable into a FOURIER series. The measure of the set of all points where there exists no second derivate is zero. Calling these curves T_2 curves, the question is whether every T_2 curve is a closed convex curve. Answer: if and only if the curvature does not change sign. It is an interesting feature of T_2 curves that the algebraic length of their evolutes is zero."

It is well known that at the outset of his scientific career FRIGYES RIESZ was also attracted to research into geometric problems. A remarkable result of his was related to the investigations of SCHOENFLIES; a converse of the JORDAN theorem, it says that every simple closed curve has a one-to-one continuous mapping onto the circle.

A one-time pupil of GYULA VÁLYI, GYULA SZŐKEFALVI-NAGY (1887—1953) started his rich and extensive scientific researches after the turn of the century. His papers numbering some 150 address the most diverse fields of geometry: the geometry of rational and transcendental functions, convex curves and surfaces, the arithmetic properties of algebraic curves and surfaces, theory of geometrical constructions, topology of curved lines, etc.

Descriptive and *projective geometry* played an important role especially in university education about the turn of the century. At the beginning of his activity spanning more than fifty years, LIPÓT KLUG (1854—1944) turned to problems studied by GYULA VÁLYI, consistently ignoring analytic methods. Later he studied more diverse fields with success. The work of BÉLA TŐTÖSSY (1854—1933) remained on a smaller scale, his most significant contribution being his work in the preparations for the second edition of the *Tentamen*. He enriched the theory of surfaces of the fourth order by examining some examples.

Descriptive geometry was given less attention in this period than other mathematical disciplines. The number of original investigations decreased and Hungarian periodicals published relatively few papers on the subject. Instead, experts of the field were content to write some good textbooks. GYULA KŐNIG raised his voice to change the situation, demanding that at least *Mathematikai és Physikai Lapok* should devote more attention to studies in descriptive geometry. That would "promote a rapprochment between this subject and mathematics for, regrettably enough, the gulf between them has been deepening."[5] But GYULA KŐNIG's efforts failed to change the situation substantially.

Mechanics and mathematical physics. Around the turn of the century these disciplines were still so closely interwined with mathematics that one cannot decide for sure whether their practitioners were physicists or mathematicians. The literary heritage of MÓR RÉTHY, GYULA FARKAS and KÁLMÁN SZILY are almost equally divided between physics and mathematics, although as professors they were considered physicists. Even such prestigeous representatives of Hungarian physics as IZIDOR FRÖHLICH, LORÁND EÖTVÖS and GYŐZŐ ZEMPLÉN were wont to use mathematical tools and were not averse to independent mathematical research, either. And conversely, many of our outstanding mathematicians like GYULA KŐNIG, LIPÓT FEJÉR, MANÓ BEKE had an impact on the development of physics in Hungary.

We have already mentioned this aspect of GYULA KŐNIG's work. MANÓ BEKE developed an intriguing theory in one of his papers [6]. According to the variational principle of the least action by MAUPERTUIS, the value of the integral

$$\int_{I}^{II} \sum_{i=1}^{n} m_i v_i \, ds_i$$

[5] *Math. és Phys. Lapok*, 1, 1892, p. 109.

formed from the masses and velocities of the moving points and the elementary path is minimal. Relying on the surface theory of GAUSS, MANÓ BEKE arrived at a geometrical interpretation of this principle by identifying the minimum of the integral with the arc length of a geodesic of a hypersurface.

LIPÓT FEJÉR's virtuosity in making apparently complicated things simple and turning the abstract into visual is well known. This gift of his is obvious from his papers on physical subjects as well. Let us mention for illustration a less-known paper in which he gave the mathematical equivalent of the so-called OSTWALD variational principle [21]. The habilitation lecture of LIPÓT FEJÉR [22] also partly belonged to physics: joining to the researches of KNESER and LYAPUNOV, he solved some problems in lability and stability.

One of the greatest personages of Hungarian theoretical physics and an inventive user of mathematical tools was GYULA FARKAS. His essays on the conditions of stable thermodynamic equilibrium and his article of 1895 in which he obtained the so-called CARATHÉODORY principle of thermodynamics 14 years ahead of CARATHÉODORY himself stirred an international sensation.[6] He began a new study of the nearly forgotten FOURIER principle; the articles he devoted to this subject contain several original mathematical results as well [16, 17]. For a better understanding of the course he followed we explain some physical concepts.

According to the usual definition, an infinitely small displacement at infinite velocity satisfying the constraints is said to be *virtual*. Although from the physical point of view this definition may seem laboured, it is useful because with its help virtual work can be defined, the D'ALEMBERT principle stated, and the conclusions drawn from the latter comply with experience. Let us now define the components of virtual displacement in the usual manner as the difference of the displacements permitted by the constraints $(\partial x_i, \partial y_i, \partial z_i)$ and those actually taking place (dx_i, dy_i, dz_i):

$$\delta x_i = \partial x_i - dx_i; \quad \delta y_i = \partial y_i - dy_i; \quad \delta z_i = \partial z_i - dz_i.$$

Keeping to this notation, the FOURIER principle can be written in the form

$$\sum_{i=1}^{n} \{(m_i\ddot{x}_i - X_i)\delta x_i + (m_i\ddot{y}_i - Y_i)\delta y_i + (m_i\ddot{z}_i - Z_i)\delta z_i\} \geq 0,$$

where m_i is the mass of the i-th point while $\ddot{x}_i$, $\ddot{y}_i$, $\ddot{z}_i$ and X_i, Y_i, Z_i are the components of the acceleration and free force, respectively.

LAGRANGE only considered the case of equality in the above relation, so the FOURIER principle is more general as it includes the case of inequality as well. For mechanics this means the following. If the constraint is such that the mass point can only move on the surface of a rigid body (or on a curve), then the constraint is described by equalities; if, on the other hand, the point may equally move on or above the surface (e.g. an aeroplane), then also inequalities are needed to describe the constraint.

On the basis of some deep theorems, GYULA FARKAS arrived at equations corresponding to but more general than LAGRANGE's equations of motion. But his results

[6] In any neighbourhood of any state of a thermically homogeneous system there are states that cannot be attained in a reversible-adiabatic way.

reach far beyond physics together with the researchers of HERMANN MINKOWSKI, ALFRED HAAR and others; they are fundamental in the theory of linear equalities and inequalities, and today they are major theorems of *linear programming*. Arising from the study of physics but used in operation research today, FARKAS's *lemma* can be stated as follows (the "or" being exclusive): the m-dimensional vector b can either be written as a linear combination with non-negative coefficients of the column vectors of the $m \times n$ matrix A, or there exists an m-dimensional vector which forms an acute angle with vector b and at most a right angle with the column vectors of matrix A.

According to the celebrated *theorem* of GYULA FARKAS derived from the above lemma, the polar of the polar of the finite set of vectors H coincides with the convex cone spanned by H [17].

The special literature on linear programming gives ample information on FARKAS's relevant researches and on the chronology of his and MINKOWSKI's results in that field.

MÓR RÉTHY at the beginning of his scientific activity stirred a sensation with his results in the theory of the diffraction of light, but his propositions are partly outdated now. His most attractive and certainly classic investigations are concerned with the flow of incompressible fluids, and are also mathematically remarkable. RÉTHY searched for the laws of free flow and flow between walls and dams of different shapes [46]. KIRCHHOFF showed in 1876 that the problem could conveniently be treated with the help of conformal mapping — the theory that grew out of his idea is now an integral part of textbooks on complex functions as well.

MÓR RÉTHY started from the following function

$$\left(\frac{\zeta^2-1}{\zeta^2+1}\right)^2 = k^2(1-e^w) \qquad\qquad (k>1,\text{ real})$$

This function maps the quadrants of unit or infinite radius in the ζ plane in a one-to-one and conformal manner onto a strip of width π in the w plane. With the help of this function, RÉTHY was able to describe several flows important for practice; they include, e.g., the flow of finite width encountering a weir with straight or rectangular cross section.

The Hungarian physicist and prolific writer IZIDOR FRÖHLICH made ample use of mathematical tools. At the beginning of his career, in the 1870s, he was intrigued by the diffraction of light, later by the polarization of light. He was sometimes joined by MÓR RÉTHY in his researches. FRÖHLICH established by several thousand experiments and later also theoretically, the law that the total intensity of light after diffraction equals the intensity of light falling on the openings. It is noteworthy that in the course of solving a differential equation of diffraction, FRÖHLICH applied FOURIER series for the first time in Hungary.

IZIDOR FRÖHLICH's two-volume handbook for theoretical physics published in 1890 was a source of knowledge for a long time. A separate chapter of it was devoted to the mathematical tools frequently used in physics, which greatly promoted the application of mathematics in physical research.

Undoubtedly, KÁLMÁN SZILY's investigations in physics, especially thermodynamics, are far more significant than his mathematical ideas. He was a mechanist, i.e. he professed that the phenomena belonging to various fields of physics could acceptably be explained only by theories based on mechanical principles. In keeping with this approach, he directly continued the trend initiated by RANKINE, BOLTZMANN, CLAUSIUS and HELMHOLTZ. SZILY was seeking an answer to the question from which variational principle of mechanics the second law of thermodynamics could be deduced. CLAUSIUS started from D'ALEMBERT's principle. According to SZILY, replacing change of heat by the change of total energy, and absolute temperature by total kinetic energy, the HAMILTON *principle* might be a suitable point of departure [55]. Though fiercely disputed by CLAUSIUS, SZILY's ideas achieved international success. The *Encyclopädie* (5.1, p. 148) also wrote of SZILY's work in appreciative terms and there was a time when the 2nd law of thermodynamics was called the CLAUSIUS—SZILY *theorem*.

A great theoretical physicist was Győző ZEMPLÉN, who died at a young age. His studies on the viscosity of fluids and gases are fundamental. He also wrote a lengthy article about the problem for the Encyclopädie (4.3, pp. 281—323). ZEMPLÉN was among the first who realized the value of the special theory of relativity and he offered the following explanation for the failure of the MICHELSON experiment: light emitted from a moving source travels in different directions at different velocities.

Astronomy, geodesy. Hungarian astronomy, already having an impressive past, gained new momentum from the multiplication of private observatories (Herény, Kalocsa, Ógyalla) and their improving equipment. Busy work was going on especially at the Ógyalla observatory founded by MIKLÓS KONKOLY-THEGE in 1871. It was initially confined to observations and the publication of the data obtained. Many people excelled in this work; moreover, the data enabled some Hungarian astronomers to calculate orbits.

A task of astronomy, interesting also from the viewpoint of mathematics, was to determine the geographical position of the observatory. Namely, the first, inaccurate, data relied on the geographical position of the observatory on Gellért Hill, and this had not been obtained directly with astronomical tools but by transferring the data of the Vienna observatory geodetically. KONKOLY-THEGE worked out a new method for the determination of geographical position [32], and KÁROLY BRAUN constructed an instrument called trigonometer to make computations easier [9].

Some astronomical studies also contained new mathematical elements. E.g., ÁRMIN KOBOLD proposed an approximation formula for the period of the Jupiter's rotation around its axis by integrating certain differential equations and using data concerning the displacement of spots [31]. RADÓ KÖVESLIGETHY employed deep mathematical tools in both astronomical and geodetic researches. His inquiries about the spectra of heavenly bodies and the many-body problem are characterized by an expert use of differential equations. What brought him world fame, however, was his achievements in seizmology: they are based on the hypothesis that the laws of refraction for seimic and light waves are the same. Of salient importance is his suggestion to apply the so-called BERTRAND problem in seizmology.

BERTRAND had raised the question as to which central forces cause a heavenly body to describe a closed path, whatever its initial position, direction and speed were. KÖVESLIGETHY adapted this problem to seizmology as follows. Let us consider the isotropic elastic media with spheric stratification (for these, the index of refraction — i.e. the reciprocal of the velocity of waves — is a function of the distance r measured from the centre of stratification); we have to find those functions $f(r)$ for which the seismic trajectories for all initial points and directions are closed curves. KÖVESLIGETHY found two essentially different solutions of the problem and later, in 1917, JENŐ EGERVÁRY determined all solutions.

It would be a fascinating task to give a detailed account of RADÓ KÖVESLIGETHY's endeavours from the aspect of mathematics. Let us illustrate his inventiveness and acumen with but a few examples here. In the 1840s, Britain's HOPKINS developed a theory in which the investigation of the inner structure of the earth was based on precession. It was, however, realized soon that his method could not produce reliable results: KÖVESLIGETHY's mathematically and geologically well-founded method relies on earthquakes [34]. For earthquakes are easy to describe in mechanical terms and the frequent occurrence provides sufficiently many empirical data. His theory built on the calculus of variations furnished more precise values of the velocity of seismic waves and the elasticity of the earth's crust.

Mathematical tools were used by *orometry*, a discipline founded by the Austrian SONKLAR and engaging the attention of our scientists for quite some time. The discipline was concerned with determining the cubic content of mountains from certain measurable data. The first theories, however, assumed the knowledge of data which were technically difficult to measure. KÖVESLIGETHY introduced the notion of *oroid* or "level surface" whose equation could be obtained from more easily measurably orometric elements [33]. When the oroid is known, the location of peaks, valleys, ridges and cols reduces to the calculation of extreme values, while the slope can be obtained simply by differentiation. KÖVESLIGETHY used orometry successfully also in examining the craters of the moon.

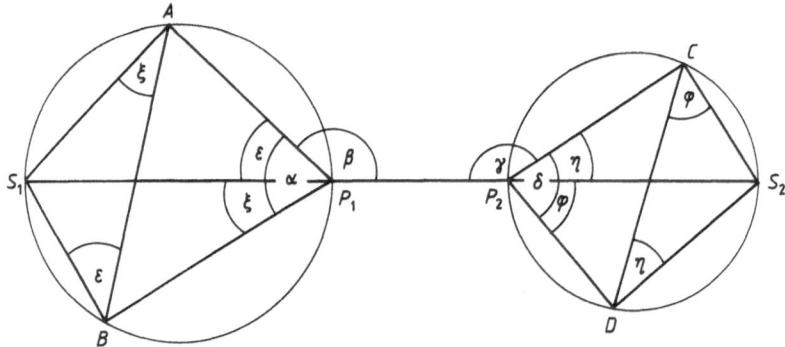

Figure 59

Of researches connected with triangulation, a method called "countersection" (Gegenschnitt) devised by the head of the Hungarian Triangulation Office, JÁNOS MAREK, needs to be mentioned [40]; later it was independently rediscovered by others. In narrow valleys and under bad visibility conditions the so-called MAREK method is commonly used: in the knowledge of four geographical points, it makes possible the determination of another two points. From point P_1 we can see the known points A and B as well as P_2; from P_2 we can see the known points C and D as well as P_1 (*Figure 59*). From P_1 we adjust to points A, B and P_2, and measure the angles α and

β. Similarly, from P_2 we adjust to C, D and P_1, and measure the angles γ and δ. Constructing the points S_1 and S_2 with the help of the circles ABP_1 and DCP_2, the coordinates of P_1 and P_2 can be computed by using simple planimetric relations.

*

The last chapter of my writing is but an introduction to the history of Hungarian mathematics in the 20th century hallmarked by many a great achievement. To work it out in detail is a hard but noble and obligatory task of the future.

Bibliography

Without aiming at completeness, we list some major treatesies published around the turn of the century by Hungarian mathematicians and after the sign (+), their collected works as well as writings about them.

[1] BAUER, MIHÁLY: On the theory of skew-symmetric substitutions. *Math. és Phys. Lapok*, 1, 1892, pp. 356—359 (In Hung.)

[2] — On the theory of characteristic equations. *Math. és Phys. Lapok*, 3, 1894, pp. 293—298. (In Hung.)

[3] — Remark to a theorem of Dirichlet. *Math. és Phys. Lapok*, 3, 1894, pp. 368—372; 4, 1895, pp. 331—336. (In Hung.)

[4] — Number theoretic theorems. *Math. és Phys. Lapok*, 5, 1896, pp. 149—160; 265—272.

[5] — On the theory of geometric constructions. *Math. és Phys. Lapok*, 12, 1903, pp. 251—255. (In Hung.)

(+) RÉDEI, LÁSZLÓ, *Matematikai Lapok*, 4, 1953, pp. 241—262. (In Hung.)

SERES, IVÁN, *Műszaki nagyjaink*, Vol. 3, 1967, pp. 327—336. (In Hung.)

[6] BEKE, MANÓ: The principle of least action on the basis of Gauss' theory of curvature. *Math. és Term. tud. Értesítő*, 2, 1883—84, pp. 133—162. (In Hung.)

[7] — Die Irreducibilität der homogenen linearen Differentialgleichungen. *Math. Annalen*, 45, 1894, pp. 278—294.

[8] — Zur Gruppentheorie der homogenen linearen Differentialgleichungen. *Math. Annalen*, 49, 1898, pp. 573—580.

[9] BRAUN, KÁROLY: The trigonometer, an instrument to resolve spheric triangles. *Math. és Term. tud. Értesítő*, 1, 1882—83, pp. 313—317. (In Hung.)

[10] CSILLAG, VILMOS: Graphic solution for systems of linear equations. *Math. és Phys. Lapok*, 7, 1898, pp. 157—170. (In Hung.)

[11] DÁVID, LAJOS: Theorie des Gauss'schen verallgemeinerten u. speziellen arithmetisch-geometrischen Mittels. *Math. u. Naturw. Berichte aus Ungarn*, 25, 1907, pp. 153—171.

[12] — Zur Theorie der Schapiraschen Iteration. *Journal für die reine u. angew. Math.*, 135, 1908, pp. 62—74.

(+) SZÉNÁSSY, BARNA, *A Debreceni Kossuth Lajos Tudományegyetem Évkönyve*, 1980—81, pp. 61—62. (In Hung.)

FILEP, LÁSZLÓ: *Ganita Bhāratī (India)*, 3, 1981, pp. 65—70.

[13] DEMECZKY, MIHÁLY: Résolution des systèmes des congruences linéaires. *Comptes Rendus*, 88, 1879, pp. 1311—1313.

[14] FARKAS, GYULA: Sur l'application de la théorie des sinus, etc. *Comptes Rendus*, 90, 1880, pp. 1542—1545.

[15] — Sur la théorie des sinus des ordres supérieurs. *Comptes Rendus*, 91, 1881, pp. 209—211.

[16] — On the algebraic foundations of applying the Fourier principle of mechanics. *Math. és Phys. Lapok*, 5, 1896, pp. 49—54. (In Hung.)

[17] — Über die Theorie der einfachen Ungleichungen. *Journal für reine u. angew. Math.*, 124, 1902, pp. 1—27.

[18] — The theory of vectors and simple inequalities. Kolozsvár (no date). (In Hung.)

(+) ORTVAY, RUDOLF, *Math. és Phys. Lapok*, 34, 1927, pp. 3—25. (In Hung.)

ORTVAY, RUDOLF, *Az MTA elhunyt tagjai felett tartott Emlékbeszédek*, 21, 1933, No. 15. (In Hung.)
FILEP, LÁSZLÓ, *Matematikai Lapok*, 29, 1977—1981, pp. 231—244. (In Hung.)
PRÉKOPA, ANDRÁS, *American Math. Monthly*, 87, 1980, pp. 165—191.

[19] FEJÉR, LIPÓT: Sur les fonctions bornées et intégrables. *Comptes Rendus*, 131, 1900, pp. 984—987.

[20] — Investigations in the theory of Fourier series. *Math. és Phys. Lapok*, 11, 1902, pp. 49—68, 97—123. (In Hung.)

[21] — On the Ostwald principle of mechanics. *Math. és Phys. Lapok*, 15, 1906, pp. 24—48. (In Hung.)

[22] — Stability and lability studies in the mechanics of mass systems. *Math. és Phys. Lapok*, 15, 1906, pp. 152—172. (In Hung.)

[23] — Sur le calcul des limites. *Comptes Rendus*, 143, 1906, pp. 957—959.

(+) *Gesammelte Arbeiten. 1—2*. Hrsg. und mit Kommentaren versehen: PÁL TURÁN. Budapest—Basel—Stuttgart, 1970.
TURÁN, PÁL, *Matematikai Lapok*, 1, 1950, pp. 160—169. (In Hung.)
TURÁN, PÁL, *Matematikai Lapok*, 11, 1960, pp. 8—18. (In Hung.)
ACZÉL, J., *Publ. Math. Debrecen*, 8, 1961, pp. 1—24. (In Hung.)
TANDORI, KÁROLY, *Matematikai Lapok*, 29, 1977—1981, pp. 7—11. (In Hung.)
KAHANE, J. P., *Matematikai Lapok*, 29, 1977—1981, pp. 21—31.

[24] FEST, VILMOS: Traffic works and lines. *Értekezések a Math. Tud. Köréből*, 2, 1872—1873. (In Hung.)

[25] GHYCZY, GÉZA: Determination of interest rates in computing annuities. *Report of the Budapest Academy of Commerce for 1889—90*. (In Hung.)

[26] HAAR, ALFRÉD: *Zur Theorie der orthogonalen Funktionensysteme*. Inaugural-Dissertation. Göttingen, 1909.

(+) Collected works. (Ed.: SZŐKEFALVI-NAGY, B.) Budapest 1959, with biography on pp. 9—10, obituary on p. 659.

[27] JORDAN, KÁROLY (— FIEDLER, RAYMOND): Contribution à l'étude des courbes convexes fermées, etc. Paris, 1912.

(+) RÉNYI, ALFRÉD, *Matematikai Lapok*, 3, 1952, pp. 111—121. (In Hung.)
GYIRES, BÉLA, *Alkalmazott Matematikai Lapok*, 1, 1975, pp. 275—298. (In Hung.)
TAKÁCS, L., *The Annals of Math. Stat.*, 32, 1961, pp. 1—11.

[28] KISFALUDI LIPTHAY, SÁNDOR: On the question of fares in connection with the profitableness of railways. *Értekezések a Math. Tud. Köréből*, 15, 1892, No. 1. (In Hung.)

[29] KLUG, LIPÓT: *A synthetic treatment of third-order spheric curves*. Pozsony, 1881. (In Hung.)

[30] — *Projective geometry*. Budapest, 1903. (In Hung.)

(+) ZIGÁNY, FERENC, *Mat. és Fiz. Lapok*, 50, 1943, pp. 205—222. (In Hung.)
KÁRTESZI, FERENC, *Matematikai Lapok*, 24, 1973, pp. 219—223. (In Hung.)

[31] KOBOLD, ÁRMIN: Data on the elements of rotation of Jupiter. *Értekezések a Math. Tud. Köréből*, 8, 1881. (In Hung.)

[32] KONKOLY-THEGE, MIKLÓS: The geographical longitude of the observatory of the royal Hungarian central institute of meteorology and seismography. *Math. és Term. tud. Értesítő*, 10, 1891, pp. 63—67. (In Hung.)

[33] KÖVESLIGETHY, RADÓ: Über eine neue Methode der Morphometrie der Erdoberfläche. *Math. u. Naturw. Berichte aus Ungarn*, 13, 1894—1895, pp. 365—378.

[34] — A new geometric theory of seismic phenomena. *Math. és Term. tud. Értesítő*, 13, 1895, pp. 363—407. (In Hung.)

[35] KÜRSCHÁK, JÓZSEF: Über dem Kreise ein- und umgeschriebene Vielecke. *Math. Annalen*, 30, 1887, pp. 578—581.

[36] — The history and theory of cyclometry. *Math. és Phys. Lapok*, 1, 1892, pp. 30—51, 130—142, 251—264; 2, 1893, pp. 297—310, 341—355, 373—387; 3, 1894, pp. 102—110, 170—180, 230—239. (In Hung.)

[37] — On the regular 12-gons. *Math. és Phys. Lapok*, 7, 1898, pp. 53—54. (In Hung.)

[38] — Das Streckenabtragen. *Math. Annalen*, 55, 1902, pp. 597—598.

[39] — Über Limesbildung und allgemeine Körpertheorie. *Journal f. die reine u. angew. Math.*, 142, 1913, pp. 211—253.

(+) RADOS, GUSZTÁV, *Az MTA elhunyt tagjai felett tartott Emlékbeszédek*, 22, 1934, No. 7. (In Hung.)
STACHÓ, LAJOS, *Műszaki nagyjaink*, Vol. 3, Budapest, 1967, pp. 242—279. (In Hung.)

[40] MAREK, JÁNOS: *Technische Anleitung zur Ausführung der trigonometrischen Operationen* ... Budapest, 1875.

300

[41] PETZVAL, JÓZSEF: Integration der linearen *Differentialgleichungen*, 1—2, Wien, 1853, 1859.

 (+) ERMÉNYI, LAJOS, *P. J.'s life and merits*. Budapest, 1906. (In Hung.)
RUMANOVSKY, IVAN, *P. J.* Ružomberok, 1957.

[42] RADOS, GUSZTÁV: Zur Theorie der Congruenzen höheren Grades. *Math. u. Naturw. Ber. aus Ungarn*, 1, 1882, pp. 266—278.

[43] — Zur Theorie der Determinanten. *Math. u. Naturw. Berichte aus Ungarn*, 8, 1890, pp. 60—64.

[44] — Zur Theorie der Raumkurven. *Math. u. Naturw. Berichte aus Ungarn*, 8, 1890, pp. 90—98.

[45] — Die Theorie der adjungierten Substitutionen. *Math. Annalen*, 48, 1892, pp. 417—424.

 (+) GYIRES, BÉLA, *Műszaki nagyjaink*, Vol. 3. Budapest, 1967, pp. 283—304. (In Hung.)

[46] RÉTHY, MÓR: Fluid currents. *Értekezések a Math. Tud. Köréből*, 15, 1894, No. 4. (In Hung.)

[47] RIESZ, FRIGYES: Position-geometrical analysis of point configurations on spheric curves of the fourth order and first class. *Math. és Phys. Lapok*, 11, 1902, pp. 293—309; 13, 1904, pp. 191—204. (In Hung.)

[48] — Sur les systèmes orthogonaux de fonctions. *Comptes Rendus*, 144, 1907, pp. 615—619.

[49] — Stetigkeitsbegriff und abstrakte Mengenlehre. *Atti del IV. Congresso Internazionale Matem. Roma*, Vol. 2, 1908, pp. 18—24.

 (+) Anonymous, *Matematikai Lapok*, 7, 1956, pp. 1—9. (In Hung.)
SZŐKEFALVI-NAGY, BÉLA: *Matematikai Lapok*, 29, 1977—1981, pp. 1—5. (In Hung.)
HALMOS, P. R., *Matematikai Lapok*, 29, 1977—1981, pp. 13—20. (In Hung.)

[50] SCHLESINGER, LAJOS: Handbuch der Theorie der linearen *Differentialgleichungen*, 1—2, Leipzig, 1895, 1897—98.

[51] — *Einführung in die Theorie der gewöhnlichen Differentialgleichungen auf funktionentheoretischer Grundlage*. Leipzig, 1900 and other dates.

[52] SZŐKEFALVI-NAGY, GYULA: Über ein Theorem von Jacobi und seine Verallgemeinerung. *Jahresb. d. Deutsch. Math. Ver.*, 18, 1909, pp. 4—7.

[53] — Über arithmetische Eigenschaften algebraischer Kurven. *Math. u. Naturw. Ber. aus Ungarn*, 26, 1910, pp. 168—195.

[54] — Zur arithmetischen Theorie der ternären Gleichungen von höherem Geschlechte. *Math. Annalen*, 73, 1913, pp. 230—240.

 (+) SZŰCS, ADOLF, *Mathematikai és Physikai Lapok*, 33, 1926, pp. 1—11. (In Hung.)
OBLATH, RICHÁRD, *Matematikai Lapok*, 5, 1954, pp. 189—243. (In Hung.)

[55] SZILY, KÁLMÁN: Das Hamilton'sche Prinzip und der zweite Hauptsatz der mechanischen Wärmetheorie. *Ann. d. Phys. u. Chemie*, 145, 1870, pp. 295—302.

[56] — On the sum of the squares of binomial coefficients. *Math. és Phys. Lapok*, 2, 1893, pp. 289—296. (In Hung.)

[57] ZEMPLÉN, GYŐZŐ: An elementary theorem on the decomposition of integers. *Math. és Phys. Lapok*, 8, 1899, pp. 135—137. (In Hung.)

[58] — A contribution to the theory of interpolation and partial fractions. *Math. és Phys. Lapok*, 9, 1900, pp. 386—404. (In Hung.)

25. Biographies

The following pages deal with Hungarian scientists of old times up to the beginning of the twentieth century.

János Apáczai Csere
(Apáca, 1625—1659, Kolozsvár)

Was born in a small village in Transylvania. His parents died when he was very young and practically nothing is known about them. He grew up as an orphan without means and had to interrupt his studies in Kolozsvár and Gyulafehérvár twice because of financial problems. In 1648 his great master BISTERFELD recommended Bishop ISTVÁN GELEJI KATONA to provide APÁCZAI CSERE a scholarship to study abroad. He attended universities mostly in Holland and was conferred the degree of doctor of theology at Harderwijck in 1651. In Utrecht APÁCZAI dealt with DESCARTES's philosophy in detail and wrote an encyclopedia, the very first one in the Hungarian language, which was published in Utrecht in 1655. After five years' stay abroad he was asked to return to Hungary, and Prince GYÖRGY RÁKÓCZI II appointed him to teach at Gyulafehérvár College. APÁCZAI happily accepted the assignment and addressed himself to the task enthusiastically. His inauguration speech in 1653, a declaration of his belief in puritanism, a religious trend starting from England, and of his belief in DESCARTES's teachings, however, brought about resentment in the ruling circles. Their pressure caused the Prince of Transylvania to transfer APÁCZAI, after his master BISTERFELD died in 1655, and in spite of the protest of the students in Gyulafehérvár, to be the head of the college in Kolozsvár, which was practically in ruins. APÁCZAI worked zealously, and the college soon became so famous that the students from Gyulafehérvár herded there. ZSUZSANNA LORÁNTFFY and Prince ÁKOS BARCSAY gave a lot of acknowledgement to APÁCZAI's activity, but his *Encyclopedia* was not a real success. Death reaped an extremely active young life when APÁCZAI died on 31st December 1659.

József Arenstein
(Pest, 1816—1892, Stuppach)

First he was a teacher at a secondary school of the Piarist order, then at the Technical School in Pest. In 1847 he became member of the Hungarian Scholarly Society in the 1850s he was assigned to teach at a secondary school in Vienna, and he gave up that job in 1865. ARENSTEIN wrote several books on mathematics and essays on agriculture, industry, navigation both in Hungarian and German.

János Bedőházi
(Szászvesszős, 1853—1915, Marosvásárhely)

BEDŐHÁZI studied at the Technical University of Budapest and at the University of Sciences in Kolozsvár where he graduated as a teacher of chemistry and physics. He went to teach at the College of Marosvásárhely in 1881 and became a director there in 1887. Besides teaching, he actively participated in the literary life of Transylvania and wrote the first comprehensive book about the two BOLYAIS (Marosvásárhely, 1897). The readable book has an engaging style, but as far as the professional side is concerned, it is of low standard, and data of the biographies are wrong. BEDŐHÁZI could not get over the humiliation and the unjust judgement of the local public.

Károly Bein
(Mágocs, 1853—1907, Budapest)

A teacher at the College of Rabbinical Studies and lecturer at the Academy of Commerce. The author of several essays on financial mathematics, natural science and pedagogy.

Henrik Berzevitzy
(Hámbor, 1652—1713, Besztercebánya)

Few data are available concerning his life. He taught mathematics at the University of Nagyszombat in 1691 and 1692. When FERENC RÁKÓCZI invaded the town in 1704, BERZEVITZY, deputy director of the university, was captured and taken prisoner. After being released he went to teach in Graz, then returned to Besztercebánya and continued teaching there. His book entitled *Practica Arithmetica* was a Hungarian version of the work of JULIUS CAESAR OF PADUA, and was first published in either 1682 or 1687.

József Beszédes
(Magyarkanizsa, 1787—1852, Dunaföldvár)

Son of an illiterate village mayor. He went to secondary school in Szeged and Temesvár and took tutorship to finance his studies. He wanted to be a priest and went to study theology in Eger, but realized before he was ordained that the career was not for him. He chose to go to Pest and became an engineer. His activity at the improvement of riverways proved extremely useful. In 1831 he became corresponding member of the Academy of Sciences. Most of his essays were published in *Tudományos Gyűjtemény*.

Lajos Bitnicz
(Ják, 1790—1871, Nagykanizsa)

A mass-priest in Szombathely, then a titular bishop. One of the founding members of the Academy. He taught mathematics and Hungarian literature in Szombathely. Most of his scientific activity concerns Hungarian linguistics; BITNICZ was a pioneer in this field. Two school-books of mathematics and a couple of essays written by him were published. The latter ones testify that BITNICZ was a man of wide reading. In 1847 he was chosen honorary member of the Academy of Sciences, the only one in mathematics before the Compromise of 1867.

Samu Bogyó
(Paptamási, 1857—1928, Budapest)

A teacher at the Academy of Commerce and a mathematician at several institutes of finance. He wrote studies of great importance on political arithmetic and actuarial mathematics. An active member of the Council of Public Education.

Farkas Bolyai
(Bolya, 1775—1856, Marosvásárhely)

The Bolyai family has a long historical past. Some of them are remembered as fighters against the Turks, others actively participated in Transylvania's political life. The family's long history of struggles ran its course to impoverishment. GÁSPÁR BOLYAI (1732—1804), FARKAS BOLYAI's father, had only a small estate left at Bolya, near Nagyszeben. GÁSPÁR BOLYAI's wife KRISZTINA PÁVAI VAJNA inherited another small farm at Domáld, near Marosvásárhely.

FARKAS BOLYAI was born in Bolya, on 9th February 1775. His learned father taught him until he was six, then he went to the famous Calvinist school in Nagyenyed. His exceptional talent — which first appeared at learning languages and numerical calculation — was accompanied by diligence and soon distinguished him from his schoolmates. The teachers readily attributed him the dubious role of infant prodigy, which was an obvious pedagogical mistake and carried the danger of turning his spiritual development into a wrong direction. Fortunately when FARKAS BOLYAI was twelve, he went to Baron KEMÉNY's estate in Marosvécs and became the private tutor of SIMON KEMÉNY (1779—1826), four years younger than FARKAS BOLYAI.

FARKAS BOLYAI's turn of life proved favourable. The cultured KEMÉNY family helped his development, and SIMON KEMÉNY became his faithful friend till the end.

The tutor and his pupil went to study at the Calvinist school of Kolozsvár in 1790. During the five years FARKAS BOLYAI spent in Kolozsvár, he went through the laps of ideological changes which were due to conflicts between the religious education at school and the ideas of the Enlightenment spreading in Transylvania, as well as the awakening Hungarian national ambitions against Vienna's oppressive policy. FARKAS BOLYAI's future plans changed according to the impact of those objective influences and his manifold gifts. MIHÁLY SZATHMÁRI PAP, BOLYAI's professor of philosophy and winner of several foreign competitions in philosophy, tried to school him in religious fanatism and strongly advised him not to deal with mathematics. GYÖRGY MÉHES (1746—1809), BOLYAI's teacher of mathematics, worked diligently and enthusiastically, but his slightly superficial professional knowledge was hardly able to balance SZATHMÁRI's influence.

For some weeks FARKAS BOLYAI tried the actors' career as well, but in the autumn of 1795 he decided together with his friend SIMON KEMÉNY to go abroad on a study trip, which proved to be a turning point of his career. BOLYAI's trip was delayed by an unexpected and long-lasting illness, so he could travel to Jena only in the spring of the next year.

The few months FARKAS BOLYAI spent in Jena were of essential importance for his future because it was then that he started dealing with mathematics systematically and thoroughly. The asthenopia he developed during an illness while he was young restrained him from excessive reading, but did not hold him back from thinking about the axioms of mathematics on his long and lonely walks.

The two friends' next station after Jena was Göttingen where they arrived in September 1796. They both registered at the university there, which provided them an opportunity to study within the frame of established education.

FARKAS BOLYAI's specialization in mathematics was determined by the years in Göttingen. He made a lot of friends and established scientific relationships with a great many people, among them was SEYFFER (1762—1822), a very young teacher of astronomy who intensively dealt with the axioms of geometry. There was KÄSTNER

(1719—1800), a professor of mathematics, who gave a detailed historical analysis of EUCLID's eleventh axiom and listed quite a number of books and papers dealing with the problem. Also GAUSS was attending the University of Göttingen then.

FARKAS BOLYAI was particularly impressed by the friendly conversations and discussions with GAUSS. It was then that a mathematical system began to take shape in his thinking, and it must have been those conversations that impelled him later to deal with the Fifth Euclidean Postulate. Several essays written by FARKAS BOLYAI indicate his appreciation of GAUSS' friendship and how important it was for his scientific investigations. Their encounters must have been mutually profitable, and when they parted, they followed their independent scientific paths. The two careers were essentially different. GAUSS received a lot of acknowledgement from the very beginning, together with financial independence and ideal conditions for serious work. FARKAS BOLYAI, however, soon experienced financial straits, rejection and the petty-minded, philistine atmosphere of the small town of Marosvásárhely.

FARKAS BOLYAI had a very strange reason to stay in Göttingen much longer than he intended to. After completing his studies, SIMON KEMÉNY JR left for home in the autumn of 1798, but they had not enough money for FARKAS BOLYAI to return, since SIMON KEMÉNY SENIOR, referring to financial difficulties, withdrew support from BOLYAI, who was forced to stay far from his country till he was redeemed. That additional year must have seen a lot of misery, starvation and poverty. FARKAS BOLYAI had to live on charity but he always remembered that period with great affection. That was the very last time in his life he had the opportunity to imbibe knowledge and had partners to exchange thoughts with people who understood and appreciated his ideas.

Finally a supporter of his sent him enough money to clear his debts and FARKAS BOLYAI could leave for his home in Transylvania on foot in July 1799.

In his will, FARKAS BOLYAI divides his long life into three periods as follows: "... Until then it had been a morning with the prospect of beautiful days which, after some days laden with fire and ice, turned into raining from the permanently overcast sky until this recent snowfall."

Those days laden with fire and ice began after BOLYAI returned home. The fire was a case of love in young BOLYAI's life and his enthusiasm for science, the ice was a sickly wife in a troubled home along with much disappointment in, and neglect of, his scientific activity.

From his estate in Bolya, where he settled down after arriving home, he would go to Kolozsvár where he took a tutorial job once again. In Kolozsvár he met ZSUZSANNA BENKŐ (1780—1821) whom he married in 1801. They had their son JÁNOS BOLYAI of that marriage.

There was hardly anything remarkable in his career until a Calvinist consistory decided to offer FARKAS BOLYAI a job at the new mathematics–physics–chemistry department of the Marosvásárhely college. Yielding mostly to his father's arguments, FARKAS BOLYAI accepted the job and went to work and live at Marosvásárhely until his death.

The countless daily routines of teaching and the poor financial situation gave FARKAS BOLYAI little chance to do research. The low college salary forced him to try and find various other sources of income. In addition to doing gardening he acquired the right of running the college pub for some years and realized some witty technical innovations that were quite well-known in the area. His designs of tilestoves were popular all over Transylvania, and his stoves made of iron were cast at remote foundries of the country.

FARKAS BOLYAI wrote six dramas, which were published. The abundance of poetic devices in those dramas is remarkable but their dramatic structure is less successful.

FARKAS BOLYAI's splendid talent was being chipped away in the large field between literature and engineering, so research in mathematics was only occasional relaxation for him where he would always return after he "wandered through the whole world of human spirit".[1] BOLYAI's basically restless and rebellious disposition and ideology were restricted by the social order, which he was forced to serve by the nature of his office, but he refused to make compromises and was never willing to give up his conviction in return for some cheap compensation.

His teaching career did not award him the pleasure of well-done and successful work. The level of teaching — just like anywhere else in the country — was low, and very few of the students understood the substance of the ambitious professor's lectures, so he did not receive much recognition. BOLYAI's essays and what colleagues and students said about him show that he had modern didactic methods. He was aware of both the teoretical and practical significance of mathematics and tried to apply individual methods at teaching.

In his seclusion in Marosvásárhely FARKAS BOLYAI tried to develop his own system of mathematics. Almost completely isolated and without the atmosphere of science, in 1832—33 he published *Tentamen*, the result of much contemplation of a scientist who could lean on nothing else but a couple of source-books. A supplement to the first volume was an essay by JÁNOS BOLYAI entitled *Appendix*, which appeared also separately — before Tentamen — on 20th June 1831.

Unfortunately, FARKAS BOLYAI's activity in mathematics did not receive due attention at the time. He was chosen corresponding member of the Hungarian Scholarly Society on 9th March 1832, before *Tentamen* was published, but not so much for his achievements in mathematics as for his contributions to Hungarian literature. BOLYAI's contemporaries gave more attention to his articles, actually of minor importance, on natural sciences, forestry, ethnography, as well as dramatic works and translations than to his achievements in mathematics. Appreciation of his activity there was scarce and reserved, and showed incomprehension. ANTAL VÁLLAS (see below), later a professor of mathematics at the University of Pest, had a classical education and ample mathematical knowledge to comprehend FARKAS BOLYAI's theory had he studied it carefully. In 1837, long after *Tentamen* was published, VÁLLAS gave the following short criticism: "Bolyai has certain linguistic merits, but oddity prevails when it comes to his scientific views".[2] How could the public have conceived the deep and abstract thoughts of *Tentamen*, let alone those of *Appendix*, if one of the most competent experts could describe them as odd?

[1] BEDŐHÁZI, JÁNOS: *The Two Bolyais' lives and work*. Marosvásárhely, 1897, 47. (In Hung.)
[2] VÁLLAS, ANTAL: Recent Hungarian Literature in math., etc. *Tudománytár*, 1837, p. 152. (In Hung.)

Letters written by FARKAS BOLYAI to his friends testify how deeply his feelings were hurt with that criticism and the numerous ill-minded remarks of the "cackling world" that reached him.

The ill-treatment of his achievements in mathematics had consequences in BOLYAI's life. He never tried again to write anything as significant as *Tentamen* was, but rather ventured into fields that promised success. His short-lasting second marriage (1824—1833) with TERÉZ NAGY (1797—1833) was poor compensation for an unfortunate first marriage. There were only two things that could really brighten up his days. One of them was the letters — of thinning occurrence, though — he still received from GAUSS, his long-time friend. The other was his son JÁNOS's interest and achievements in mathematics. Father-and-son relationship was somewhat disturbed when in 1846 JÁNOS moved to Marosvásárhely, settled down there, and lived a life free of conventions and unusual enough to be the subject of gossip.

In 1848 the breakout of the Hungarian War of Independence stirred the aging FARKAS BOLYAI's sympathy, but with the caution of people approaching grave, he warned his nation against bloodshed. In 1851 he decided to retire. With that, all his ties to public life broke. The old teacher who had once set out with great expectations was preparing to meet with death, which eventually ensued on 20th November 1856, after successive attacks of cerebral haemorrhage.

There are numerous sources where authentic accounts of FARKAS BOLYAI's life story can be found so details are disregarded here. A great proportion of the literature about him, especially the works with novel-like tendencies, however, have one common feature that is unacceptable; some authors fabricate sensation stories or emphasizes unimportant episodes to make the father-and-son relationship more interesting. Those authors seem to be informed of ceaseless antagonism in the BOLYAI family, including fights and duels, and that father and son were jealous of each other's scientific achievements. These widespread rumours have practically no grain of truth. No case can be proved when their disagreement went beyond one usual between parent and child, and no case suggests that either of the two mathematicians should morally be condemned.

János Bolyai
(Kolozsvár, 1802—1860, Marosvásárhely)

One of the most original minds of mathematics was born in his mother's parents' house in Kolozsvár on 15th December 1802. A memorial tablet was erected on the house at the centenary of his birth. The extremes and vicissitudes, as LAJOS SCHLESINGER put it, made JÁNOS BOLYAI's life a really great tragedy.

JÁNOS BOLYAI's father, FARKAS BOLYAI, took special care of his son's physical and intellectual education. First it was the physical education that he paid more attention to, so that the intellect should have a sound body at disposal. The nimble and healthy child showed extreme intellectual capacities from the very beginning. He was endowed with an extremely observant mind and a deep sense of justice. JÁNOS BOLYAI was nine years old when his father decided to send him to school, but by then he had acquired serious knowledge of the various subjects and excelled at the exact sciences. For instance, when he was four, he could distinguish certain geometrical figures, knew about the sine function and could identify the best-known constellations. By the time he was five JÁNOS BOLYAI had learnt, practically by himself, to read. He was well above the average at learning languages and music. At the age of seven he took up playing the violin and made such good progress that he was soon playing difficult concert pieces.

FARKAS BOLYAI had his more talented students teach his son several subjects, but reserved the teaching of mathematics to himself. According to a letter written by FARKAS BOLYAI to GAUSS, he always wanted his son to be a mathematician. Unfortunately not only the talent of JÁNOS BOLYAI showed very early but several of his detrimental features as well, which he had probably inherited from his mother. He was not practical at all and tended to be tyrannizing. The child's warm affection was often replaced by sudden whims. One thing in which JÁNOS BOLYAI remained consequent throughout his life was his sense of justice, whether scientific achievements or any other matter be concerned, even if the judgement was unfavourable for himself.

At the age of twelve JÁNOS BOLYAI became a regular student of the Calvinist College of Marosvásárhely. Skipping the first three years, he started in the fourth. (In Hungary today it would be about the eighth year at school.) It often occurred, however, that he attended lectures intended for senior students. Although he did not learn very hard, he always had top marks. He qualified excellent when he sat for the exam called "rigorosum" on 30th June 1817, which entitled him to study the Latin classics.

At this point, father and son arrived at crossroads as to where and how JÁNOS should continue his studies. One solution could be that he attend the first two years of the Latin classics course, but that seemed to be a waste of time since JÁNOS had a better knowledge of most subjects than the students there. Another and more advantageous arrangement was to ask GAUSS, FARKAS BOLYAI's long-time friend, to give JÁNOS accommodation and help him explore mathematics further. FARKAS BOLYAI had written a letter to GAUSS in 1816 asking him that favour, but unfortunately GAUSS dismissed the request. So the BOLYAIS had to find some other solution. FARKAS BOLYAI was decidedly against sending his son to the university in Pest or Vienna, both of them of low standard. So after a lot of consideration they decided on a career in military engineering and chose the Academy of Engineering in Vienna. FARKAS BOLYAI took special care of preparing his son for the entrance exam, as the results accounted for which year of the seven-year studies one was admitted to. Lack of money forced JÁNOS to stay and study at the department of philosophy in Marosvásárhely for one year after rigorosum before he could leave for Vienna in August 1818, after receiving financial support form several people.

Certainly it was the narrow circumstances that made the BOLYAIS decide in favour of the military academy, but it might have proved more yielding for JÁNOS to continue his mathematical studies under his father's guidance. The military academy in Vienna was one of great tradition, and mathematics carried great importance there. The main subject in the third, fourth and fifth years was pure mathematics; applied mathematics was emphatically dealt with in the sixth year. Military subjects were taught in detail in the seventh year only. In summer the cadets had military training. The most outstanding students were awarded one more year of studies after the seventh year to receive special training in military architecture and fortification.

The military career, however, was alien to JÁNOS BOLYAI's character, to his rebellious and contemplating disposition, and he was unable to get used to military discipline and servile drill.

JÁNOS BOLYAI's performance at the entrance exam allowed him to start his studies in the fourth year the highest possible the regulations permitted. In the first year he was not an outstanding student, but in the second year he was the second best in his class. He had top marks of everything except drawing and calligraphy. At a visit that year Archduke JOHANN VON HABSBURG, chief-commander of the military academy and superintende of the Engineers had direct experience of JÁNOS BOLYAI's mathematical talent. The Archduke took pains to send a message to FARKAS BOLYAI expressing his recognition and his conviction that JÁNOS might expect fast advance at the military career if he continued to work diligently. Later the Archduke actually helped JÁNOS.

At the college, JÁNOS BOLYAI was making good progress not only in the special subjects but also in music and sport. He was well-known for his skill at fencing and playing the violin.

When he was in the seventh year at college, his mother died on 18th September 1821. The wife of FARKAS BOLYAI, a wretched woman of hysterical disposition, suffered from her increasingly serious illness as much as from being separated from her son, and turned her husband's and everybody's life around her miserable. Her death deeply shocked FARKAS BOLYAI. Although JÁNOS was bereaved of her mother, he was able to carry on with his studies as usual. JÁNOS BOLYAI completed his academic studies on 6th September 1822, but since his results were the second at the top that year, he was not sent to detachment service but, together with six other cadets, was granted additional year of studies to get trained for services requiring high technological knowledge.

While JÁNOS BOLYAI was staying in Vienna, he showed special interest in certain fields of mathematics, EUCLID's fifth postulate in particular. His interest had been roused by his father, who had unselfishly passed his splendid knowledge to his son and laid the ground for the wonderful achievements described in the *Appendix*. This was one of the indisputable merits of FARKAS BOLYAI. During his years at college, JÁNOS BOLYAI further deepened his knowledge of the subject. His ambition was increased by the inspiring interest of JOHANN WOLTER VON ECKWEHR (1791—1851), who was his mathematics professor, and the enthusiasm of KÁROLY SZÁSZ (see later), a Hungarian tutor in Vienna.

The aim JÁNOS BOLYAI set himself was to prove the fifth postulate in an indirect way. His discussions with KÁROLY SZÁSZ resulted in the recognition that to assume that the circle of infinite radius is a straight line is equivalent to EUCLID's axiom of parallelism. When SZÁSZ and BOLYAI parted, they promised each other that if either of them achieved serious results at proving EUCLID's axiom of parallelism, they would declare it a joint success. Later JÁNOS BOLYAI stressed in his writings that the agreement only involved proving the fifth postulate, but made no provision for the case of creating a new system of geometry.

In September 1823 JÁNOS BOLYAI was commissioned a sub-lieutenant and sent to service at the Board of Fortification in Temesvár. Soon after that, on 3rd of November he wrote his famous letter to his father saying that he had "created a new, another world out of nothing".

The next station in JÁNOS BOLYAI's troubled military career was Arad, where he was sent in 1826 and was promoted to lieutenant. In 1830 he was ordered to Lemberg. Before travelling there, JÁNOS BOLYAI visited his father who asked him to write a Latin summary of his research in geometry so that it could be published in *Tentamen* within a short time.

. The military career interfered with JÁNOS BOLYAI's life once again when he was sent to Olmütz in 1832. He was promoted to captain, the highest rank he achieved. Unfortunately, the unfavourable reception of *Appendix*, published in the meantime and sent to various places, and particularly GAUSS' laconic and ambiguous appreciation made JÁNOS BOLYAI irritable and turned him into a misanthrope. It became harder and harder for him to get on with others. He kept himself aloof from the other officers and because of his failing health he was unable to properly carry out his duty. In spite of Archduke JOHANN VON HABSBURG's patronage, JÁNOS BOLYAI's situation was untenable in the army. He handed in his pension claim and retired on 16th June 1833. Documents testify that his commanding officers were sad at JÁNOS BOLYAI's leaving office and hoped that the highly qualified officer would return to service, which BOLYAI never did.

A new era began in JÁNOS BOLYAI's life, but by no means happier than before. For a short time he stayed in his father's house at Marosvásárhely, then he moved to their estate in Domáld where he lived his colourless life till 1846, hardly ever leaving the place. From 1834 on, he lived together with a woman called ROZÁLIA KIBÉDI ORBÁN. As they could not deposit the caution money required, they were unable to get married until 18th May 1849 when — after the Debrecen dethronement of the Habsburg–Lotharingian dynasty — the decrees of Vienna provisionally become ineffective in the region of Marosvásárhely. One of FARKAS BOLYAI's letters, however, testifies that the emperor never endorsed the marriage. The pension and the poorly run estate did not bring income enough for the family which grew with their two children born.

JÁNOS BOLYAI dealt with problems of mathematics also in that period of his life. He began to work on the creation of a general geometrical system based on axioms. The work, however, was never completed although the preface was written as early as 1834. He also dealt with problems (algebraical solution of algebraic equations of any degree, the rendering of any integral in a closed form, etc.) that had been solved before; he could have saved the trouble if he had had easy access to information. In 1837 BOLYAI heard of a call by the JABLONOWSKY Society of Leipzig to work out constructions involving complex numbers. Both JÁNOS and his father sent in their papers, which, unfortunately, were hurriedly composed although carefully thought out. Neither JÁNOS nor FARKAS received any recognition for the competition papers.

JÁNOS BOLYAI's ambition was further deteriorated by the failure, his irritableness increased, and he became quarrelsome with his family. He did not care to look after his estate at Domáld and the letters to his father convey more and more nervousness. Financial disagreement, scientific disputes and petty intrigues disturbed the relationship between father and son.

In 1846 JÁNOS wound up the estate at Domáld and moved to Marosvásárhely. For a time the nearness worsened the atmosphere between the two BOLYAIs. They hardly

37. The tomb of the two Bolyais in Marosvásárhely

met because the rumours surrounding JÁNOS strongly disturbed FARKAS BOLYAI who was held in great respect.

1848 brought two events in JÁNOS BOLYAI's humdrum life. One was that FARKAS BOLYAI managed to acquire a copy of LOBACHEVSKY's fundamental treatise entitled *Geometrische Untersuchungen zur Theorie der Parallellinien*, so that also JÁNOS could read it. JÁNOS studied the work throughly, pointing out its shortcomings as well as its merits. He developed his remarks into an essay which was published in German and Hungarian by PAUL STÄCKEL and JÓZSEF KÜRSCHÁK (1900 and 1902).

The other event that impressed the ailing JÁNOS BOLYAI was the Hungarian War of Independence. His notes show that although he had anxieties about the success of the cause of his nation, he deeply sympathized with it. When the imperial army occupied Szászrégen, JÁNOS BOLYAI allegedly participated in a select meeting of Hungarian officers that discussed the chances of expelling them with the help of the Székely volunteers. Some biographers of JÁNOS BOLYAI seem to know that BOLYAI also put forward his expert strategic plans, but they were no accepted. For lack of reliable data SAMU BENKŐ, a BOLYAI specialist, expressed strong doubts about that story.

At the end of 1852, JÁNOS BOLYAI broke off with ROZÁLIA ORBÁN, who used to be the main source of disagreement between father and son. FARKAS and JÁNOS BOLYAI gradually came to respect and like each other, so that when the bells told for the old professor of the college, "who had written countless x-es during the eight X-es of his own decades", JÁNOS became an orphan both as a son and a scientist by losing his most understanding fellow scientist.

In the last period of his life JÁNOS BOLYAI hardly dealt with mathematics. Feeling so unfortunate all through his life, he finally "set out to establish a practical system of all former knowledge and experience for the benefit of mankind" (BENKŐ). The draft entitled "Salvation theory" follows a clear logic throughout the manuscript, and the bold ideas — especially in the field of linguistics and sociology — promise the establishment of a really original system. Unfortunately BOLYAI had no power of the mind or the body left for a detailed and systematic elaboration. The state of decline was soon followed by the end. The possibly greatest genius of Hungarian science died on 27th January 1860, without anybody sheding a tear of feeling pity for him. Most probably, the immediate cause of death was pneumonia.

No authentic picture of him has been found. The one in various publications, including stamps, is not trustworthy.

The two BOLYAI's rich legacy of manuscripts is kept at the Teleki–Bolyai Library of Marosvásárhely and at the archives of the Library of the Hungarian Academy of Sciences.

Sámuel Brassai
(Torockó, 1800—1897, Kolozsvár)

One of the most famous polyhistors of 19th century Hungary. Having graduated from the College of Kolozsvár, BRASSAI became a teacher there, then in 1837 he was chosen member of the Hungarian Scholarly Society. For his participation in the Hungarian War of Independence, BRASSAI was disgraced, and between 1848 and 1859 he was

Mátyás Butschány
(Zólyom, 1731—1796, Hamburg)

Having studied in Göttingen, he became Privatdozent there, then he lived as a person of independent means. His work entitled *Anfangsgründe der Algebra, nebst derselben Anwendung auf die Rechenkunst* was published twice (Göttingen, 1761; Vienna, 1767).

Elek Cörver
(Torna, 1714?—1747, Pest)

A priest and teacher of the Piarist order. He taught in Nyitra and Pest. He is the author of the book entitled *Compendium elementorum geometriae practicae fundatum* (Buda, 1746).

Dániel Csányi
(Nagybánya, 1820—1867, Debrecen)

Having studied in Máramarossziget, Debrecen and Kassa, CSÁNYI graduated from the Technical University of Vienna. After the Hungarian War of Independence was defeated, he had to languish in prison for a long time for having the fortification of Komárom built. While in captivity, he wrote introductions to arithmetics and calculus, for which in 1863 — after his release — he became a member of the Academy. Csányi published the legacy of FERENC KEREKES (see later).

László Csernák
(Pápa, 1740—1816, Deventer)

His father GYÖRGY CSERNÁK, a judge of the County Court of Veszprém, died very young, and his mother sent the 13-year-old boy to study in Debrecen. Having completed his studies CSERNÁK went through all the stations on the way to professorship between 1759 and 1766. He was a library attendant, deputy director of the library, an instructor of rhetoric, the president of the faculty of theology and metaphysics, and finally the senior of the College of Debrecen. In a letter CSERNÁK mentioned ISTVÁN HATVANI (see later) — one of his teachers in Debrecen — with affection.

In 1767 CSERNÁK was sent abroad on a study trip. The permission to stay abroad and his scholarship were extended by JOSEPH II several times. CSERNÁK attended the universities of Zurich, Basel, Torino, Utrecht and Göttingen, and stayed for a long time in Groningen where he published several articles on medicine and physics.

CSERNÁK had already spent eight years abroad when in 1775 a secondary school of the Dutch town Deventer, on the right bank of the river Ijssel, invited him to teach philosophy and mathematics. That is where CSERNÁK worked until his retirement in 1816. Between 1779 and 1782 he was also rector of the school.

In 1781 the board of the College of Sárospatak invited CSERNÁK to the chair of history and foreign languages, but he refused it. As one of his letters reveals, he turned down the offer because of his wife. Nor did he accept the invitation to the chair of philosophy at the University of Groningen, probably also for his wife's sake. Deventer was the place where CSERNÁK got married and his home made that little town very dear to him. His wife came from a well-to-do family there. CSERNÁK never spoke perfect Duth in his life. He died shortly after retirement in 1816. His single piece of work in mathematics, exemplary in size, appearance and precision, is the table of prime numbers published in Deventer.

János Decsi Czimor
(Decs, 1560—1601, Marosvásárhely)

DECSI CZIMOR studied in Debrecen, then in Wittenberg between 1587 and 1589, but attended school also in Italy. He was the director of the College of Marosvásárhely from 1593 until his death. His scientific activity was manifold; he dealt with history, Roman law, philosophy, and translated works by Latin authors into Hungarian.

János Dubovszky
(Höszlin, 1654—1710, Kassa)

The only fact known about him is that he was a Jesuit priest and teacher, and taught mathematics at the University of Nagyszombat between 1694 and 1697. Together with FERENC SZÉKELY (see later) DECSI CZIMOR was probably a co-author of the first trigonometric table (*Canon sinuum* ... etc., Nagyszombat, 1694).

András Dugonics
(Szeged, 1740—1818, Szeged)

A well-known figure also of Hungarian literature at the end of the 18th century. He was of Dalmatian origin and came from a family of merchants who had immigrated to Hungary not long before DUGONICS was born. He joined the Piarist order when he was 18, and after taking the monastic vow he worked in Vác, Medgyes and Nyitra as a secondary school teacher. After the Jesuit order was suppressed, applications were invited to several chairs at the university of Nagyszombat, and DUGONICS decided to compete for the chair of mathematics. The support of PÁL MAKÓ (see later) helped him to gain the chair of elementary mathematics,[3] and DUGONICS started his career at the University of Nagyszombat in 1774. After the university had moved to Pest, DUGONICS lived in Buda, then in Pest, until he retired in 1808. The last ten years of his life were spent in Szeged, his hometown, where he died in 1818.

DUGONICS's part in the history of Hungarian mathematics is an issue of strong disputes. The literature on the problem exhibits eclectic opinions. A separate investigation of his activities as a university professor, an ardent supporter of the Hungarian language and a mathematician, might help us see clearer. From the mathematical point of view, DUGONICS's activity as a professor is unimportant. Neither a rising standard of teaching mathematics, nor outstanding students mark his excellence, although — accordingly to some records — he dealt with the more gifted students individually. A few lines by DÉNES KATONA (see later), a pupil of his describe DUGONICS as a professor:

[3] The single problem in mathematics appearing among the philosophical questions of the concourse was very simple: to evaluate the height of an unaccessible mountain from suitable data.

"At lectures of mathematics DUGONICS was in the habit of sending one of the students to the blackboard before he explained the proposition, and asked the student to explain it. Then DUGONICS gave the student a mark depending on the performance. As soon as DUGONICS himself set out to explain something, everything was told clear enough to be printed. It was still extremely hard for us to answer in Latin as our text had been written in Hungarian."[4]

Other documents, however, say — and judging by DUGONICS's book, it seems very likely — that DUGONICS's lectures were of a mediocre level, and at examinations he was characterized by extreme severity and hair-splitting.

The students filled with the fever of ideas of the Enlightenment adored their professor. His ardent patriotism, which DUGONICS tried to plant in his students, the embracement of the cause of the Hungarian language gave him a leading part, and the noble minds of the students clustered around him. DUGONICS was disliked by his colleagues for his outspoken style, stubborn character, occasionally vulgar language and oppositionary behaviour. He was simply unapt to arouse affection or cherish friendship. He bore particular disaffection to the *ex-Jesuits* and strongly resented the servile imitation of the Vienna schooling system in Hungary. That provided ground to DUGONICS's enemies to accused him of connexions with the Hungarian Jacobins. All he had to do with the Jacobins, however, was his friendship with JÓZSEF HAJNÓCZY, whereas he condemned their movement, and several of his articles attacked MARTINOVICS.

DUGONICS's activity in literature and mathematics formed a part of the increasing resistance against the Germanization policy of JOSEPH II. Only one piece of his rich lifework was written in Latin (*Argonauticon*, published in Pozsony and Kassa, 1778). Widely popular was his novel entitled *Etelka* (first published in 1788), which criticized the reign of JOSEPH II, and in the frame of an old story refuted the view proclaimed by the Vienna Court that the Hungarian language was underdeveloped to convey scientific and literary material.

Gyula Farkas
(Sárosd, 1847—1930, Pestszentlőrinc)

Born on 28th March 1847, at a manor called Sárosd in Fejér County where his father was the steward of the estate, FARKAS attended secondary school at the Benedictines in Győr. Then he went to Pest to study las and music. He realized before long that he was not interested in law and was not gifted enough to become a musician. After working for some time as a private teacher, FARKAS — on ÁNYOS JEDLIK's advice — registered at the arts department where he graduated as a teacher of physics and chemistry. The Modern School of Székesfehérvár had chosen FARKAS to teach there. In 1874 he left that school and went to work as a mentor at Polgárdi. The Count of

[4] PERÉNYI, JÓZSEF: Dénes Katona's life and works. *Roman Catholic Secondary School's Bulletin*, 1895/96, 11, Sárospatak. (In Hung.)

Polgárdi GÉZA BATTHYÁNY had a well-equipped laboratory of physics for his children, which provided FARKAS a splendid opportunity for research. Study trips abroad also enriched his knowledge.

In 1880 FARKAS moved to Pest, where he was appointed Privatdozent of function theory, chiefly for his articles published in *Comptes Rendus*.

FARKAS's career took an undisturbed rise from then on. In January 1887 he was appointed associate professor at the University of Kolozsvár, and in a year he became ordinary professor. FARKAS worked in Kolozsvár until 1915 as a highly respected professor of the university. His assignments of Dean and Rector of the University greatly contributed to the development of the University of Kolozsvár.

He was given several honorary titles as Honorary Doctor of the University of Padua, Corresponding (in 1898) (later: Ordinary in 1914) Member of the Hungarian Academy of Sciences, Honorary Member of the Eötvös Loránd Mathematical and Physical Society in 1924.

FARKAS dealt with problems of mathematics particularly at the beginning of his career. It is his merit that vector algebra and vector analysis became known in Hungary. Later FARKAS was engaged in problems of theoretical physics, in which he applied also original methods now belonging to linear programming. His essays are laconic, which makes them very hard to understand.

FARKAS's worsening eye-disease forced him — similarly to the case of GYULA VÁLYI (see later) — to relinquish professorship in 1915.

FARKAS married twice, but he survived both of his wives, so he had to live a solitary life after his retirement in Budapest. A few months before his death FARKAS moved in the flat of a relative and died there on 27th December 1930.

Vilmos Fest
(Yaroslav, 1815—1879, Sopron)

Worked as an engineer at the regulation of the River Danube from 1834 on, later at the construction work of the Chain Bridge in Budapest. FEST became member of the Hungarian Scholarly Society in 1845. After 1850 he worked in Kassa and for the Ministry of Transportation in Pest. FEST wrote several essays in engineering, and his single essay in mathematics won him the second prize of the Academy.

Zoárd Geőcze
(Budapest, 1873—1916, Budapest)

He was born on 23rd August 1873. His father was a teacher at the Military Academy, one of the best experts in military science in the period of the Compromise, who greatly contributed to the development of Hungarian military language.

ZOÁRD GEŐCZE's marks at the secondary school — Modern Schools in the 8th and 4th District of Budapest — were not better than the average. He went to study at the Arts Faculty of Budapest University for a Master's degree in teaching. The seminars held by GYULA KŐNIG (see later) greatly inspired GEŐCZE to do research. GEŐCZE submitted his results in a competition paper but the work did not meet KŐNIG's satisfaction. The hot-tempered GEŐCZE answered his professor's criticism in a vulgar language, which caused KŐNIG to stop support of GEŐCZE. Since the majority of mathematicians then depended on GYULA KŐNIG for starting their careers — KŐNIG had the decisive word in professional and educational matters — the incident proved to be extremely unfortunate.

After graduation GEŐCZE took a job at the secondary school of Podolin. In Podolin he soon married IRMA LIPPÓCZY, an ideal partner and devoted helper of her husband through the difficulties of life.

From Podolin, GEŐCZE went to teach at Ungvár in 1899.

Their marriage was blessed with seven sons and a daughter, and GEŐCZE adored his children, although the happiness of the family was overcast with grave financial difficulties. Nonetheless there was only one thing that really embittered GEŐCZE, namely that his research in the fast developing field of real functions ran repeatedly stranded for lack of literature. GEŐCZE's first paper was published in the 1905 yearbook of the junior modern school of Ungvár, and another followed it one year later concerning surface measurement, GEŐCZE's exclusive subject of research later on. LAJOS SCHLESINGER, professor of the University of Kolozsvár, noticed the brand-new ideas in GEŐCZE's essay and asked him to write a summary of his results. The note was published in *Comptes Rendus* in 1907, and on its basis the Ministry of Culture granted GEŐCZE a one-year scholarship in Paris, the then capital of research in real functions. In Paris GEŐCZE had opportunity to study the latest achievements of French mathematicians such as BAIRE, BOREL, LEBESGUE and POINCARÉ. GEŐCZE paid special attention to LEBESGUE's activity, and he sent LEBESGUE his first paper written in Paris asking him to report on it. Unfortunately, at the beginning he was as unlucky with LEBESGUE as with KŐNIG, because LEBESGUE sent back the paper unread. LEBESGUE's answer attached to the essay says he was unable to get down to GEŐCZE's subject for its crabbed style and the large number of terms and notations introduced. But documents also prove that when LEBESGUE later on became familiar with GEŐCZE's achievements, he appreciated them very much.

The ample but unfinished scholarship studies made it justified for GEŐCZE to return to Paris in 1910, after one year's stay at home. The doctor's degree of Sorbonne praises GEŐCZE's accomplishment there. GEŐCZE's rising fame and widening activities caused him to leave Ungvár and move to Budapest, the centre of Hungarian scientific life. He was teaching at the modern school of the 5th district of Budapest till the outbreak of the World War I. In 1913 GEŐCZE became Privatdozent in "the theory of manifolds and functions of real variables" at the Budapest University of Sciences. LIPÓT FEJÉR says that the references had great difficulties in understanding GEŐCZE's papers.

In 1913 the Hungarian Academy of Sciences conducted a competition in compiling a comprehensive book on mathematics. On the basis of his project entitled *Geometrical applications of set theory*, the Academy commissioned GEŐCZE to do the job. Unfortunately only a fragment of that work has been written. Initially his teaching obligations hindered GEŐCZE later his front service made it possible to complete the pioneering work which would have required thorough preliminary studies.

Soon after the outbreak of the World War GEŐCZE was called up. First he served at the Serbian front and participated in the retreat along the River Drina. The military operation demanded incredible sacrifice, and GEŐCZE's regiment was dispersed. The surviving troops and officers were gathered in a battalion and taken to the northern front. GEŐCZE's commander wanted to keep him away from the tribulations of the war, and sent him to take care of, and supervise, an electric network branching off a power station. It was also a tiring job, but after work GEŐCZE had some time left to deal with mathematics. He sent his essays home by military post, and the proof-reading was done by his colleagues in Budapest.

But the exerting work was too much even for GEŐCZE's strong physique. Seriously ill, he was taken from Chernovits to Vienna, but he was homesick and was eventually taken home. In Budapest he had to be taken to hospital with a serious cold and exhausted nerves in the spring of 1916. Oedema coupled with heart disease killed GEŐCZE at the peak of his working powers on 26th November 1916.

On 7th December 1916, at the meeting of the Mathematical and Physical Society, LORÁND EÖTVÖS commemorated ZOÁRD GEŐCZE — together with war dead Győző ZEMPLÉN — with words of deep reverence:

"I also have to announce an other tragic loss to you. Of a disease developed at the front the great mathematician, our dear colleage ZOÁRD GEŐCZE, who used to inform us on his sensational results in the theory of surfaces at this place, died on the 26th last month. No difficulties in his home, no loudness of his children, no horror of the trenches or thunder of the guns were able to disturb ZOÁRD GEŐCZE at concentrating on the solution of his favourite problem and making efforts to widen and deepen our knowledge of the subject. Everlasting respect and gratitude go to those who, like ZOÁRD GEŐCZE, set grand aims to strive for and remain faithful to those aims to the end."

Sándor Győry
(Tarján, 1795—1870, Pest)

Studied in Pozsony, Nagykőrös, Debrecen, and graduated as an engineer in Pest in 1825. Member of the Academy from 1832 to his death. An enthusiastic participant in the activities of the mathematical section who — besides giving communications in mathematics and questions of education — dealt with music, philosophy and practical science.

Károly Hadaly
(Nagysziget, 1743—1834, Pest)

A doctor of philosophy and law. A teacher of mathematics at the University of Nagyszombat, then at the Academy of Győr, Pécs and Pozsony. The professor of mathematics at the University of Pest after 1809. An outstanding expert on water-supply engineering and member of several scientific associations for his knowledge in that field. Among his written works there are some, mediocre, textbooks of mathematics in Latin and German.

István Hatvani
(Rimaszombat, 1718—1786, Debrecen)

All over Hungary, the best-known professor of the College of Debrecen. Posterity surrounded the "Hungarian Faust" by a mysterious reputation. His interesting and colourful character stirred the imagination of countless writers and poets who invented

stories and misintegrated HATVANI's scientific experiments in order to attribute him magic power, and make his personality grand and attractive beyond reality. The lots of written works on HATVANI fail to properly deal with HATVANI's scientific activity.

HATVANI was born in Rimaszombat in 1718. The parents wanted the weak and sickly child to become a priest. He went to comprehensive schools in Rimaszombat, Losonc, Kecskemét and Komárom. The schools HATVANI attended aroused his interest in various fields, but his teachers were impressed mostly by his deep inclination to philosophy. He started his studies at Debrecen College, but an epidemic of plague, which interfered also in MARÓTHI's life (see later), forced him to leave the town. He went to Losonc, where he worked as a tutor till 1741. Between 1741 and 1745, again, his name appeared in the annals of the college.

After graduating from Debrecen College, HATVANI continued his studies in Basel, Switzerland, with the support of the town of Debrecen and the counties of Nógrád and Szabolcs. Besides the faculty of theology, he also attended the faculty of medicine, and he soon became a doctor of both faculties in 1748. During his stay in Basel, HATVANI had the opportunity to attend lectures of JOHANN BERNOULLI (1667—1748) and his son DANIEL BERNOULLI (1700—1782) on mathematics, physics and medicine.

On the influence of the outstanding masters, HATVANI became an expert scientist, for whom the universities of Heidelberg, Marburg and Leyden were competing. HATVANI, however, "preferred poverty to splendour", and — refusing the favourable offers — went to work at the faculty of mathematics, physics and philosophy of Debrecen College in 1748.

During his long career as a teacher, HATVANI lectured on mathematics, geometry, philosophy, astronomy. Particularly famous were his lectures on physics with demonstrative experiments. It was HATVANI who in 1750, introduced ordinary lectures on chemistry into Hungary.

HATVANI actively participated in the administration of the college. One of the tasks involved in his job was the medical care at the college, so HATVANI can be regarded as the first medical superintendent in Hungary.

During the reign of JOSEPH II, HATVANI was the greatest authority of the struggle for the religious and educational autonomy of the Protestant Church. As an acknowledgement of his successful activity, HATVANI was chosen judge of the County Court in 1783. He retired on 11th February 1786, and died soon after his retirement, on 16th November 1786.

Sámuel Hegedüs
(Torda, 1781—1844, Szászváros)

Studied at the College of Nagyenyed and became a teacher of foreign languages at the college in 1805. In 1807 he was sent to Göttingen, there he got acquainted with GAUSS, with whom he later had a correspondence.

Having returned to Hungary, HEGEDÜS worked in Nagyenyed, Kolozsvár, Torda and Szászváros as a teacher and priest. HEGEDÜS wrote several books on mathematics and compiled a table of logarithms, but they were not printed for lack of publishers. HEGEDÜS was chosen member of the Hungarian Scholarly Society in 1832.

Miksa Hell
(Selmecbánya, 1720—1792, Vienna)

A Jesuitical teacher who worked — among other places — at Lőcse and Kolozsvár. In 1755 he was appointed director of the observatory of the Imperial Court in Vienna. In that position HELL was commissioned by CHRISTIAN VII, the King of Denmark, to observe from the Isle of Vardő, Norway, the *Venus transition* of 1769. On the successful expedition HELL was accompanied by JÁNOS SAJNOVICS, a well-known linguist. HELL's book on arithmetics and algebra was published several times, but he gained fame for his studies in astronomy.

Gáspár Heltai, Jr. (?)

Hardly anything is known about his life. He was mayor of Kolozsvár, and between 1582 and 1601 he ran the printing office started by his father. HELTAI rewrote the *Arithmetic of Debrecen* (1577) and published it in 1591. He changed the original so much that the *Arithmetic of Kolozsvár* can be regarded as a new piece of work.

Ignác Hertl
(Szakolca, 1702—1775, Kőszeg)

A Jesuitical priest and teacher who spent his novitiate in Vienna, and graduated after a three-year course from the faculty of philosophy at Nagyszombat. The stations of his teaching career were Sopron, Nagyszombat, Buda, Kassa, Kőszeg. HERTL's *Arithmetic* (published in Kassa, 1758) was one of those better-known and much used.

Ernő Hollán
(Szombathely, 1824—1900, Budapest)

Graduated from the Academy of Military Engineering in Vienna, the one JÁNOS BOLYAI had attended. During the War of Independence HOLLÁN excelled in military engineering so much that KOSSUTH appointed him chief engineer and commissioned him to build the fortification of Pétervárad. After the capitulation HOLLÁN returned to his native town, where he was arrested and held in captivity. While in custody, HOLLÁN wrote a book entitled *Analytical geometry*, for which he was chosen member of the Hungarian Academy of Sciences in 1858. Following stormy years, HOLLÁN was appointed secretary of state for transportation affairs in 1867, and wrote a number of books on the subject. Later HOLLÁN returned to the military career and was a lieutenant general when he retired. HOLLÁN was one of the most enthusiastic fighters of the cause of Hungarian independence and progressive ideas.

János Horváth
(Kőszeg, 1732—1799, Buda)

Joined the order of the Jesuits in 1751, and worked as a professor in Nagyszombat and Buda also after the order was dissolved. HORVÁTH's books on physics were extremely popular. He was chosen the Dean and the Rector of the University several times. His mathematics texts written in Latin were published also abroad.

Jenő Hunyady
(Pest, 1838—1889, Budapest)

Born on 28th April 1838. HUNYADY's father was a well-to-do physician in Pest, who did not spare money and effort from a good education of his son. The elevated atmosphere in the family greatly contributed to the spiritual development of the child.

HUNYADY went to secondary school in Pest, and excelled in mathematics from the very beginning. His knowledge of music was also remarkable. The HUNYADY's house was one the centres of musical life in the Hungarian capital. After leaving secondary school HUNYADY studied at the Technical College of Pest, where one of his teachers was ISTVÁN KRUSPÉR (see later), who had especially great influence on him. Although HUNYADY's interest in mathematics further deepened at the Technical College, he could not get into contact with the latest results and trace new problems for research in Hungary of those years of oppression. So in 1857 HUNYADY set out on a study-trip and spent eight years abroad. The first station was Vienna, then he visited the greatest mathematical centres of contemporary Europe. At the beginning lectures of KUMMER and KRONECKER made particular impression on HUNYADY, and KRONECKER greatly valued HUNYADY's first results in research. HUNYADY, however, chose to mainly deal with problems set in lectures of ALFRED CLEBSCH and OTTO HESSE and works by STEINER, CHASLES and MÖBIUS.

Some of HUNYADY's shorter notes were published while he was a university student. His doctoral thesis was published in Göttingen in 1864. GYULA KŐNIG (see later) said about the thesis:

"... a significant treatise of a real expert. Based on source studies, it gives an original systematization of the general theory of algebraic curves — a theory now treated in textbooks —, enriches that theory with several minor details and finally gives a new classification of curves of the third and fourth degree by their asymptotes." (KŐNIG, GYULA: In memory of Jenő Hunyady. *Akadémiai Értesítő*, 2, 1891, pp. 3—4.)

With ample knowledge and considerable results HUNYADY returned to Hungary in 1865 and qualified for the title of Privatdozent at the Technical University. In 1867 he became corresponding member of the Hungarian Academy of Sciences, and when SÁNDOR KOMNENOVICS (see later) died in 1869, HUNYADY was promoted Professor of Elementary Mathematics and Political Arithmetic at the Technical University, where he worked practically until his last day.

At the beginning of HUNYADY's teaching career he was extremely busy having to give lectures on disciplines of mathematics that fell far out of his interest. After a hard struggle in 1873 the Technical University achieved to have another department of mathematics. Afterwards HUNYADY had to give lectures on geometry only.

HUNYADY was a very successful teacher. Hundreds of future engineers and teachers attended his carefully devised lectures, which were free from anything that did not belong to the matter. HUNYADY never sought popularity. His students knew that their professor was an expert of algebraic geometry, a fast developing branch of mathematics and that he interwove his lectures with the results of his own research. The seminars held by HUNYADY and KŐNIG by turns brought many would-be teachers to like mathematics and led them to acquire the methods of individual research.

HUNYADY was generally considered a person of kindly disposition. He liked to live in an affectionate and friendly atmosphere and enjoyed witty conversations while having a few glasses of wine. He was quite the opposite of the dry, retiring type of scholars who held themselves aloof from the students and colleagues. His serene humour was not broken when in 1873 he was reduced to poverty after a financial crash.

HUNYADY took an active part in starting and editing the short-lived *Műegyetemi Lapok*. In 1883 he was chosen — together with LORÁND EÖTVÖS — member of the Hungarian Academy of Sciences. At the same time he was awarded the first prize of the competition of mathematical papers published between 1876 and 1882; his winning essay was entitled *Various forms of the equation for six points to lie on a conic*.

At that time, however, HUNYADY was seriously ill. His inherited heart disease became worse and worse because of the exhausting work and haphazard life. But his love for work carried on, and he worked hard to realize his old plan to build up analytical geometry by his own concept and have it published. HUNYADY tried hard but the work was never completed. The unfinished manuscript has never been published.

HUNYADY was one of the initiators of a union for Hungarian mathematicians. At the meetings of the *Mathematical Society*, established in 1885, he was a regular lecturer. But he could not live to see the realization of his long-cherished dream, the start of a periodical on physics and mathematics. In two instalments — April and October 1889 — HUNYADY held his inaugural address at the Academy about orthogonal substitutions, but that was his swan song. His declining constitution was too weak to overcome a cold, and he died on 26th December 1889.

János Iváncsics
(Komárom, 1722—1784, Nagyszombat)

A mentor in Vienna for a period of time, then he attended the faculty of theology. IVÁNCSICS wrote — probably together with ANTAL REVICZKY (see later) — a textbook of mathematics in three volumes between 1752 and 1755, which was the guide for teaching mathematics at the university. IVÁNCSICS worked as a teacher at the universities in Nagyszombat and Vienna. After the dissolution of the order of the Jesuits he was the abbot of Esztergom and Siklós.

Miklós Jánosi
(Kolozsvár, 1700—1741, Nagyszeben)

A teacher of the Jesuitical order lecturing on mathematics and philosophy in Vienna and at the College of Kolozsvár.

Dénes Mihály Katona
(Dercsika, 1782—1874, Sátoraljaújhely)

A Piarist teacher at the secondary school of Léva. One of the best-known naïve circle squarers and angle trisectors.

Ferenc Kerekes
(Erdőhegy, 1784—1850, Balatonfüred)

A child of poor parents. "No rank or goods and chattels did I inherit from my ancestors" — says one of his letters. His strong social consciousness — palpable in all his writings — can be attributed to his youth full of tribulations. KEREKES went to school at the colleges of Debrecen and Keszthely, where he had a long experience of the hard way of living that the "servant students" had to share.

"The life of a servant student" — KEREKES writes — "is worse than that of a beast of burden in a lot of ways. He is scolded and beaten by his lord, then the public preceptor continues in the same fashion, and so do the inspector and the excitator who have that big club in their hands to thrash the fellow-servants with."[5]

It is certainly a sad description of college life at those times, especially as the description goes on telling about the unusual way of corporal punishment, which KEREKES so frequently protested against:

"I still remember how it astonished me when I first went to invocation. As soon as we got to the 'amen', they started butchering the children and make them howl, and when they had enough of that, we sang one more hymn and then we left."[6]

[5] Quoted from NAGY, SÁNDOR: *The History of Debrecen College.* Debrecen, 1940, p. 89. (In Hung.)
[6] *Ibid.*, p. 82.

After graduation, the versatile and original young man was sent abroad. The first station on his long and colourful journey was Vienna. It is known that KEREKES began translating VERGIL's *Georgicon*, but only parts of it were published in *Magyar Kurír*.

The history of literature has recorded an interesting episode of KEREKES's stay in Vienna. In 1815 KEREKES had MIHÁLY FAZEKAS's extremely popular epic poem *Matthias of the Geese* published in Vienna, without letting the author know about his plan or having his consent. KEREKES wrote a prologue of 28 hexameters with the definite makings of a poet.

During his trip abroad, KEREKES was mostly engaged in his studies of chemistry, which is testified also by his letters, diary and scientific notes of that period and shows that KEREKES kept in touch with several of the outstanding chemists. He apparently did not deal much with mathematics then.

The University of Saint Petersburg offered KEREKES a job, but he accepted the invitation of the College of Debrecen instead. His knowledge of languages, manifold education and his gift of teaching gave him great authority in Debrecen.

Hungarian science remembers versatile KEREKES — with some reason — mainly as a mathematician. KEREKES dealt with the problems of teaching mathematics and some of the then current topics of mathematical research. He gave a detailed description of the material he thought should be taught in secondary and higher education on the principle of practice. Why should the students' heads be crammed — KEREKES asks at some place — with Roman numbers, which have no function whatsoever, and why should time be wasted on learning things like that instead of dealing with something useful? He claimed that material be properly compiled, there should be good texts available and good teachers to teach. "How could he teach who does not know? Nemo dat, quod non habet." KEREKES is known to have been the first teacher who — seeing the difficulties of teaching mathematics — wanted the teachers to be helped by means of methodology.

"Why do you not commission some competent Hungarian to write instructions for the teachers how to teach algebra efficiently?"[7]

There is an unpublished manuscript of about 700 pages written by KEREKES that goes through the subject of college mathematics in a form of questions and answers.[8] The exchanges between teacher and students are presented in dramatic form, interwoven with subtle observations of methodology and — here and there — with the golden thread of humour.

[7] Library of the College of Debrecen, manuscript. R. 608/53, p. 42.
[8] *Ibid.*, manuscript. R. 608, p. 45.

Pál Kerekgedei Makó
(Jászapáti, 1724—1793, Pest)

No scientist in Hungary had as great an effect on the pedagogy of the Hungarian enlightenment, and on the teaching of mathematics in particular, as PÁL KEREKGEDEI MAKÓ had. The greatest authority of public education at his age, he had a decisive word in the problems of educational issues, and praiseworthy is the work he did to have the Hungarians' interest considered within the narrow limits set by the Imperial Court in Vienna.

MAKÓ came from the lower class of the nobility. After joining the order of the Jesuits in 1741, he worked at Trencsén and Győr. Between 1744 and 1747 he attended the faculty of philosophy at the University of Nagyszombat. A short period of teaching in Ungvár and Nagyszombat was followed by a lecturing job at the University of Vienna, but MAKÓ also went to Graz and Besztercebánya during his varied career, and undertook teaching at the University of Nagyszombat once again. He started his literary activity at Nagyszombat. His elegies written in Latin were so successful and well-known that they were read at schools in Switzerland. Those poems called the attention of the Habsburg Court to him, and MAKÓ was invited to work at the faculty of mathematics, physics and mechanics at *Teresianum* in Vienna, established by MARIA THERESA in 1746, where he worked for ten years. At the same time MAKÓ was teaching logic and methaphysics at the University of Vienna. He wrote his books on mathematics during his stay in Vienna.

After the order of the Jesuits was dissolved, MAKÓ went to the diocese of Vác, but according to MARIA THERESA's measures he had to stay in office in Vienna, and together with JÓZSEF ÜRMÉNYI and DÁNIEL TERSTYÁNSZKY he worked intensively on the preparation of *Ratio Educationis*. When the queen relocated the university from Nagyszombat to Buda, MAKÓ returned there and was appointed dean of the faculty of arts, where he worked till his unexpected death in 1793.

Mózes Kézy
(Fehérgyarmat, 1781—1831, Sárospatak)

A teacher at the College of Sárospatak and the author of several books on physics and mathematics.

István Király
(Debrecen, 1669?—1726, Debrecen)

Studied in Holland and in Halle. After returning home, he was a teacher of philosophy at the College of Debrecen.

Sándor Komnenovics
(Körmöcbánya, 1813—1869, Pest)

Worked as an engineer between 1847 and 1852 at the regulation of the Tisza, then — from 1852 — he taught mathematics at the Technical College, and also at the Technical University till his death. KOMNENOVICS had no significant results in mathematics, but his text books in Hungarian were popular.

Gusztáv Kondor
(Hercegszántó, 1825—1897, Budapest)

Went to secondary school in Baja and Szabadka, then he attended the College of Arts in Szeged. After two years of preparatory course KONDOR studied engineering in Pest, but his studies were interrupted by the Hungarian War of Independence. He fought as a first-lieutenant and experienced a lot of harassment after the capitulation at Komárom. In 1850 KONDOR graduated as an engineer, then for two years he studied astronomy at the University of Pest, and mathematics in Vienna under the guidance of JÓZSEF PETZVAL (see later). The letter circumstance resulted in his expertise in differential equations, and his book on that subject published in 1861. In 1865 KONDOR began working at the University of Pest, and was appointed professor of elementary mathematics. In 1896 the Institute of Mathematics started at the University of Budapest on KONDOR's initiative, and he was its first director. His activity in astronomy was more significant than that in mathematics.

Sámuel Köleséri
(Szendrő, 1663—1732, Nagyszeben)

Studied at the College of Debrecen and got his doctorate of philosophy in Leyden, then the doctorate of theology in Franeker. For a short time he returned to Debrecen but left again for Leyden, where he also graduated as a physician. In Nagyszeben, where he finally settled down, KÖLESÉRI worked as a doctor of the province and a chief-supervisor of the mining industry. In 1729 he was chosen member of the British Royal Society. KÖLESÉRI was one of the strongest advocates of the Enlightenment in Hungary.

Gyula Kőnig
(Győr, 1849—1913, Budapest)

It is almost impossible to give a brief account of GYULA KŐNIG's life and activity. If Hungary is considered a great power of mathematics, it is accounted for the results of a *great number* of mathematicians instead of the achievements of just a *few* outstanding ones. The tradition of handing down fields of research and methods to later generations, the acknowledgement of efforts, without which enthusiasm may go blunt, knowledge — essential to research —, and finding appropriate research problems for scientists at the beginning of their careers, all these contributed to the achievements of Hungarian mathematics. GYULA KŐNIG was one of the first and greatest at establishing those conditions in Hungary. Adverse circumstances made it impossible for the two BOLYAIS to give a *direct* impetus to mathematics in Hungary. In that respect

GYULA KŐNIG's activity is of far greater significance. The accomplishment of Hungarian mathematics up to this date goes back to his merits.

GYULA KŐNIG was born in Győr on 16th December 1849, and he went to primary and secondary school there. At the beginning of his studies he excelled in literature. His memoirs and articles on general topics still capture the reader with their polished style, the abundance of poetic images, and descriptiveness.

At the age of 16, right after leaving secondary school, KŐNIG went to study at the Medical Department of the University of Vienna and attended lectures on mathematics. After a short stay in Vienna, his attachment to natural sciences took him to Heidelberg, where outstanding authorities like BUNSEN, KIRCHHOFF and HELMHOLTZ were working. In 1869 also KÖNIGSBERGER became professor at Heidelberg. Although KŐNIG was more attracted by HELMHOLTZ, KÖNIGSBERGER had a greater influence on his scientific career.

KŐNIG's first essay treated a medical subject *(Beiträge zur Theorie der elektrischen Nervenreizungen)*, its topic had been suggested by HELMHOLTZ, and it was published in the medical gazette of the Vienna Academy of Sciences (1870). KŐNIG's doctoral thesis already dealt with mathematics, namely the theory of module functions, a topic recommended by KÖNIGSBERGER. KŐNIG got his doctor's degree at Heidelberg in June 1870.

Then KŐNIG spent half a year in Berlin and attended lectures of WEIERSTRASS and KRONECKER. The latter had specially great influence on KŐNIG, who turned out to be one of the most outstanding students of the excellent German mathematician.

After KŐNIG returned to Hungary, he soon received the various scientific degrees deserved for his qualification and results. He became a Privatdozent of the University of Budapest at the age of 22 and was appointed professor of the Teachers' College in 1873. Then, in 1874, he became professor of the Third Department of the Technical University. The magnetic personality of KŐNIG was present at the training of teachers and engineers and in the whole intellectual life of Hungary for about forty years. It is almost impossible to survey the far-flung areas which attracted his interest and in which he played a leading role. Instead of going into details it may be a sufficient verification to refer to the countless obituaries that appeared in the various newspapers and journals when this great man of the nation died.

His lectures at the university — some of them being attended by engineering and arts students together — dealt with modern as well as classic chapters of mathematics. KŐNIG gave as thorough a rendition to classic and abstract algebra, analysis, differential geometry, number theory as to the then-emerging theories of real functions, sets, probability and further to political arithmetic and the history of mathematics. KŐNIG's lectures were not based on books written by others, but followed his individual arrangement of the material observing possible simplifications, deficiencies and further ways of research. Many of the problems treated in KŐNIG's papers emerged through his lectures, but there are several times as many works on subjects which KŐNIG recommended or which were based on KŐNIG's passing remarks. Particularly stimulating were his seminars of practice, which KŐNIG held by turns first with JENŐ HUNYADY, later with JÓZSEF KÜRSCHÁK. Many footnotes in the mathematical journals of that time say that the problem in question originated with GYULA KŐNIG.

KŐNIG was extremely conscientious and he expected the same of his students. He went into details whatever the subject be and analyzed everything thoroughly even for students of the Technical University. He emphasized the importance of the quality rather than the quantity of knowledge, and tried to get his students to acquire an exact mathematical thinking.

KŐNIG's teaching activity was not confined to the university. He held the view that the professors of universities should always be in touch with secondary education, and they should guide and help work there. The algebra part of the 1879 programme and directions for secondary schools were worked out by KŐNIG. He did not only declare expectations concerning secondary education, but wrote textbooks to help realize those expectations. His texts — later rewritten by MANÓ BEKE — were used in secondary education for a long time.

KŐNIG established the education of commerce in Hungary. In the 1890s there were four kinds of commercial schools in Hungary with almost the same programme but different qualifications of the teachers. In 1895 the minister of culture GYULA WLASSICH asked KŐNIG to supervise those schools as a ministerial commissioner to do away with the prevailing differences. KŐNIG did his fruitful work with much competence and tact, and encouraged teachers personally to pass the necessary examinations. That work always filled KŐNIG with love and pleasure, and whenever he had the possibility, he undertook to preside at final exams every year so as to get an inside view of the life of those schools. Later on KŐNIG, as president of the Hungarian section, participated in the reformation project of the international committee of teaching mathematics.

Similarly manifold and extensive was KŐNIG's activity in other fields. He was among the founders and most enthusiastic editors of *Műegyetemi Lapok* in existence for three years altogether. Together with others, KŐNIG believed in the 1870s that a new era was dawning for Hungarian mathematics, and he was deeply disappointed when the publication of the journal ceased for lack of money.

KŐNIG was one of the founders of the *Mathematical Society* which was the first of its kind in Hungary and had not statutes, and when the *Mathematical and Physical Society* started in 1891, KŐNIG was chosen deputy chairman. He became a member of the Hungarian Academy of Sciences in 1889, and for 19 years he was member of the board of directors and secretary of the Third Department there. Characteristic of KŐNIG's many-sidedness is that he helped the editors of the high-standard periodical *Budapesti Szemle* with his advice, and that he was a member of the board of directors at *Franklin*, then largest publishing house in Hungary, from 1894 and acted as general director from 1904 to his death.

KŐNIG took part in all activities of the Technical University. He was the dean of the faculty of engineering and architecture three times, and Rector of the University also three times. His authority was decisive in the University Board. "A significance achievement of KŐNIG's activity as Rector", said GUSZTÁV RADOS, "was that he kindled the idea of giving the Technical University a new building, kept the cause alive and ardently urged to have it realized until the government, in principle, decided in favour of the construction".

During his extraordinarily active life KŐNIG received ample acknowledgement from the authorities as well as from scientific institutions in Hungary and abroad. KŐNIG was often spent sunny hours with his loving wife and gifted sons (DÉNES an outstanding mathematician of ill fate, and GYÖRGY, a doctor of law and historian of literature). His intimate home provided KŐNIG the atmosphere that is essential to a creative life. For relaxation and recreation KŐNIG liked reading. His favourite was GOETHE's *Faust*.

In 1905, because of his manifold obligations, KŐNIG retired, but he continued to lecture at the Technical University. In fact he gave his most important special courses — related to his own investigations — after retirement. KŐNIG remained active practically until the last moment of his life: towards the end, he worked exclusively on his vast monograph on the foundations of mathematics, published posthumously. He was working on the last chapter when a sudden attack killed him on 8th April 1913.

In 1914 the Academy of Sciences commemorated GYULA KŐNIG with a due ceremony. In 1918 the *Kőnig Prize* was established; it was awarded twelve times between 1922 and 1944, but after 1930 — because of the depreciation of foundation — only in the form of a medal.

István Kruspér
(Miskolc, 1818—1905, Budapest)

First he studied law, then graduated from the Technical University of Vienna. A professor at the Technical College of Buda from 1850 and then at the Technical University. Corresponding member of the Hungarian Academy of Sciences from 1858 and ordinary member from 1869. His most famous work is *Geodesy*, and the measuring instruments constructed by him received great appreciation. KRUSPÉR was Hungary's delegate at the Paris conference of 1870 which worked out the metric system. His modern lectures on mathematics and mechanics greatly elevated the standard of engineers' training.

Mihály Lipsicz
(Óvár, 1703—1765, Győr)

Studied mathematics in Vienna in 1737, worked as a teacher at Kolozsvár and one year later at Kassa, where his book *Algebra* was published in 1738. He also worked at Nagyszombat and Győr and dealt with astronomy and technical problems.

Nándor Lutter
(Bér, 1820—1891, Budapest)

A Piarist priest-teacher, he taught in several schools of his order. For his participation in the 1848 War of Independence he was suspended from teaching for a long time. In 1858 he was appointed director of a secondary school in Pest. LUTTER wrote a number of textbooks on mathematics and compiled a five-figure table of logarithms, which was widely used and brought him corresponding membership in the Academy of Sciences. After the Compromise he took an important part in public education and became inspector of an educational district in 1884.

Master György of Hungary — Georgius de Hungaria
(?, 1422?—1502?, Rome)

Nothing is known for sure about his life. He is supposed to have lived in Utrecht or nearby and was a priest who was also a master of arithmetics. His book of twenty pages entitled *Arithmeticae summa tripartita Magistri Georgii de Hungaria* was published probably in Utrecht in 1499.

Zsigmond Maksay
(Maksa, 1850—1896, Pécs)

A secondary-school teacher at Nagykálló and Pécs. Published several papers on mathematics in yearbooks of his schools, in *Mathematikai és Physikai Lapok*, and in *Középiskolai Mathematikai Lapok*. In LIPÓT FEJÉR's writings, MAKSAY is mentioned several times in terms of great respect.

György Maróthi
(Debrecen, 1715—1744, Debrecen)

Little is known of his life, though he was one of the most famous professors of the College of Debrecen. His lifework is clearly shown by his papers, but little is known of his life story. MARÓTHI was born in 1715 in Debrecen.

His father was first a town counsellor, then a chief justice. The boy acquired his elementary education in the lower classes of the College of Debrecen where he studied until he was sixteen. Then MARÓTHI went to study abroad for seven years (1731—1738). The main stations of his trip were Basel, Zurich, Bern and Amsterdam.

After returning to Hungary, in 1738 he was appointed teacher of the College of Debrecen. MARÓTHI's career started amid sorrowful events: a disastrous epidemic of plague that had ravaged through Transylvania reached Debrecen in 1738 and 1739, decimating the population. Everyone who could afford it ran away from Debrecen, and the number of students at the college reduced to some thirty. MARÓTHI withstood the tribulations and immediate danger. The six short years he spent there proved highly successful. Stremous work, however, undermined MARÓTHI's weak health and he died in 1744.

The most reliable data on MARÓTHI, the teacher and the man, come from the funeral address held at MARÓTHI's grave by SÁMUEL SZILÁGYI. He describes a remarkable man of knowledge, character and appearance who gained fame in Hungary and abroad during his short career. He was held in great respect as a polymath, an eloquent lecturer, a thorough expert of music, in engineering matters a reliable adviser of the town, the author of several pieces of literature and excellent textbook of mathematics. MARÓTHI certainly became a legendary figure and his students Professor ISTVÁN HATVANI, Professor FERENC VARJAS and others remembered him with warm appreciation. Two household expressions involving his name show what an authority MARÓTHI was: "Maróthi claims that it is so" and "According to Maróthi two by two is four". (These sayings were translations of German ones involving GEMMA FRISIUS.)

While MARÓTHI travelled abroad he could compare the cultural level of foreign countries with that of Hungary, and the result was rather disadvantageous for his native country. That recognition made him write two drafts (*Idea*, 1740; *Opiniones*, 1741) soon after he returned home putting forward some ideas concerning primary and secondary education. First of all, MARÓTHI urged the introduction of teaching in the Hungarian language, and set up a hierarchy of the school subjects. Besides the classical subjects he considered natural science equally important. For example, he regarded a *detailed* account of the elements of mathematics — within the subject of philosophy — as necessary, and suggested that afternoon playing on Wednesdays and Saturdays be replaced by arithmetics. MARÓTHI had ideas also on the desirable programme of mathematics at school, and for the realization of his ideas he proposed that a book arithmetic in Hungarian be published.

His greatest achievement was that he wrote that book himself and it turned out to be the first Hungarian arithmetic European standard. The impact of the book was significant, and it greatly contributed to the development of natural sciences in Hungary. There is evidence that at the beginning of the 19th century MARÓTHI's book was still one of the most popular guides for teaching elementary arithmetics in Hungary.

MARÓTHI's concept of school organization and methodology, further that he adopted the cause of the college library, inspired teachers to learn permanently and develop their skills, and emphasized the importance of acquiring special books and periodicals from abroad, all that gives a picture of a highly qualified teacher who, though not a researcher, had great merits in raising the standard of mathematical education in Hungary.

Lajos Martin
(Buda, 1827—1897, Kolozsvár)

After leaving the Roman Catholic Secondary School of Buda, MARTIN attended the university of arts, then became a student of the engineering department. In 1848 he volunteered into the army and fought heroically as an artilleryman in the War of Independence. After the fall of the war he was hiding for a while but was captured and imprisoned. Later on he was enlisted into the Austrian army, where his qualification in mathematics was discovered, and he was asked to take the courses of the military academy. In 1859 MARTIN left the army, graduated as a teacher and worked all over the country as an engineer, teacher and director of a telegraph office. In 1872 he was appointed professor at the department of advanced mathematics of the University of Kolozsvár, where he worked till his death.

His activity in mathematics was not significant. In the 1860s he wrote some texts for secondary-schools and several studies on the theory of the best propeller and windmill. In 1896 a bicycle-driven mechanism which had been constructed by him and looked something like a helicopter was said to have ascended about two or three meters above the ground. MARTIN prophetically foretold the future importance of flying at traffic, war and commerce. In Hungary he was a laughing stock for his research and experiments. The importance of the mechanisms MARTIN built was, however, recognized abroad, buth he refused to sell the patents there, no matter how tempting the offers were. Several recent studies have dealt with MARTIN's inventions.

Ignác Martinovics
(Pest, 1755—1795, Buda)

A teacher, originally a Franciscan, who turned an atheistic philosopher, and the leader of the Hungarian Jacobin Movement. Due to MARTINOVICS's role in Hungarian history, a number of books and studies have given detailed accounts of his life and activity. Therefore, we restrict attention to some — less known — facts related to his work in mathematics.

MARTINOVICS got his doctor's degree in theology and art at the University of Buda, and mostly dealt with philosophy and natural sciences. He started his career at church schools but his ambition made him do more. In 1780 MARTINOVICS applied for the post of the head of the mathematics section at the College of Nagyvárad, but he was rejected. ANDRÁS DUGONICS says in his *Notes*[9] that MARTINOVICS was found unsuitable for the post at the interview held in Pest.

"I began questioning him about that branch of science, but to his bad luck I asked questions he failed to answer, but he was talking about the glory of mathematics in so enthusiastic a voice that all their Excellencies went over to his support. I stood up against them."

In 1781 MARTINOVICS left the Franciscan order and was appointed priest in the army at Chernovits where he worked also as a mathematician of the Engineers. From there

---

[9] DUGONICS, ANDRÁS: *Notes*. Olcsó Könyvtár, Nos 401—402, p. 25. (In Hung.)

he went to Paris where he got in touch with the enlightened circles, among them a sectet leftist organization. In 1783 MARTINOVICS was appointed to teach at the department of natural sciences at the College of Lemberg, which post he held also after the college became a university.

Philosophical and political problems increasingly engaged MARTINOVICS, and he made plans to establish a republic in Hungary. In pamphlets published anonymously he turned against the Hungarian clergy and aristocracy and sent classified information about them to the court in Vienna.

As far as natural sciences are considered in MARTINOVICS's life, most typical was what happened to him in the spring of 1791, when he applied for a post vacant at the Department of Physics at the University of Pest. The committee of nomination did not choose MARTINOVICS among the first three eligible candidates. His appeal was rejected after the authorities at the University of Pest gave the matter a thorough investigation by asking the most competent professors for expert opinion about MARTINOVICS's activity in chemistry, physics and mathematics. The three reports were equally critical. JÁNOS PASQUICH, for example, had the following opinion (originally written in Latin) of MARTINOVICS's activity in mathematics:

"His *General theory of equations of any degree explained by new formulae* ... (Buda, 1780) ... lacks a thorough and necessary knowledge of the subject altogether", ... "the general theory of equations of any degree and the formulae that would explain them cannot be found in this short essay."

"*A mathematical essay on some properties of the circle.* Those properties are of the kind which not even a middling mathematician would possibly consider important."[10]

The expert opinions entitled the University of Pest to give the post to the original candidate, and the emperor appointed the former Jesuitical priest JÓZSEF DOMIN (1754—1819) head of the department. DOMIN's name is mentioned in three books on physics and LORÁND EÖTVÖS remarks[11] appreciatively that he used electricity in medical treatment.

MARTINOVICS's life, rich of unexpected turns, took a tragic direction then. In 1792 he fell out of confidence with the Imperial Court of Vienna and that made him establish an organization of Hungarian reformers in 1794. MARTINOVICS was arrested in Vienna the same year and — together with the other four principal defendants — was executed in Buda for high treason on 20th May 1795.

[10] That manuscript written by Martinovics in Latin has recently been found at an archive in Leningrad. See the review in *Mathematikai Lapok*, Vol. 20, 1969, pp. 57—62. (In Hung.)

[11] EÖTVÖS, LORÁND: The Memory of Ányos Jedlik. *Természettudományi Közlöny*, Vol. 29, 1897, pp. 387—402. (In Hung.)

Sámuel Méhes
(Kolozsvár, 1785—1852, Kolozsvár)

His father, GYÖRGY MÉHES, was a teacher of mathematics and philosophy at the College of Kolozsvár, where SÁMUEL MÉHES also went to school. In 1806 he went abroad with a letter of recommendation from FARKAS BOLYAI to GAUSS.[12] MÉHES studied mainly medical sciences during his stay abroad, but in 1809 he was appointed teacher of physics and mathematics at the College of Kolozsvár. In 1817 he had his father's *Mathematics* and his own *Algebra* published. MÉHES was elected member of the Hungarian Scholarly Society in 1836. MÉHES's significance lies in his unrelenting struggle for a greater work of natural sciences at school.

Ferenc Mentovich
(Nagydebrek, 1819—1879, Marosvásárhely)

Studied art and law, then attended universities in Vienna, Berlin and Switzerland studying mathematics, physics, chemistry and mineralogy. In Göttingen he visited GAUSS, who called MENTOVICH's attention to LOBACHEVSKY's *Geometrische Untersuchungen*. MENTOVICH gave an account of the visit in a Hungarian magazine. He started his teaching career at Nagykőrös, then from 1856 he worked at the College of Marosvásárhely till his death. MENTOVICH dealt also with fiction but more significant was his role in philosophy. He was the most outstanding materialist among Hungarian natural scientists in the 19th century.

Ferenc Menyői Tolvaj (?)

All that is known about him is that he was a teacher at Gyöngyös and Losonc, and died about 1710.

Sámuel Mikoviny
(Ábelfalva, 1700—1750, somewhere near Trencsén)

Studied at the University of Jena, an architect at the Imperial Court and a teacher at the Academy of Selmecbánya. He had great merits in geodesy, regulation of rivers, cartography and architecture. For years, MIKOVINY gave lectures in several theoretical

---

[12] Briefwechsel GAUSS–BOLYAI. p. 97.

subjects under the collective name "mathematica", but only two of his essays dealt with mathematics proper. The Hungarian traditions in geodesy, cartography and hydraulics are due to MIKOVINY. In 1749, the royal castle of Buda was begun according to his plans. He was a member of the Academy of Science of Berlin.

József Mitterpacher
(Bellye, 1739—1788, Pest)

A priest teacher at Nagyszombat, then the professor of mathematics at the University of Pest. His *Analysis and mechanics* was published by JÁNOS PASQUICH (see later) in Leipzig in 1790.

Károly Nagy
(Komárom, 1797—1868, Paris)

Studied in Vienna and worked at an observatory there. The economic adviser of LAJOS KÁROLYI and KÁZMÉR BATTHYÁNY after 1845, which role led to his imprisonment in 1849. After being released he had to sacrifice his modest fortune to get a passport and go to Paris, where he died in 1868.

NAGY wrote several basic geometries and arithmetics, but he mostly dealt with astronomy. He compiled the calendars published by the Academy between 1837 and 1843. NAGY was one of those who urged the introduction of the metric system in Hungary, and he was a delegate at the conference in Paris. A member of the Academy from 1832.

Lajos Naszluhácz
(Nagykanizsa, 1815—1877, Székesfehérvár)

An engineer and director of the Hungarian Lowlands–Fiume Railways.

Sándor Nékám
(Pest, 1827—1885, Budapest)

A teacher and physician, and the professor of elementary mathematics at the University of Pest between 1863 and 1870. He is considered a pioneer of statistical geodesy in Hungary, but had no essential results in mathematics.

István Nyiry
(Átány, 1776—1838, Sárospatak)

Brought up and educated at Sárospatak, later a teacher at the college there. He began to passionately deal with fine arts and mathematics in his early childhood. Hungarian became the official language of education at Sárospatak in 1810, just when PÁL SIPOS (see later) left there. NYIRY, the successor of SIPOS, was the first lecturer of philosophy in the Hungarian language. NYIRY's first essay was written on mathematics (*Prima elementa matheseos intensorum*, Kassa, 1821), but he was increasingly interested in philosophy. NYIRY became member of the mathematics department of the Hungarian Scholarly Society in 1831 and worked there actively: collected mathematical terms, reviewed textbooks and competition papers. NYIRY's most ambitious plan was to write an encyclopedia of several volumes, but only the first three volumes were completed.

János Onadi (?)

All that is known about his life is that he was a school-attendant at the Calvinist College of Kassa. His *Arithmetic* — carrying a Latin title but written in Hungarian — presents the rules of counting in the form of poems.

János Pasquich
(Vienna, 1753—1829, Vienna)

A teacher of advanced mathematics at the University of Buda from 1788, and the director of the observatory there. The author of several books on mathematics. One of the most enthusiastic pioneers of Hungarian astronomy; the rich collection of his correspondence on the subject is a most valuable source on the history of Hungarian astronomy.

Ferenc Pethe
(Büdszentmihály, 1762—1832, Szilágysomlyó)

A teacher at the Georgicon of Keszthely. In 1812 he published in Vienna a two-volume Hungarian school-book of elementary mathematics, a most useful piece of work in respect Hungarian of mathematical terminology.

József Petzval
(Szepesbéla, 1807—1891, Vienna)

PETZVAL spent only his first 31 years in Hungary, so he is one of those who displayed their scientific activities abroad.

PETZVAL was born in the family of a poor cantor teacher on 6th January 1807. He went to primary school at Késmárk and to secondary school at Podolin. All his biographers point out that PETZVAL did very poorly at mathematics before a school-book aroused his interest toward the subject. After attending the College of Kassa and working as a tutor — which ha was forced to do for financial reasons — he studied engineering in Pest from 1826. After graduation PETZVAL worked for the city of Pest for seven years as a most gifted expert of flood-prevention and drainage. From 1832 he lectured on mathematics and mechanics at the university. In 1835 he was appointed professor at the department of advanced mathematics.

PETZVAL's career in Hungary was brief for in 1837 he was invited to teach at the department of advanced mathematics at the University of Vienna. The vigorous, eccentric and romantic PETZVAL became a well-known figure of great authority in Vienna.

344

His life was composed of the quiet investigations of thorough scientist, the surprising and important technical discoveries — especially in the field of optics — of an experimentalist, and the countrywide outstanding results of a splendid sportsman.

Of all branches of mathematics, PETZVAL was most interested in the theory of differential equations, and wrote essays also on ballistics. In one paper he fronted fencing, and in another the mathematical theory of horsewalk. His international reputation, however, was due to his theoretical and practical results concerning lenses.

Students, liked PETZVAL, and considered him a true, firm and helpful friend, but his colleagues suffered much from his strong criticism.

PETZVAL was elected member of the Academy of Sciences of Vienna in 1851. He died on 17th September 1891.

Several recent Hungarian, Slovak and Austrian studies have dealt with PETZVAL's life and his achievements in physics.

Ottó Petzval
(Szepesbéla, 1809—1883, Budapest)

Brother of JÓZSEF PETZVAL. An engineer and a doctor of philosophy. After graduation he worked as an assistant engineer at the Engineers' Training Institute for several years. When his brother left his post, OTTÓ PETZVAL was appointed professor of advanced mathematics at the university, where he worked almost till his death. When the Engineers' Training Institute assumed the name József Technical College and became independent of the University of Art and Science in 1850, OTTÓ PETZVAL worked there for several years as a deputy director. He chose the topics of his lectures from various fields of technology. Several textbooks — lithographed and printed — bore his name. Today OTTÓ PETZVAL's works on mathematics seem to be of the secondary-school level, and there had been more essential approaches to the subject than his *Advanced Mathematics* (1850). He was elected to member of the Academy of Sciences in 1858.

Kristóf Pühler
(Siklósd, 1500?—1583?)

Little is known about his life. He studied at the University of Vienna and probably belonged to the Augustine order. In various documents his surname has several forms (Puehler, Puchler, Püchler, etc.). His place of birth is also uncertain; most likely it was Sigless, now in Austria.

Ferenc Rausch
(Prellenkirchen, 1743—1816, Pozsony)

A teacher of applied mathematics at the University of Buda, and the author of several insignificant text books of mathematics.

Mór Réthy
(Nagykőrös, 1848—1925, Budapest)

He went to secondary school at Nagykőrös and attended the technical universities of Buda and Vienna. After graduation he was an assistant at the University of Buda, then went to teach at the Modern School at Körmöcbánya. A state scholarship took him to Göttingen and Heidelberg. His first paper, on the diffraction of light, was presented at Göttingen. After RÉTHY returned to Hungary in 1874, he was appointed extraordinary professor at the department of theoretical physics and mathematics at the University of Kolozsvár. His seminars in mathematics elliptic functions, complex functions, determinants — together with his investigations into JÁNOS BOLYAI's *Appendix* and the "end-like equality of areas" defined by FARKAS BOLYAI, gave a new colour and momentum to Hungarian mathematics.

In 1886 RÉTHY was offered a job at the Technical University of Budapest. First he lectured on geometry there, but gradually he turned to theoretical problems of physics and mechanics. He had the post of a dean several times, became corresponding member of the Academy of Sciences in 1878 and ordinary member in 1900. RÉTHY was extremely active at movements of mathematicians and physicists and was one of the founding members of the *Eötvös Loránd Mathematical and Physical Society.*

Endowed with extreme modesty and working capacity, RÉTHY was a person worthy of affection and an ardent servant of all Hungarian causes. His most significant essays deal with the legacy of the two BOLYAIs and the development of some of their results. Internationally, however, better known are RÉTHY's investigations in physics. His results concerning the shape of jets of incompressible fluids — where he applied deef tools of function theory — were the basis for many a further investigation. Also well known is RÉTHY's research into the *Ostwald theory* and into the classic "principles" of mechanics. RÉTHY died on 16th November 1925.

346

His ancestors' name was KOVÁTS, but it was Germanized during the reign of JOSEPH II. SCHMIDT studied at the universities of Vienna and Munich, and he graduated as an engineer. He worked as an architect in Temesvár and from 1869 in Pest.

SCHMIDT's absolute devotion and sympathy towards the two BOLYAIS — even though he was not a mathematician — drove him to do his utmost to gain due acknowledgement for the two unrecognized Hungarian scientists. His untiring and successful activity brought SCHMIDT to be considered the nestor of Hungarian mathematicians around the turn of the century, and made him a great authority also at the *Mathematical and Physical Society*.

János György Schmidt
(Pest, 1765—1848, ?)

Undoubtedly one of the professor of greatest knowledge at the Engineers' Training Institute of Pest between 1800 and 1837; many of his students became outstanding engineers. As for as we know, he did not deal with mathematics but certainly applied advanced mathematical tools in treating problems of technology.

Jakab Schnitzler
(Nagyszeben, 1636—1684, Nagyenyed)

Studied at the University of Wittenberg, then went to teach at Nagyszeben and Nagyenyed. From 1668 to his death SCHNITZLER was a priest at Nagyenyed.

Ágoston Scholtz
(Kotterbach, 1844—1916, Veszprém)

One of the pioneers of linear algebra in Hungary. Studied at the universities of Vienna and Berlin, then went to teach at the secondary school at Igló. In 1871 he was appointed director of the Lutheran Secondary School of Pest, and in 1884 professor of the Technical University of Budapest where he worked till 1909. SCHOLTZ obviously received the first impetus toward mathematics from HUNYADY whose investigations are closely related to those of SCHOLTZ.

János András Segner
(Pozsony or Szentgyörgy, 1704—1777, Halle)

One of the first scientists from Hungary who is known and recognized by the world history of mathematics. His family was of German origin. The ancestors came from Steiermark in the 15th century, and soon took public offices in Hungary and were raised to noble rank.

JÁNOS ANDRÁS SEGNER's devotion to Hungary is indicated by the fact that when he lived in Germany his interest in Hungarian affairs never diminished, and Hun-

garians visiting Germany were always welcome in SEGNER's home. A statue and memorial tablet have been erected in Debrecen in his honour.

SEGNER went to school in Pozsony and Győr, and most probably spent a year at the College of Debrecen in 1724. In 1725 he left for Jena to study medicine, but he also studied natural sciences and mathematics. His talent showed very young, and the first attempts at scientific activity come from the period of his university studies. His first essay was on mathematics and was soon followed by several papers and books on philosophy, physics, chemistry, astronomy, mathematics and medicine. After graduating as a physician in 1729, SEGNER returned to Pozsony and most probably started negotiations with the authorities in Debrecen who offered him a doctor's job in November 1730. SEGNER accepted the offer and began to work in Debrecen, where he spent only one year and a half. In 1732 he went to Jena once again, most probably because he was in love with somebody there, but also his thirst for knowledge inspired him to go abroad where he saw better chances to study and he wanted to get a magister's degree in Jena.

In Germany SEGNER did very successfully and had a lot of recognition. First he was offered a job at the University of Jena, then he worked at the newly established University of Göttingen for twenty years (1735—1754) and also at the University of Halle, lecturing on physics, mathematics and medicine. In Halle SEGNER also dealt with writing textbooks of mathematics and established an observatory as well.

SEGNER experienced countless recognitions all through his life. He was elected member of several German and other foreign scientific societies and FREDERICK II of Prussia presented him several awards.

A wheel constructed on the theory of action and counteraction (1750) made SEGNER's name generally known. EULER dealt with the mathematical theory of SEGNER's wheel

349

in several of his essays, and EULER's correspondence shows that he regarded SEGNER as one of the greatest scientists of the time.[13] Also well known are SEGNER's examinations of inertia, acoustic and optics.

Pál Sipos
(Nagyenyed, 1759—1816, Tordos)

Studied at the Bethlen College of Nagyenyed. After a few years of tutoring he became rector professor at the Szászváros Division of the College of Nagyenyed. Then he took the job of a mentor once again and taught Count JÓZSEF TELEKI's son. In 1791 his patron's support made it possible for him to go abroad on a study-trip. SIPOS studied theology at the University of Frankfurt, then he went to Göttingen where KÄSTNER was one of his teachers. After several years of tutoring in Vienna he went to teach at Szászváros again, then to Sárospatak, where he became one of the most successful teachers in the history of the College. During his stay there SIPOS wrote several drafts outlining his ideas on curriculum and methods to improve the results of teaching mathematics. SIPOS's idea was to teach a detailed and thoroughly worked out portion of mathematics even at the expense of the amount of the material. The history of pedagogy shows the success of SIPOS's plan, which was the basis of the curriculum introduced at the College of Sárospatak in 1810, since it included essentially the same materials as that tought at grammar schools towards the end of the 19th century.

In 1810 SIPOS gave up his post at Sárospatak for the less tiring and more profitable job of a Calvinist minister in Tordos. He died — most probably of typhoid fever — on 15th September 1816.

SIPOS was one of the most versatile and learned Hungarians of the time. His philosophical essays dealt with KANT and FICHTE, he wrote several successful poems, and a rich correspondence of great scientific value belongs to his legacy. He also wrote theological books and studies.

Károly Szász Sr.
(Vízakna, 1798—1853, Marosvásárhely)

A teacher of physics and mathematics at the College of Nagyenyed. After FARKAS BOLYAI retired, SZÁSZ worked at Marosvásárhely from 1851. He wrote one textbook of mathematics.

---

[13] Die Berliner und die Petersburger Akademie der Wissenschaften im Briefwechsel Leonhard Eulers. Berlin, 1959.

Ferenc Székely
(Gyarmat, 1658—1715, Ungvár)

A Jesuitical priest-teacher, who worked at Kassa, Eger and Kőszeg. He lectured on philosophy and logic at the University of Nagyszombat between 1693 and 1696. Most probably, co-author — with JÁNOS DUBOVSZKY — of the first Hungarian trigonometric table.

Kálmán Szily Sr.
(Izsák, 1838—1924, Budapest)

Studied at the technical department of the universities of Budapest and Vienna. After graduation in 1860, he was teaching at József Technical College, then at its legal successor, the Technical University of Budapest. For three decades SZILY actively participated in training engineers. He essentially contributed to the modernization of the *Natural Science Association*. Till 1898 SZILY edited the journal of the Association, *Természettudományi Közlöny*, which he started in 1869. His interest in linguistics led him to deal with the development of the Hungarian terminology of natural sciences. A corresponding member of the Hungarian Academy of Sciences from 1865, and ordinary member from 1873. Became chairman of the *Natural Science Association* in 1880, and General Secretary of the Academy of Sciences in 1889. Then SZILY gave up teaching so as to be able to fully concentrate on his new job. In 1890 he launched the journal *Akadémiai Értesítő*, which he edited till 1904. In the 20th century part of his life SZILY dealt almost exclusively with linguistics. In 1904 he established the

Hungarian Linguistic Society and took on editing the periodical *Magyar Nyelv*. In 1905 SZILY gave up the post of General Secretary of the Academy for the post of chief librarian. Most significant was his research in theoretical physics. His study of the interrelation between mechanics and thermodynamics received international response. SZILY's investigations into linguistics were also useful, and all his works are written in a perfect style. Some of his mathematical papers lack the necessary thoroughness. More important are his works concerning the history of mathematics, which — although often differ from today's point of view — carry useful information about the Hungarian mathematics of old times.

Károly Taubner
(Nagyveleg, 1809—1860?, Verona)

Studied philosophy, history, mathematics and astronomy in Berlin between 1834 and 1837. He was appointed teacher of the Lutheran Secondary School of Pest in 1837, became corresponding member of the Academy in 1840. In 1844 TAUBNER went to work in Italy as an army chaplain and remained there. He wrote several mathematical texts and some essays on geometry, but more important was his philological activity. TAUBNER was one of HEGEL's most enthusiastic followers in Hungary.

István Tichy
(?, after 1750—after 1800, ?)

After receiving his diploma in engineering at the Institutum Geometricum in 1785, he lectured there for a short time before going to Kassa to become a teacher of the Academy in 1786. He was a freemason, pensioned off in 1795 for his contacts with the Hungarian Jacobin movement. He published, an anonymous highly enlightened pamphlet demanding reforms at the faculty of phylosophy of Pest University (*Philosophische Bemerkungen über das Studienwesen in Ungarn*. Pest–Ofen–Kaschau, 1792, 150 p.)

Pál Tittel
(Pásztó, 1784—1831, Buda)

Went to secondary school in Gyöngyös and Kecskemét. In 1806 he became a teacher at the College of Eger. His favourite branch of science was astronomy, and the observatory in Eger — out of operation for a long time — was put back to function on TITTEL's initiative. TITTEL frequently visited foreign countries and he was a close friend of GAUSS. In 1824 he became the director of the observatory in Buda and lectured on astronomy at the university there. TITTEL did not deal with mathematics intensively but emphasized the importance of the subject in many of his articles. For his activity in astronomy, he was elected member of the Hungarian Scholarly Society in 1830. On the bicentenary of TITTEL's birth a commemorating conference was arranged in Budapest.

Béla Tőtőssy
(Billéd, 1854—1923, Budapest)

Studied geometry intensively at the University of Zurich under the guidance of FIEDLER, an outstanding geometrician. In 1882 he began to work at the Technical University of Budapest where he was appointed professor of projective geometry in 1889. Corresponding member of the Academy of Sciences in 1899. His papers on projective geometry were published in Hungarian periodicals and in *Math. Annalen*. His greatest contribution was the preparatory work for the second edition of *Tentamen* and *Appendix*.

Antal Vállas
(Pest, 1809—1869, New-Orleans)

The son of a poor coachman. He joined the Piarists' order, but left it before he was ordained a priest, and went to study in Kassa and Pest. FERENC KAZINCZY soon noticed VÁLLAS's talent for writing and encouraged him to deal with literature seriously, but VÁLLAS — after writing a couple of poems — chose natural sciences as a career. In 1830 he moved to Vienna, where he got acquainted with ETTINGSHAUSEN, a well-known mathematician. In 1836 VÁLLAS returned to Hungary and accepted a teaching job at the economic college of Rohonc, and later at the University of Pest. In 1850 the absolute government fired him, so in 1851 VÁLLAS decided to leave Hungary. After a lot of adventures and difficulties he settled down in New Orleans and died there. His death was announced four years later by JÁNOS ARANY, general secretary of the Academy of Sciences, who described VÁLLAS as "an excellent champion of exact sciences". VÁLLAS actively participated in the life of the Academy and held lectures at the meetings not only on mathematics but also on physics, engineering and astronomy. He wrote several textbooks of algebra and geometry, and his paper on the solution of equations of higher degree was published also in English. VÁLLAS had progressive ideas and firm character; he greatly contributed to the cause of Hungarian natural sciences and mathematics.

Gyula Vályi
(Marosvásárhely, 1855—1913, Kolozsvár)

Hardly any other professor at the University of Kolozsvár had a greater knowledge and — especially at teachers' training — a greater influence than GYULA VÁLYI. He was born on 25th January 1855 at Marosvásárhely, the town of the BOLYAIs. His father was a judge and later the curator of the Calvinist church. His mother was RÁKHEL DÓZSA, a descendant of the rebel leader GYÖRGY DÓZSA.

VÁLYI first went to school at Marosvásárhely, then he attended the University of Kolozsvár where he graduated as a teacher of mathematics and physics. MÓR RÉTHY and LAJOS MARTIN helped him to get a two-year bursary at the University of Berlin where he diligently attended the lectures of the "great three", KRONECKER, KUMMER and WEIERSTRASS. VÁLYI began his scientific activity at that time. The first steps were solutions to problems set in *Műegyetemi Lapok*. MÓR RÉTHY incited him to examine

the problem of the best propeller, and VÁLYI wrote his doctor's dissertation on the theme. On the basis of this dissertation, he was promoted to Privatdozent at the University of Kolozsvár in 1881. In 1884 he became professor of theoretical physics there, and professor of elementary mathematics in 1885. VÁLYI's scientific activity and the subjects of his lectures included analysis, geometry and number theory. He had few publications, but each was a major event in Hungarian scientific life. VÁLYI's ill health, progressive eye-trouble and homely appearance made him shy and seclusive. A good description of his personality occurs in the telegram LAJOS SCHLESINGER sent to the science department of the university on the occasion VÁLYI's death:

"The tragedy of the life which has ended now is that the frail body hindered the full exertion of intellectual power possessed in abundance. MR VÁLYI's accomplishment, however, is outstanding and lasting, achieved through serene reconciliation with the unchangeable, an attitude reminding us of the classic idols, and so characteristic of our deceased colleague."

VÁLYI had a personality and knowledge engaging enough to get the students to love mathematics. Hungary especially Transylvania, gained a large number of highly qualified and enthusiastic teachers owing to VÁLYI's activity. VÁLYI was always compassionate with his students, shared their joy and borrow. Heart-stirring commemorations tell about the countless sacrifices that his unselfishness and compassion made for his students. LAJOS DÁVID, one of his students, said about VÁLYI:

"In the lectures VÁLYI dealt strictly with mathematics, refraining from any, say philosophical, remarks. He went through the tiniest details of calculation with great care, and practically no mistakes thanks to his extraordinary skill at calculation. The amount of the material covered was always considerable, but everything was clearly arranged which, together with his splendid memory, enabled — and his bad eyes also compelled — him to speak without notes. VÁLYI's lectures did not always follow the same pattern. Of the special literature read aloud to him regularly every year he could pick up and fit in his lectures those items, which made the course more perfect. Thus much of the material gradually built up to expedient texts even in the not everywhere authentic form published by the students. There were two courses of VÁLYI that hardly changed with time. One of them presented the *Basic theory of functions*; in it VÁLYI followed his old master WEIERSTRASS and included the theory of elliptic functions. The other concerned with JÁNOS BOLYAI's *Appendix*; in that course VÁLYI, especially in the first years stuck to the order and contents of the sections of BOLYAI's masterpiece."

VÁLYI belonged to the university and Kolozsvár, the peaceful and civilized town, which befitted his character so well. Unpublished letters in his legacy tell that when GUSZTÁV KONDOR died, VÁLYI's friends in Budapest offered him professorship at the Budapest University of Art and Sciences, but he refused the honour.

VÁLYI retired in 1911, at his own request. The reason of his decision to give up lecturing shows his strong self-criticism and great character: during a lecture his memory — the only reliance because of his bad eyes — failed, and VÁLYI regarded the episode as an indication that he was no longer suitable for the job of a professor. VÁLYI spent the short time he had left of his life in modest circumstances in the home of his brother GÁBOR VÁLYI, a retired professor of statistics. GYULA VÁLYI died on 13th October 1913. A solemn commemoration was held by MÓR RÉTHY, one of VÁLYI's teachers, at a meeting of the Hungarian Academy of Sciences.

A letter in RÉTHY's legacy indicates that RÉTHY tried to have VÁLYI's collected studies and letters published, but the plan was never realized.

Pál Vásárhelyi
(Szepesolaszi, 1795—1846, Pest)

After studying at Miskolc and Eperjes, VÁSÁRHELYI graduated as an engineer in Pest. He has undying merits at the improvement of Hungarian riverways including the Danube. VÁSÁRHELYI's essays deal with transportation and river control. Two of his studies are concerned with triangulation. When solving engineering problems he willingly applied the latest achievements of mathematics. In 1838 VÁSÁRHELYI was elected member of the Hungarian Academy of Sciences.

János Ármin Vész
(Szeged, 1826—1882, Budapest)

After secondary school studies at Szeged, VÉSZ graduated as an engineering career was followed by a professional job at the university and the Technical College, then at the Technical University; he lectured on advanced mathematics. VÉSZ became corresponding member of the Hungarian Academy of Sciences in 1858 and ordinary member in 1868. He wrote several papers on actual mathematics and the calculus of variations. VÉSZ's best-known work is a two-volume textbook entitled: *The Outlines of Advanced Mathematics* (Pest, 1861—1862, in Hungarian).

János Warga
(Kovácsvágás, 1804—1875, Nagykőrös)

A teacher of the Calvinist College at Nagykőrös, and member of the Hungarian Scholarly Society from 1835. He published several books and papers, mostly on pedagogy. His activity in mathematics is insignificant.

József Wolfstein
(Károlyváros, 1773—1859, Buda)

A teacher at the College of Kassa and at the department of mathematics at the University of Art and Science in Pest. The author of several textbooks of mathematics written in Latin.

Győző Zemplén
(Nagykanizsa, 1879—1916, Asiago)

Graduated as a teacher from the Budapest University of Art and Science and began to teach physics there in 1905. Privatdozent of physics at the Technical University in 1908. Professor at the department of theoretical physics at the Technical University after 1912. At the beginning of his scientific career ZEMPLÉN wrote several articles on number theory and interpolation. Later on ZEMPLÉN dealt exclusively with research in physics and he applied advanced means of mathematics. His investigations in the problem of fluid flow, thermodynamics and the theory of relativity are of fundamental importance. ZEMPLÉN's wast encyklopaedia-article: *Besondere Ausführungen über Unstetige Bewegungen in Flüssigkeiten* (IV. Vol. 3, pp. 281—323) is frequently cited. ZEMPLÉN died at the peak of his powers — a tremendous loss to Hungarian physics — at the Italian front in World War I.

The titles of Hungarian periodicals
in this volume

Akadémiai Értesítő = Gazette of the Academy
Alkalmazott Matematikai Lapok = Papers in Applied Mathematics
Budapesti Szemle = Budapest Gazette
A Cselekvés Iskolája = The School of Activity
Csillagászati Lapok = Astronomical Journal
A Debreceni Déri Múzeum Évkönyve = Yearbook of the Déri Museum in Debrecen
A Debreceni Kossuth Lajos Tudományegyetem Évkönyve = Yearbook of the Kossuth Lajos University
   in Debrecen
Építünk = We are building
Értekezések a Math. Tud. Köréből = Studies in Mathematics
Évkönyv = Yearbook
Földméréstani Közlemények = Surveying Bulletin
Középiskolai Matematikai Lapok = Mathematical Journal for Secondary Schools
Közlemények a Debreceni Tudományegyetem Matematikai Szemináriumából = Bulletin of Mathematics
   of the Debrecen University of Sciences
Magyar Académiai (Akadémiai) Értesítő = Hungarian Academic Gazette
A Magyar Filozófiai Társaság Közleményei = Bulletin of the Hungarian Society of Philosophy
Magyar Könyvszemle = Hungarian Bookreview
Magyar Középiskola = Hungarian Secondary Schooling
Magyar Kurír = Hungarian Curier
Magyar Mérnök- és Építész-egylet Közlönye = Bulletin of the Hungarian Society of Engineers and Architects
Magyar Nyelv = Hungarian Language
Magyar Pedagógia = Hungarian Pedagogy
Magyar Tudomány = Hungarian Science
A Magyar Tudományos Akadémia Évkönyve = Yearbook of the Hungarian Academy of Sciences
A Magyar Tudós Társaság Évkönyve = Yearbook of the Hungarian Scholarly Society
Az MTA Évszázada = A Century of the Hungarian Academy of Sciences
Az MTA Értesítője = Gazette of the Hungarian Academy of Sciences
Az MTA Matematikai és Fizikai Tudományok Osztályának Közleményei = Bulletin of the Mathematical
   and Physical Division of the Hungarian Academy of Sciences
Az MTA III. Osztályának Közleményei = Bulletin of the 3rd Division of the Hungarian Academy of Sciences
Mathematikai Műszótár = Dictionary of Mathematics
Mathematicai Pályamunkák = Mathematical Competition Papers
Matematikai Lapok = Mathematical Journal
Math. és Phys. Lapok = Mathematical and Physical Journal
Math. és Term. tud. Értesítő = Gazette of Mathematics and Natural Sciences
Math. és Term. tud. Közlemények = Bulletin of Mathematics and Natural Sciences
Mennyiségtani és Természettudományi Didaktikai Lapok = Journal of Arithmetics and Didactic Natural
   Sciences
Műegyetemi Lapok = Journal of the Technical University

Műszaki nagyjaink = Great Engineers of Hungary
Pannonhalmi Szemle Könyvtára = Library of the Pannonhalma Gazette
Századok = Centuries
A Szent István Akadémia Évkönyve = Yearbook of the St. Stephan Academy
Természettudományi Közlöny = Bulletin of Natural Sciences
Tudományos Gyűjtemény = Scientific Collection
Tudománytár = Science Digest
Az ungvári reáliskola Értesítője = Gazette of the Secondary School of Ungvár
TK = Teleki—Bolyai Library

Geographical glossary

Ábelfalva, see Ábelová CS
Ábelová CS
Aiud (Nagyenyed) RO
Alba Iulia (Gyulafehérvár) RO
Apáca, see Apaţa RO
Apaţa RO
Arad RO
Átány H

Baia Mare (Nagybánya) RO
Baja H
Balatonfüred H
Banská Bystrica (Besztercebánya) CS
Banská Štiavnica (Selmecbánya) CS
Bardejov CS
Bártfa, see Bardejov CS
Bellye, see Bilje YU
Bér H
Besztercebánya, see Banská Bystrica CS
Biled (Billéd) RO
Bilje (Bellye) YU
Billéd, see Biled RO
Bolya, see Buia RO
Braşov (Brassó) RO
Brassó, see Braşov RO
Bratislava (Pozsony) CS
Breslau, see Wrocław PL
Brezovička (Hámbor) CS
Brîncoveneşti (Marosvécs) RO
Buda, from 1872 Budapest H
Büdszentmihály, today: Tiszavasvári H
Buia (Bolya) RO
Burbach (Peyerbach) A

Celldömölk H
Chemnitz (Karl-Marx-Stadt) D

Cluj, today: Cluj-Napoca (Kolozsvár) RO
Cluj-Napoca (Kolozsvár) RO

Danzig, see Gdańsk PL
Debrecen H
Debric (Nagydebrek) RO
Decs H
Dercsiká, see Jurová CS
Domáld, see Viişoara RO
Dunaföldvár H

Eger H
Eperjes, see Prešov CS
Erdőhegy, see Pădureni RO
Esztergom H

Fehérgyarmat H

Gdańsk (Danzig) PL
Gheorgheni (Gyergyószentmiklós) RO
Güssing (Némétújvár) A
Gyarmat H
Gyergyószentmiklós, see Gheorgheni RO
Gyömrő H
Gyöngyös H
Győr H
Gyulafehérvár, see Alba Iulia RO

Hámbor, see Brezovička CS
Hercegszántó H
Hortobágy H
Höszlin H
Hurbanovo (Ógyalla) CS

Igló, see Spišská Nová Ves CS
Izsák H

359

Ják H
Jászapáti H
Jurová (Dercsika) CS
Jur pri Bratislave (Szentgyörgy) CS

Kaliningrad (Kőnigsberg) SU
Kalocsa H
Karl-Marx-Stadt (Chemnitz) D
Karlovac (Károlyváros) YU
Károlyváros, see Karlovac YU
Kassa, see Košice CS
Kecskemét H
Késmárk, see Kežmarok CS
Keszthely H
Kežmarok (Késmárk) CS
Kolozsvár, see Cluj-Napoca RO
Komárno (Komárom) CS
Komárom, see Komarno CS
Komárom H
Kőnigsberg, see Kaliningrad SU
Körmöcbánya, see Kremnica CS
Košice (Kassa) CS
Kőszeg H
Kotterbach, see Rudnany CS
Kovácsvágás H
Kremnica (Körmöcbánya) CS

Lemberg, see Lvov SU
Leningrad (Saint Petersburg) SU
Léva, see Levice CS
Levice (Léva) CS
Levoča (Lőcse) CS
Lőcse, see Levoča CS
Losonc, see Lučenec CS
Lučenec (Losonc) CS
Lvov (Lemberg) SU

Mágocs H
Magyarkanizsa, see Stara Kanjiža YU
Magyaróvár H
Maksa, see Moacşa RO
Máramarossziget, see Sighetul Marmaţiei RO
Marosvásárhely, see Tîrgu-Mureş RO
Marosvécs, see Brîncoveneşti RO
Medgyes, see Mediaş RO
Mediaş (Medgyes) RO
Ménhárd, see Vrbov CS
Miskolc H
Moacşa (Maksa) RO

Nagybánya, see Baia Mare RO
Nagydebrek, see Debric RO
Nagyenyed, see Aiud RO
Nagykálló H

Nagykanizsa H
Nagykőrös H
Nagyszeben, see Sibiu RO
Nagysziget, see Velký Ostrov CS
Nagyszombat, see Trnava CS
Nagyvárad, see Oradea RO
Nagyveleg H
Németpróna, see Nitrianske Pravno CS
Németújvár, see Güssing A
Nitra (Nyitra) CS
Nitrianske Pravno (Németpróna) CS
Nyitra, see Nitra CS

Óbuda, from 1872 Budapest H
Ocna Sibiului (Vízakna) RO
Odorheiu Secuiesc (Székelyudvarhely) RO
Ógyalla, see Hurbanovo CS
Olmütz, see Olomouc CS
Olomouc CS
Oradea (Nagyvárad) RO
Orăştie (Szászváros) RO
Óvár, see Starý Hrádok CS

Pădureni (Erdőhegy) RO
Pannonhalma H
Pápa H
Paptamási, see Tămăşeu RO
Pásztó H
Pécs H
Pest, from 1872 Budapest H
Pestszentlőrinc (Pestlőrinc), Budapest H
Pétervárad, see Petrovaradin YU
Petrovaradin (Pétervárad) YU
Peyerbach (Burbach) A
Pişcolt RO
Piskolt, see Pişcolt RO
Podolin, see Podolinec CS
Podolinec (Podolin) CS
Polgárdi H
Pozsony, see Bratislava CS
Prešov (Eperjes) CS

Rechnitz (Rohonc) A
Régen, see Reghin RO
Reghin (Régen) RO
Rimaszombat, see Rimavská Sobota CS
Rimavská Sobota (Rimaszombat) CS
Rimetea (Torockó) RO
Rohonc, see Rechnitz A
Rózsahegy, see Ružomberok CS
Rudnany (Kotterbach) CS
Ružomberok (Rózsahegy) CS

Saint Petersburg, see Leningrad SU
Sárosd H

Name index

GÜNTHER, Siegmund 94, 261
GYIRES, Béla 228, 293, 300, 301
GYŐRY, Sándor 124, 129, 199, 202—204, 206, 207, 209, *322*, 347

HAAR, Alfréd 219, 272, 290, 291, 296, 300
HABSBURG, Johann von 310, 311
HADALY, Károly 76, 77, *322*
HADAMARD, Jacques 242
HADWIGER, H. 157
HAJNÓCZY, József 318
HAJÓS, György 169, 227
HALLEY, Edmond 84
HALMOS, P. R. 301
HALPHEN, Georges 275
HALSTED, Georg Bruce 263—26ᴜ
HAMBURGER, Hans 250
HAMILTON, William Rowan 137, 189, 190, 248, 255, 297, 301
HANKEL, Hermann 140, 247
HARNACK, Carl 275
HARRIOT, Thomas 89, 94
HÁRS, János 21, 51
HATVANI, István 81—86, 208, 316, *322—324*, 337, 347
HAUSDORFF, Felix 139, 236
HAVAS, Miksa 292
HEGEDŰS, Sámuel 124, 128, *324*
HEGEL, Georg 129, 352
HELL, Miksa *324*
HELLEBRANT, Árpád 70
HELMHOLTZ, Hermann 143, 222, 252, 297, 333
HELTAI, Gáspár Jr. 51, *324*
HENSEL, Kurt 243, 287
HEPPES, Aladár 93
HERMAN, Ottó 16
HERMITE, Charles 222, 247, 278
HERON 270
HERSCHEL, John 205
HERTL, Ignác *324*
HERTZ, Heinrich Rudolph 251
HESS, E. 274
HESSE, Otto 224, 231, 326
HILB, Ernst 290
HILBERT, David 136, 157, 233, 236, 238, 243, 263, 266, 267, 293
HJELMSLEV, Johannes 293
HÖLDER, Otto 279
HOLLÁN, Ernő 199, *325*
HOLYWOOD (HALIFAX), John, see SACROBOSCO
HOPKINS 298
HOPPE, E. 274
HORÁNYI, Alexius 51, 119
HORNER, William George 207

HORVÁTH, Ádám, see PÁLÓCZI HORVÁTH, Ádám
HORVÁTH, János *325*
HORVÁTH, Róbert 85
HOÜEL, Guillaume-Jules 130, 172, 257
HUMBERT, Georges 275
HUMPHREYS 126
HUNFALVY, Pál 14
HUNYADY, Jenő 218, 220—222, 224—232, 257, 259, 272, *325—327*, 333, 348
HUYGENS, Christian 81, 92

IGEL, B. 227
IMRE, Sándor 219
IMSCHENETZKY, V. G. 249
ISENKRAHE, Kaspar 261
ISTVÁN I 22
IVÁNCSICS, János *327*, 347

JACOBI, Carl G. J. 215, 240, 248, 249, 256, 265, 301
JÁNOSI, Miklós 33, 36, *327*
JAUSZ, Béla 51
JEAN DE MEURS (JOHANNES DE MURIS) 23
JEDLIK, Ányos 108, 318, 340
JELITAI, József 79, 109, 117, 119
JENTSCH, Werner 89
JOHANNES DE MURIS, see JEAN de Meurs
JOHANNES DE SACRO BOSCO, see SACROBOSCO
JOHN OF HOLYWOOD, see SACROBOSCO
JORDAN, Camille 139, 192, 278
JORDAN, Károly 292—294, 300
JOSEPH II 73, 76, 78, 79, 316, 318, 324, 348
JULIUS CAESAR OF PADUA 33, 35, 36, 44, 52, 303
JÜRGENS, Enno 233

KAGAN, V. F. 157, 164, 183, 185, 186, 196, 263
KAHANE, J. P. 300
KALMÁR, László 196, 234, 237, 238, 252, 254
KANT, Immanuel 135, 137, 185, 350
KÁNTOR, Sándor 278
KAPTEYN, Willem 269
KÁRMÁN, Mór 217
KÁRMÁN, Tódor 217, 289
KÁROLYI, Lajos 342
KARSTEN, W. J. G. 87
KÁRTESZI, Ferenc 9, 161, 174, 195, 196, 300
KÄSTNER, Abraham 87, 128, 305, 306, 350
KATONA, Dénes 112, 119, 317, 318, *327*
KATONA, István, see GELEJI KATONA, István
KAZINCZY, Ferenc 100, 102, 109, 353
KEERSEBOOM 84
KELETI, Károly 292
KEMÉNY, Simon Sr. 305, 306
KEMÉNY, Simon Jr. 305, 306

365